AF361739

# Advances in the Processing and Application of Polymer and Its Composites

# Advances in the Processing and Application of Polymer and Its Composites

Editors

**Wei Wu**
**Hao-Yang Mi**
**Chongxing Huang**
**Hui Zhao**
**Tao Liu**

MDPI • Basel • Beijing • Wuhan • Barcelona • Belgrade • Manchester • Tokyo • Cluj • Tianjin

*Editors*
Wei Wu
City University of Hong Kong
China

Hao-Yang Mi
Zhengzhou University
China

Chongxing Huang
Guangxi University
China

Hui Zhao
Guangxi University
China

Tao Liu
Shenzhen University
China

*Editorial Office*
MDPI
St. Alban-Anlage 66
4052 Basel, Switzerland

This is a reprint of articles from the Special Issue published online in the open access journal *Polymers* (ISSN 2073-4360) (available at: https://www.mdpi.com/journal/polymers/special_issues/ Advances_Processing_Application_Polymer_its_Composites).

For citation purposes, cite each article independently as indicated on the article page online and as indicated below:

LastName, A.A.; LastName, B.B.; LastName, C.C. Article Title. *Journal Name* **Year**, *Volume Number*, Page Range.

**ISBN 978-3-0365-5413-6 (Hbk)**
**ISBN 978-3-0365-5414-3 (PDF)**

# Contents

*Article*

# Study of the Preparation and Properties of Chrysin Binary Functional Monomer Molecularly Imprinted Polymers

Long Li [1,2,3,4], Lanfu Li [1], Gege Cheng [1,2,3,4], Sentao Wei [1], Yaohui Wang [1,2,3,4], Qin Huang [1,2,3,4], Wei Wu [5], Xiuyu Liu [1,2,3,4,*] and Guoning Chen [6,*]

[1] School of Chemistry and Chemical Engineering, Guangxi Minzu University, Nanning 530006, China; lilong19980227@163.com (L.L.); a17876072393@163.com (L.L.); ggcheng2022@163.com (G.C.); a1184306866@163.com (S.W.); wwangyh970218@163.com (Y.W.); huangqin@gxun.edu.cn (Q.H.)
[2] Key Laboratory of Chemistry and Engineering of Forest Products, State Ethnic Affairs Commission, Nanning 530006, China
[3] Guangxi Key Laboratory of Chemistry and Engineering of Forest Products, Nanning 530006, China
[4] Guangxi Collaborative Innovation Center for Chemistry and Engineering of Forest Products, Guangxi Minzu University, Nanning 530006, China
[5] Jihua Laboratory, 13 Nanpingxi Road, Foshan 528200, China; scutw.wei@gmail.com
[6] Guangxi Bossco Environmental Protection Technology Co., Ltd., Nanning 530007, China
* Correspondence: xiuyu.liu@gxun.edu.cn (X.L.); chenguonin2@126.com (G.C.)

**Citation:** Li, L.; Li, L.; Cheng, G.; Wei, S.; Wang, Y.; Huang, Q.; Wu, W.; Liu, X.; Chen, G. Study of the Preparation and Properties of Chrysin Binary Functional Monomer Molecularly Imprinted Polymers. *Polymers* **2022**, *14*, 2771. https://doi.org/10.3390/polym14142771

Academic Editors: Hao-Yang Mi, Chongxing Huang, Hui Zhao and Tao Liu

Received: 12 May 2022
Accepted: 1 July 2022
Published: 6 July 2022

**Publisher's Note:** MDPI stays neutral with regard to jurisdictional claims in published maps and institutional affiliations.



**Abstract:** Chrysin is a natural bioactive molecule with various groups, and it has been a challenge to separate and enrich chrysin from natural products. Molecularly imprinted polymers have been widely used in the extraction of natural products, but the number and type of functional monomers limits the separation effect. The synergistic action of multiple functional monomers can improve the separation effect. In this paper, molecularly imprinted polymers (Bi-MIPs) were prepared using methacrylic acid and acrylamide as binary functional monomers for the separation and enrichment of chrysin. The Bi-MIPs were characterized using thermogravimetric analyzer (TGA), Fourier transform infrared spectroscopy (FT-IR) and scanning electron microscope (SEM). The performances of Bi-MIPs were assessed, which included adsorption isotherms, selective recognition and adsorption kinetics. The experimental results show that Bi-MIPs are shaped as a uniform sphere with an abundant pocket structure on its surface. The adsorption of chrysin on the Bi-MIPs followed a pseudo-second-order and adapted Langmuir–Freundlich isotherm models. The adsorption performance of the Bi-MIPs was determined at different temperatures, and the Bi-MIPs showed excellent adsorption performance at 30 °C. The initial decomposition temperature of the Bi-MIPs was 220 °C. After five times of adsorption and desorption, the adsorption performance of the Bi-MIPs decreased by only 7%. In contrast with single functional monomer molecularly imprinted polymers (Si-MIPs), the Bi-MIPs showed excellent specificity, with an imprinting factor of 1.54. The Bi-MIPs are promising materials in the separation and enrichment of chrysin for their high adsorption capacity, low cost and being environmentally friendly.

**Keywords:** chrysin; molecular imprinting; adsorption performance; binary functional monomers

## 1. Introduction

Natural products have gained enormous popularity over the years as they are used in clinical settings [1]. They have made excellent contributions historically to drug development, and many of them have had profound effects on our lives [2]. Chrysin, chemically known as 5,7-dihydroxy flavone [3], is a kind of flavonoid compound with extensive pharmacological activity isolated from active ingredients of Chinese traditional medicine [4,5] and has great antioxidative [6] and anti-inflammatory effects [7]. Research has shown that chrysin inhibits tumor cell proliferation [8], induces tumor cell apoptosis [9], suppresses tumor angiogenesis [10] and circumvents drug resistance [11]. Therefore, studies on the extraction and identification of chrysin are of great value.

According to literature reviews, there are many methods for the separation of chrysin, including HPLC [12], column chromatography [13], solid-phase extraction (SPE) [14], gas chromatography–mass spectrometry(GC-MC) [15], macroporous resin [16] and other traditional methods. However, due to the problems of long extraction time and a large amount of solvent and waste, the instrumental methods are not suitable for industrial mass production applications. Although the traditional separation methods, such as extraction and precipitation, can be applied at a low cost, these approaches are restricted because of low recoveries and purities. At present, there is still a lack of efficient extraction methods of chrysin.

The molecularly imprinted polymer has been widely used for the separation and enrichment of active components from natural products [17–21]. Molecular imprinting is a technique for preparing polymers of desired and predetermined selectivity [22]. This polymer is a material with complementary spatial structure and functional group interaction with template molecule. The molecularly imprinted polymer has a strong affinity and recognition ability for the template molecule [23–28]. The method has excellent prospects for application in the field or market for easy and fast molecular identification. Due to their unique properties, molecularly imprinted polymers have been widely used in various applications, such as drug delivery [29], detection of viruses [30], chemical sensor [31,32], specific recognition of protein [33], chromatography [34], solid-phase extraction [35] and bioanalysis [36].

Generally speaking, most natural products have multiple active groups. Single functional monomers and template molecules are easily destroyed during hydrogen bond formation, which reduces adsorption capacity and separation factor and, thus, cannot be well separated and purified from traditional Chinese medicine. The synergistic effect of the two functional monomers is beneficial to the separation and purification of natural products. For example, Wan [37] selectively extracted myricetin from traditional Chinese medicine with an adsorption capacity of 10.58 mg/g using glycidyl methacrylate (GMA) and 4-vinylpyridine as bifocal monomers. Huan [38] used acrylamide and 2-vinylpyridine as bifocal monomers to prepare a solid-phase extraction column for the separation of rutin extract from traditional Chinese medicine, and the recovery rate was 85.93%, which was better than the traditional separation column. The calculation and practical experiments show that MIP synthesized from an acrylamide (AM) monomer has a higher specific factor and adsorption capacity [39]. Chrysin has a rigid benzene ring structure and contains both a hydroxyl group and an aldehyde group, belonging to polar flavonoids. The functional monomer methacrylic acid (MAA) was acidic and acrylamide (AM) neutral. Under the synergistic action of AM, the force of MAA on the hydrogen bonds, electrostatic and $\pi$–$\pi$ stacking of chrysin increased, which is expected to improve the adsorption selectivity of molecularly imprinted polymers. Therefore, the strategy of multifunctional monomer is a valid synthetic option to synthesize imprinted materials for the template molecules of flavonoids with polar functional groups, such as chrysin.

In this paper, Bi-MIPs are used for the separation and enrichment chrysin for the first time. The Bi-MIPs were prepared by precipitation polymerization using methacrylic acid and acrylamide as functional monomers and ethylene glycol dimethacrylate as a crosslinking agent. The adsorption selectivity of the Bi-MIPs was evaluated by preparing Si-MIPs with methacrylic acid as a functional monomer as the control group. Furthermore, we also study the feasibility of the Bi-MIPs as effective sorptive materials for the dissociation and enrichment of chrysin. The equilibrium, kinetics and thermodynamics of the adsorption process were investigated to study the adsorption mechanism of chrysin on the Bi-MIPs. Thus, Bi-MIPs are presented for their promising output application toward the extraction of chrysin.

## 2. Materials and Methods

### 2.1. Materials

Chrysin, 2,2′-azobisisobutyronitrile (AIBN) and acrylamide (AM) were purchased from Shanghai Aladdin Biochemical Technology Co., Ltd. (Shanghai, China). Methyl alcohol and acetic acid were purchased from Chengdu Cologne Chemicals Co., Ltd. (Chengdu, China). Methacrylic acid (MAA) was purchased from Sinopharm Chemical Reagent Co.,

Ltd. (Shanghai, China). Ethylene glycol dimethacrylate (EGDMA) was purchased from Alfa Aesar (Qingdao, China). Chloramphenicol and oxytetracycline were purchased from Shanghai Maclin Biochemical Technology Co., Ltd. (Chengdu, China).

*2.2. Preparation of Bi-MIPs, Bi-NIPs, Si-MIPs and Si-NIPs*

The Bi-MIPs were synthesized by the precipitation polymerization method, using chrysin as template molecule, MAA and AM as function monomers, EGDMA as cross-linker and AIBN as initiator. Chrysin (0.3 mmol), MAA (2.4 mmol) and AM (0.4 mmol) were dissolved in 50 mL of methanol in a 250 mL three-necked round-bottomed flask. The function monomers (MAA (2.4 mmol) and AM (0.4 mmol)) interacted with the template molecule (chrysin) through intermolecular force to form a stable "pre-polymerization" complex. Then, the 50 mL of mixture solution of EGDMA (9.6 mmol), AIBN (0.1043 g) and methanol was added and dispersed under ultrasound. The mixture was heated followed by mechanical agitation (50 rpm) and heat-polymerized at 65 °C for 12 h. The synthesis process is shown in Figure 1. The Bi-MIPs were collected through vacuum filtration, and the unreacted monomers were removed by methanol: acetic acid (9:1, *v/v*) extraction for 24 h and methanol extraction for 12 h. The Bi-MIPs were dried under a vacuum at 50 °C for 12 h. The binary functional monomers molecularly non-imprinted polymer (Bi-NIP) microspheres were synthesized using the same procedure as the Bi-MIPs, but without the chrysin (template molecules). The Si-MIPs were synthesized following the same procedure without adding the AM. The single functional monomer molecularly non-imprinted polymers (Si-NIPs) were synthesized following the same procedure without adding the AM and chrysin.

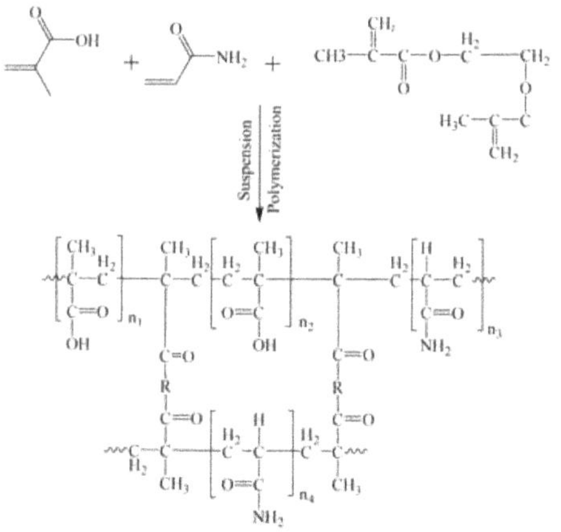

**Figure 1.** The scheme for preparing the Bi-MIPs (R is -O-CH$_2$-CH$_2$-O-).

*2.3. Characterization of Bi-MIPs, Bi-NIPs, Si-MIPs and Si-NIPs*

The analysis of the size and the morphology of the Bi-MIPs, Bi-NIPs, Si-MIPs and Si-NIPs were performed using field-emission scanning electron microscopy (SEM; Supra 55 Sapphire, Carl Zeiss, Jena, Germany). Thermogravimetric analysis (TGA) and differential thermogravimetric analysis (DTG) (STA 449 F5, NETZSCH-Gerätebau GmbH, Selb, Germany) were conducted to evaluate the thermal stabilities of the Bi-MIPs, Bi-NIPs, Si-MIPs and Si-NIPs under an N$_2$ atmosphere at a heating rate of 10 K/min. The functional groups of the Bi-MIPs, Bi-NIPs, Si-MIPs and Si-NIPs were recorded by Fourier transform infrared spectrometer (MagnA-IR55, Thermo Fisher Scientific, Massachusetts, USA) in the range of 4000–400 cm$^{-1}$. The chrysin solution was determined by UV-visible spectra (UV-2600 Shimadzu, Tokyo, Japan). The Bi-MIPs, Bi-NIPs, Si-MIPs and Si-NIPs were weighed by analytical balance with an accuracy of 0.1 mg (Practum124-1cn, Sartorius AG, Göttingen, Germany).

### 2.3.1. Scanning Electron Microscopy (SEM)

The surface morphology of the Bi-MIPs, Bi-NIPs, Si-MIPs and Si-NIPs was determined by SEM analysis (SEM, Supra 55 Sapphire, Carl Zeiss Germany, Jena, Germany). The samples were evenly coated on the conductive adhesive of the sample sheet and then sprayed with gold for 0.5 h. The surface morphology of the Bi-MIPs, Bi-NIPs, Si-MIPs and Si-NIPs after the samples were sprayed with gold was observed by SEM under low vacuum conditions.

### 2.3.2. Diameters of the Bi-MIPs, Bi-NIPs, Si-MIPs and Si-NIPs

The diameters of the Bi-MIPs, Bi-NIPs, Si-MIPs and Si-NIPs were measured from their SEM images using image analysis software (Image J). The diameter of more than 60 samples in each sample were measured. AD represents mean diameter ± standard deviation, and N represents sample quantity.

### 2.3.3. Thermogravimetric Analysis (TGA)

The thermal stability of the Bi-MIPs, Bi-NIPs, Si-MIPs and Si-NIPs was determined by thermogravimetric analysis (TGA) (STA449F5, NETZSCH-Gerätebau GmbH, Selb, Germany). The samples were heated from 30 to 800 °C for thermal degradation under nitrogen protection at a rate of 10 °C/min [40].

### 2.3.4. Nitrogen Adsorption/Desorption Analysis (BET)

The specific surface area and pore size of the samples were measured by the specific surface area and pore size analyzer (ASAP2460, Micromeritics, Georgia, USA). Nitrogen adsorption–desorption was measured at 90 °C for 12h under nitrogen protection [41].

### 2.4. Static Adsorption

### 2.4.1. Adsorption Isotherm of Chrysin on Bi-MIPs, Bi-NIPs, Si-MIPs and Si-NIPs

To estimate the adsorption properties of Bi-MIPs, the Bi-MIPs were weighed (0.02 ± 0.0002 g) with an analytical balance and placed into a round of 50 mL conical flasks, and added into 20 mL of a chrysin methanol solution with different concentrations (0.2 mg/mL–1.4 mg/mL). The adsorption process of the Bi-MIPs was carried out for 5 h in an 80 rpm constant temperature oscillator at 25 °C. After the sorption experiments, the Bi-MIPs were separated from the mixed solution by a centrifuge and the concentration in the supernatant was obtained for UV-vis absorbance. The chrysin adsorption amount of Bi-MIPs was determined based on the following Equation (1):

$$Q_e = \frac{(C_0 - C_e) \times V}{W} \tag{1}$$

where $C_0$ (mg/mL) represents the initial concentration of chrysin in the solution, $C_e$ (mg/mL) represents the equilibrium concentration of chrysin after adsorption, V (mL) represents the volume of the adsorbed chrysin solution and W (g) represents the mass of polymers in the adsorbed chrysin solution.

At the same time, to evaluate the adsorption capacity of Bi-NIPs, Si-MIPs and Si-MIPs. A total of 20 mg of Bi-NIPs, Si-MIPs or Si-MIPs was placed into a round of 50 mL conical flasks and added into 20 mL of chrysin solution with different concentrations (0.2 mg/mL–1.4 mg/mL). Under the same operating conditions of Bi-MIPs, the concentration of chrysin in the supernatant was analyzed using UV-vis absorbance. The equilibrium adsorption capacity was calculated using Equation (1).

### 2.4.2. Adsorption Kinetics

To estimate the adsorption kinetics of Bi-MIPs on chrysin, the influence of the adsorption properties over time was studied. The Bi-MIPs were weighed (0.02 ± 0.0002 g) with an analytical balance and placed into a round of 50 mL conical flasks. After adding 20 mL of chrysin methanol solution, the Bi-MIPs were used to adsorb chrysin (1 mg/mL) in an

80 rpm constant temperature oscillator at 25 °C. Samples of 0.05 mL volume were taken at 15, 30, 45, 60, 75, 90, 105,120, 150, 180, 210, 240 and 300 min, and the concentration in the supernatant was obtained for UV-vis absorbance. The chrysin equilibrium adsorption amount of the Bi-MIPs was obtained based on the following Equation (2):

$$Q_t = \frac{(C_0 - C_t)}{W} \times V \qquad (2)$$

where $C_0$ (mg/mL) represents the initial concentration of chrysin in the solution, $C_t$ (mg/mL) represents the concentration of chrysin solution at time t (min), V (mL) represents the volume of adsorbed chrysin solution and W (g) represents the mass of polymers in the adsorbed chrysin solution.

At the same time, to evaluate the adsorption capacity of Bi-NIPs, Si-MIPs and Si-MIPs. Under the same operating conditions of the Bi-MIPs, the concentration of chrysin in the supernatant was analyzed using UV-vis absorbance. The equilibrium adsorption capacity was calculated using Equation (2).

### 2.4.3. Adsorption Thermodynamics

To determine the adsorption temperature of chrysin on the Bi-MIPs, the effect of five different temperatures (10, 20, 30, 40 and 50 °C) on the adsorption capacity of chrysin by Bi-MIPs was evaluated. The Bi-MIPs were weighed (0.02 ± 0.0002 g) with an analytical balance and placed into a round of 50 mL conical flasks. After adding 20 mL of chrysin methanol solution, the Bi-MIPs were used to adsorb chrysin (1 mg/mL) in an 80 rpm constant temperature oscillator. The concentration of chrysin in the supernatant was analyzed using UV-vis absorbance. The equilibrium adsorption capacity was calculated using Equation (1). At the same time, to evaluate the adsorption temperature of chrysin on the Bi-NIPs, Si-MIPs and Si-MIPs, under the same operating conditions of the Bi-MIPs, the concentration of chrysin in the supernatant was analyzed using UV-vis absorbance. The equilibrium adsorption capacity was calculated using Equation (1).

### 2.5. Selective Adsorption

To evaluate the specific recognition ability of Bi-MIPs for chrysin, selective experiments were conducted with chloramphenicol and oxytetracycline as structural analogues. The Bi-MIPs were weighed (0.02 ± 0.0002 g) with an analytical balance and placed into a round of 50 mL conical flasks. After adding 20 mL of a solution of chrysin (1 mg/mL), chloramphenicol (1 mg/mL) and terramycin (1 mg/mL), it was adsorbed in an 80 rpm constant temperature oscillator at 25 °C. After the adsorption experiments, the Bi-MIPs were separated from the mixed solution by a centrifuge. The concentration in the supernatant was obtained by UV-vis absorbance. The selective adsorption ability of Bi-MIPs was obtained using the selectivity factors ($\alpha$), which is shown in Equation (3).

$$\alpha = \frac{Q_{\text{Bi-MIPs}}}{Q_{\text{Bi-NIPs}}} \qquad (3)$$

### 2.6. Adsorption Reusability

After 20 mg Bi-MIP was adsorbed and balanced in 20 mL chrysin solution, the saturated adsorption sample was obtained by centrifugation. The Bi-MIPs were cleaned with a methanol solution for several times, and then dried and weighed. The processed Bi-MIPs were subjected to the adsorption test again. Under the same conditions, the adsorption–desorption cycle was performed 5 times, and the adsorption amount determined the concentration after adsorption each time and calculated the adsorption amount according to Formula (1).

## 3. Results and Discussion

### 3.1. Characterization of Bi-MIPs, Bi-NIPs, Si-MIPs and Si-NIPs

#### 3.1.1. SEM

The surfaces of Bi-MIPs, Bi-NIPs, Si-MIPs and Si-NIPs were analyzed using a scanning electron microscope. The most representative SEM images of all particles including Bi-MIPs, Bi-NIPs, Si-MIPs and Si-NIPs are shown in Figure 2. The SEM results show that were differences in the polymerization process in the presence of the template or without it. From Figure 2a–d, the as-synthesized polymers products present a regular and spherical structure with a diameter size of about 1–2 μm. The pore volume and surface area of the Bi-MIPs are $4.459 \times 10^{-3}$ cm$^3$/g and $2.8488 \pm 0.1059$ m$^2$/g, respectively. The Bi-MIPs and Bi-NIPs particles are the same, but the Bi-MIPs have more surface folds, which can provide more adsorption sites [19,42]. The results show that the chrysin occupies a position on the polymer surface during the synthesis process. After elution with methanol and acetic acid, the surface of the Bi-MIPs becomes more wrinkled than that of Bi-NIPs. However, some clustered small particles remain on the surface of the Bi-MIPs, which shows that the template molecules had a remarkable effect on the shape and adsorption properties. This is due to the addition of template molecules in the synthesis of Bi-MIPs, which have a larger particle size and more surface adsorption sites due to the presence of template molecules that give the crosslinking agent and functional monomers a fuller skeleton. The Si-MIP particles are smaller than the Bi-MIP particles, which indicates that the number of functional monomers has a significant influence on the formation of particles and the final particle size. These results suggest that the Bi-MIPs are successfully synthesized.

#### 3.1.2. FT-IR

The structures of the Bi-MIPs, Bi-NIPs, Si-MIPs and Si-NIPs were investigated by FT-IR spectroscopy, and the results are shown in Figure 3. As shown in Figure 3a, the Bi-MIPs, Bi-NIPs, Si-MIPs and Si-NIPs display a stretching vibration peak of -CH$_2$- occurring at 2987 cm$^{-1}$, indicating that MAA is contained in polymerization. The stretching vibration peak of the -NH- double bond appeared at 1456 cm$^{-1}$, the single-bond vibration region of -CH appeared at 1398 cm$^{-1}$, and the stretching vibration peak of -C=O occurred at 1732 cm$^{-1}$ [42–45]. The carbonyl groups tend to have a higher vibration absorption, indicating the successful preparation of the Bi-MIPs, Bi-NIPs, Si-MIPs and Si-NIPs. The partial single-bond vibration absorption peak of -COC- was observed at 1154 cm$^{-1}$ on the Bi-MIP curve, indicating the presence of EGDMA. At the same time, in the Bi-NIP, Si-MIP, and Si-NIP curves, partial single-bond vibration absorption peaks of -COC- were observed at about 1154 cm$^{-1}$, indicating the presence of EGDMA, indicating that the polymer was successfully synthesized. The FT-IR of the Bi-MIPs and Bi-NIPs were almost the same. After removing the template molecule chrysin, the chemical structure and composition of the Bi-MIPs were the same as that of the Bi-NIPs. These results prove that the types of functional monomers have a certain influence on the synthesis of polymers, and the functional monomers play a role in the construction of adsorption pocket size.

#### 3.1.3. TGA

The thermal stabilities of the Bi-MIPs, Bi-NIPs, Si-MIPs and Si-NIPs were studied by TGA analysis, and the results are presented in Figure 3b,c. As shown in Figure 3b, the thermogravimetric analysis curves of the Bi-NIPs, Si-MIPs and Si-NIPs are the same as those of the Bi-MIPs without significant difference. The Bi-MIPs start to decompose at 220 °C, the temperature of the fastest decomposition rate occurs at 404 °C and the maximum decomposition temperature (Tmax) is 468 °C. The decomposition process is divided into two stages, including dehydration in the low-temperature zone (100–220 °C) and decomposition in the high-temperature zone (404–462 °C). The decomposition of the second stage is due to the decomposition of the crosslinker (EGDMA), indicating that the stability of the crosslinker is the main factor affecting the thermal stability of the polymers. The TGA results demonstrate that the Bi-MIPs have excellent thermal stability.

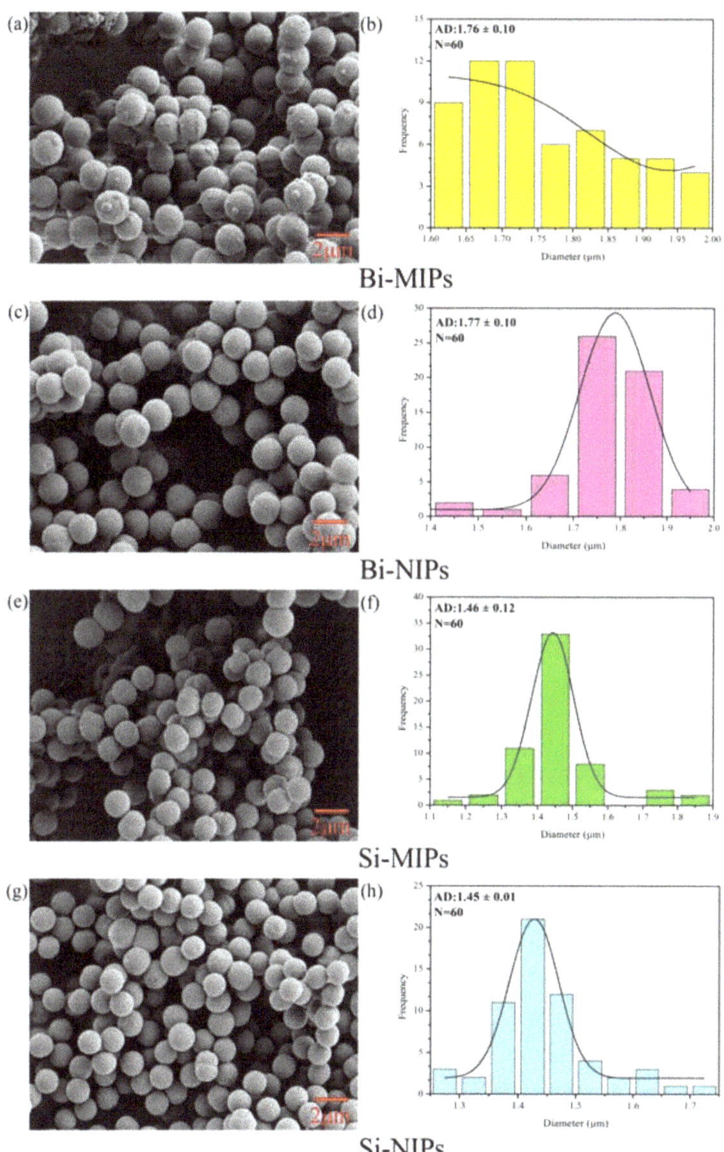



**Figure 2.** SEM, average particle size and particle size distribution images of Bi-MIPs (**a**,**b**), Bi-NIPs (**c**,**d**), Si-MIPs (**e**,**f**) and Si-NIPs (**g**,**h**).

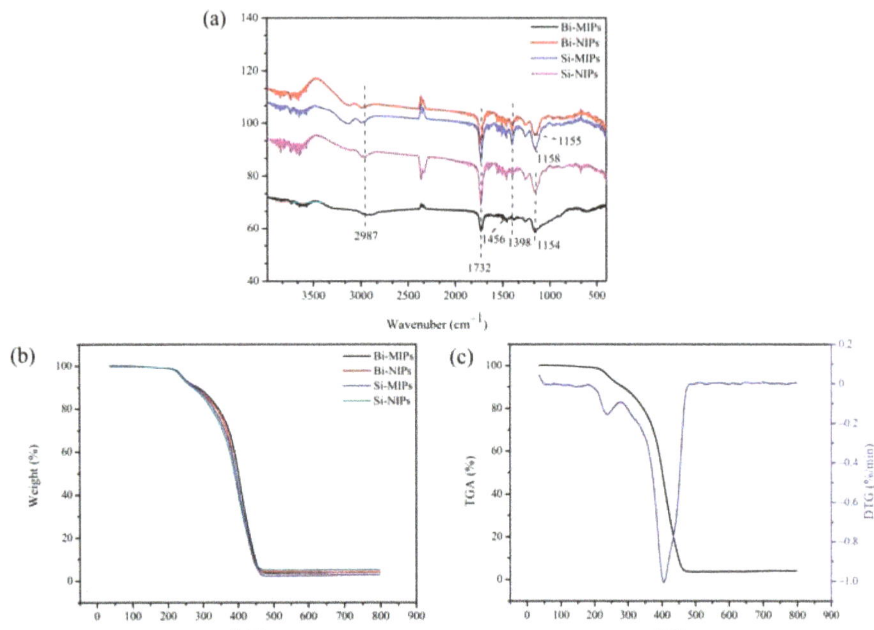

**Figure 3.** (**a**) FI-IR images of the Bi-MIPs, Bi-NIPs, Si-MIPs and Si-NIPs; (**b**) TGA images of the Bi-MIPs, Bi-NIPs, Si-MIPs and Si-NIPs; (**c**) TGA and DTG images of the Bi-MIPs.

### 3.2. Static Adsorption Experiments

#### 3.2.1. Adsorption Isotherm

The adsorption isotherms of chrysin on the Bi-MIPs, Bi-NIPs, Si-MIPs and Si-NIPs at (298 K) with chrysin concentrations of 0.2–1.4 mg/mL are presented in Figure 4. As shown in Figure 4a, the chrysin adsorption property for the Bi-MIPs, Bi-NIPs, Si-MIPs and Si-NIPs increased with increasing chrysin concentration. As the concentration increases, the adsorption difference increases. The Bi-MIPs have the highest adsorption capacity of chrysin, followed by Si-MIPs. These results indicate that the Bi-MIPs and Si-MIPs have specific cavities sizes and specific adsorption capacity for chrysin. By comparing Bi-MIPs, Bi-NIPs, Si-MIPs and Si-NIPs to the adsorption capacity of chrysin, the adsorption capacity of the binary functional monomers was better than that of the single functional monomer, which further indicates that the Bi-MIPs have an excellent application prospect.

To analyze the adsorption mechanism, Langmuir and Freundlich isotherm models were used to fit the experimental data when the adsorption process reached the adsorption equilibrium. The equations of these two models are as follows [46–49].

Langmuir isotherm equation:

$$\frac{1}{Q_e} = \frac{1}{Q_m} + \frac{1}{K_1 Q_m} \times \frac{1}{C_e} \tag{4}$$

Freundlich isotherm equation:

$$\ln Q_e = \ln k_2 + \frac{1}{n} \ln C_e \tag{5}$$

where $C_e$ represents the concentration of chrysin at the adsorption equilibrium (mg/mL); $Q_e$ represents the chrysin adsorption amount for the Bi-MIPs, Bi-NIPs, Si-MIPs and Si-NIPs at equilibrium (mg/g); $Q_m$ represents the maximum adsorption amount of monolayer coverage (mg/g); $K_1$ represents the Langmuir constant (mL/mg); $K_2$ represents the Freundlich constant; and $1/n$ represents the dimensionless Freundlich constant.

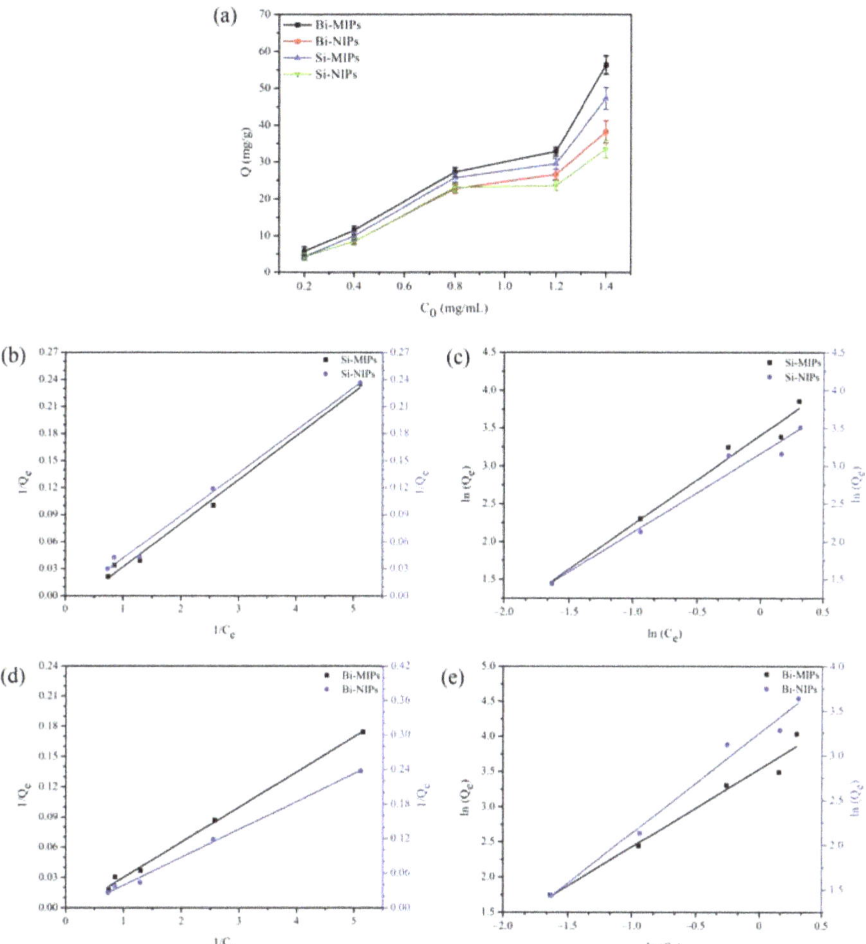

**Figure 4.** (**a**) Adsorption isotherm of the Bi-MIPs, Bi-NIPs, Si-MIPs and Si-NIPs; (**b**) Langmuir adsorption isotherm of the Si-MIPs and Si-NIPs; (**c**) Freundlich adsorption isotherm of the Si-MIPs and Si-NIPs; (**d**) Langmuir adsorption isotherm of the Bi-MIPs and Bi-NIPs; (**e**) Freundlich adsorption isotherm of the Bi-MIPs and Bi-NIPs.

According to the Langmuir and Freundlich isotherm models, the experimental data were fitted and the parameters are shown in Table 1. The Bi-MIPs are illustrated by comparing the Langmuir and Freundlich equation correlation coefficients (the Langmuir and Freundlich correlation coefficients are 0.9953 and 0.9669, respectively). The isothermal adsorption curve of the Bi-MIPs is better represented by the Langmuir model. The Bi-NIPs, Si-MIPs and Si-NIPs were compared and analyzed by the Langmuir and Freundlich equation correlation coefficients (Langmuir and Freundlich correlation coefficients are 0.9946 and 0.9788, 0.9912 and 0.9736, 0.9905 and 0.9596, respectively), indicating that the Bi-NIPs, Si-MIPs and Si-NIPs with Bi-MIPs have the same degree of compatibility with Langmuir. The isothermal adsorption curves of the Bi-NIPs, Si-NIPs and Si-NIPs are better represented by the Langmuir model. These results indicate that the adsorption of chrysin on the Bi-MIPs occurs via monolayer adsorption, which shows that the Bi-MIPs can easily adsorb chrysin.

**Table 1.** Parameters of the Langmuir and Freundlich adsorption models.

| Samples | Langmuir Isotherm | | | Freundlich Isotherm | | |
|---|---|---|---|---|---|---|
| | $K_1$ (mL mg$^{-1}$) | $R^2$ | $Q_m$ (mg g$^{-1}$) | $K_2$ (mL mg$^{-1}$) | $R^2$ | $1/n$ |
| Bi-MIPs | 0.1370 | 0.9953 | 209.64 | 34.20 | 0.9669 | 1.106 |
| Bi-NIPs | 0.1935 | 0.9946 | 106.72 | 25.77 | 0.9788 | 1.116 |
| Si-MIPs | 0.1193 | 0.9912 | 60.350 | 30.20 | 0.9736 | 1.183 |
| Si-NIPs | 0.3417 | 0.9905 | 176.99 | 23.75 | 0.9596 | 1.046 |

### 3.2.2. Adsorption Kinetics

Kinetic experiments were carried out on the Bi-MIPs, Bi-NIPs, Si-MIPs and Si-NIPs with a chrysin solution with an initial concentration of 1.0 mg/mL and the results are shown in Figure 5. The adsorption of chrysin on the Bi-MIPs showed excellent characteristics of the adsorption kinetics. The adsorption capability is increased with the increase in the adsorption time, and the adsorption rate decreased gradually with increasing adsorption time. At any time, the Bi-MIPs have the highest adsorption capacity of chrysin compared with the Bi-NIPs. As for the Bi-MIPs, the adsorption process can be divided into the fast adsorption stage (0–105 min) and the slow adsorption stage (105–240 min). The adsorption capacity in the fast adsorption stage accounted for 76% of the equilibrium adsorption capacity. As for the Si-MIPs, the adsorption capacity in the fast adsorption stage (0–75 min) accounted for 64.7% of the equilibrium adsorption capacity. The chrysin absorption capacity of Bi-MIPs reached equilibrium after 240 min, which indicates that the specific cavities of the adsorbent formed by binary functional monomers promote the adsorption effect. At the same time, the chrysin absorption capacity of the Bi-MIPs, Bi-NIPs, Si-MIPs, Si-NIPs and MIPs exceeds the NIPs, which further indicates that Si-MIPs and Bi-MIPs have successfully synthesized imprinted pores.

The pseudo-first-order (PFO) and pseudo-second-order (PSO) kinetic models were used to investigate the kinetic behaviors of the Bi-MIPs, Bi-NIPs, Si-MIPs and Si-NIPs during chrysin adsorption. The kinetic data of different initial melanoidin concentrations were fitted using the following models [50–54].

PFO kinetic equation:

$$\ln(Q_e - Q_t) = \ln Q_e - \frac{k_3}{2.303} \tag{6}$$

PSO kinetic equation:

$$\frac{t}{Q_t} = \frac{1}{k_4 Q_e^2} + \frac{1}{Q_e} \tag{7}$$

where $k_3$ (1/min) represents the rate constant of PFO kinetic adsorption, $k_4$ (mg/(g·min)) represents the rate constant of PSO kinetic adsorption, $Q_t$ represents the amounts of chrysin adsorbed (mg/g) at time t, and $Q_e$ represents the amounts' equilibrium time.

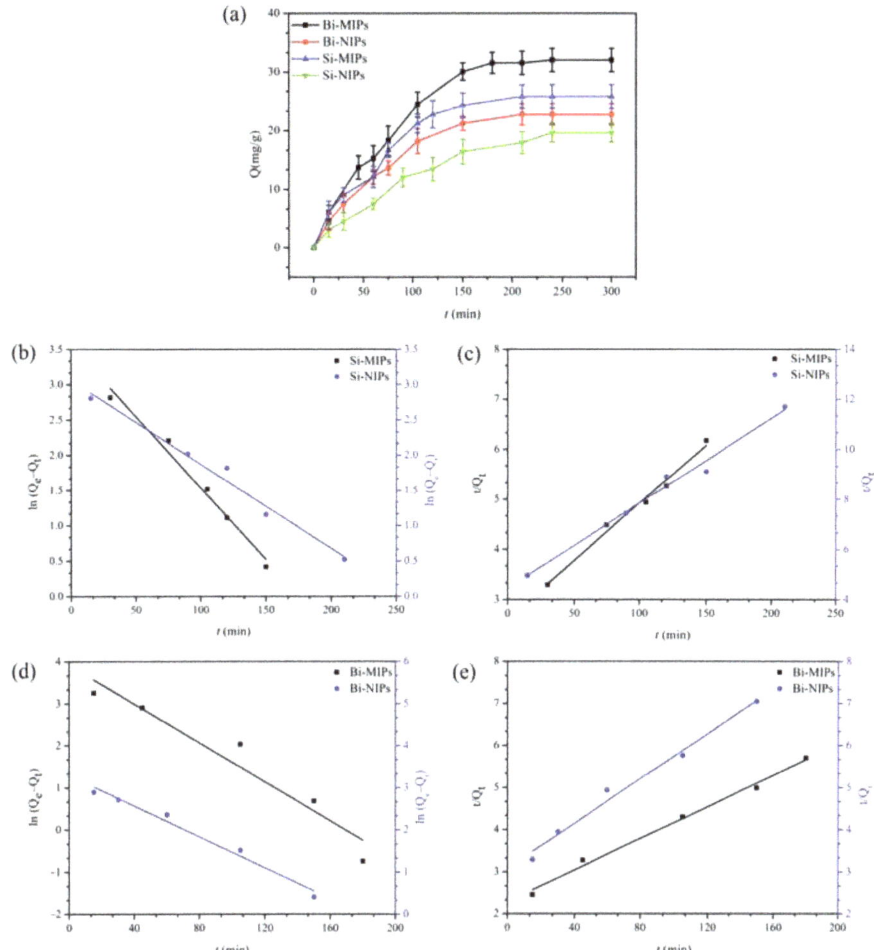

**Figure 5.** (**a**) Adsorption kinetics curves of the Bi-MIPs, Bi-NIPs, Si-MIPs and Si-NIPs; (**b**) PFO kinetic mode of the Si-MIPs and Si-NIPs; (**c**) PSO kinetic mode of the Si-MIPs and Si-NIPs; (**d**) PFO kinetic mode of the Bi-MIPs and Bi-NIPs; (**e**) PSO kinetic mode of the Bi-MIPs and Bi-NIPs.

The experimental data were fitted to the PFO kinetic and PSO kinetic to obtain the corresponding fitting curves (Figure 5) and kinetic parameters (Table 2). It is known from Table 2 that the correlation coefficient ($R^2$) values of the PSO and PFO kinetic models of the Bi-MIPs are $R^2 = 0.9903$ and $R^2 = 0.9153$, respectively. The PSO kinetic model creates better experiments with the adsorption behavior of chrysin onto the Bi-MIPs than the PFO kinetic model; this phenomenon also indicates that chemisorption is the principal mechanism involved in the sorption process. The results indicate that the Bi-MIPs are beneficial to the adsorption of chrysin, and these results further prove the potential applications in the separation of chrysin by the Bi-MIPs.

**Table 2.** Kinetic data of the PFO and PSO kinetic models.

| Samples | PFO Kinetic | | PSO Kinetic | |
|---|---|---|---|---|
| | $K_3$ (min$^{-1}$) | $R^2$ | $K_4$ (g mg$^{-1}$ min$^{-1}$) | $R^2$ |
| Bi-MIPs | 0.0533 | 0.9153 | 0.00015 | 0.9903 |
| Bi-NIPs | 0.0421 | 0.9746 | 0.00023 | 0.9816 |
| Si-MIPs | 0.0466 | 0.9751 | 0.00020 | 0.9841 |
| Si-NIPs | 0.0275 | 0.9750 | 0.00026 | 0.9808 |

### 3.2.3. Adsorption Thermodynamics

Adsorption isotherm experiments were performed at the same initial chrysin solution and at different temperatures of 10, 20, 30, 40 and 50 °C. The fitting results of the adsorption isotherms are presented in Figure 6. As shown in Figure 6a, the adsorption ability of chrysin by the Bi-MIPs is the highest, followed by the Si-MIPs. This is due to the specific cavities of the Bi-MIPs and Si-MIPs, which have specific adsorption on chrysin.

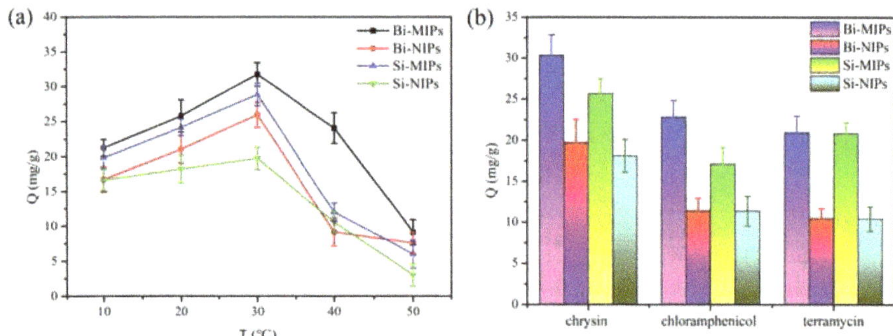


**Figure 6.** (**a**) Thermodynamic curves of the Bi-MIPs, Bi-NIPs, Si-MIPs and Si-NIPs; (**b**) Selective adsorption of the Bi-MIPs, Bi-NIPs, Si-MIPs and Si-NIPs.

With the temperature increase from 10 to 30 °C, the absorption capacity of the adsorbents increased, which due to increases in temperature accelerate the movement of the molecules in a methanol solution. Therefore, the probability of albumin binding to the adsorbent adsorption site is enhanced. As the temperature increased from 30 to 50 °C, the $Q_e$ of chrysin on the adsorbents decreased. Due to the hydrogen bonds of the adsorbents being broken with the increase in temperature, the adsorption capacity of chrysin by the adsorbents is weakened. From the point of view of economic and industrial applications, 30 °C is chosen as the optimum temperature. The energy consumption of the adsorption process is reduced, which builds the foundation for the large-scale extraction and separation of the binary functional monomer polymers in natural products.

### 3.3. Adsorption Selectivity

To further investigate the selectivity of the Bi-MIPs for chrysin, chloramphenicol and oxytetracycline were chosen as comparative substrates in the selective adsorption test. The results are shown in Figure 6b; the Bi-MIPs have the highest adsorption capacity for chrysin, followed by the Si-MIPs. This is due to the specific pockets of Bi-MIPs and Si-MIPs, which have specific adsorption on chrysin. The Bi-MIPs and Si-MIPs have electrostatic adsorption on chloramphenicol and oxytetracycline. The imprinting factor ($Q_{Bi-MIPs}/Q_{Bi-NIPs}$) of the Bi-MIPs is 1.54, and the imprinting factor ($Q_{Si-MIPs}/Q_{Si-NIPs}$) of the Si-MIPs is 1.42. These results indicate that, under the synergistic action of AM, the force of MAA on the hydrogen bonds, electrostatic and π-π stacking of chrysin increased [37–39], and the adsorption selectivity of molecularly imprinted polymers is improved. The selectivity of binary functional monomers in the separation and extraction of chrysin further proves that the Bi-MIPs have broad application prospects in the extraction and separation of natural products.

*3.4. Adsorption Reusability*

To evaluate the capacity of the Bi-MIPs to be regenerated and reused, the adsorption performance after repeated cycles was investigated. Multiple adsorptions and desorption experiments were carried out using a polymer chrysin solution, and the results are shown in Figure 7. After five times of adsorption and desorption, the remaining solid mass of the sample is 0.0402, 0.0356, 0.0356, 0.0317, and 0.0302. The solid loss rate is 24.88%. After being recycled and reused, the Bi-MIPs only lost 4.92% of adsorption capacity. This decrease may be attributed to the reduction in active binding loci following the regeneration and inadequate desorption of the adsorbed chrysin molecules. The results suggest that the Bi-MIPs exhibited excellent adsorption capability in all five cycles. At the same time, we should further consider how to improve the recovery efficiency of the Bi-MIPs.

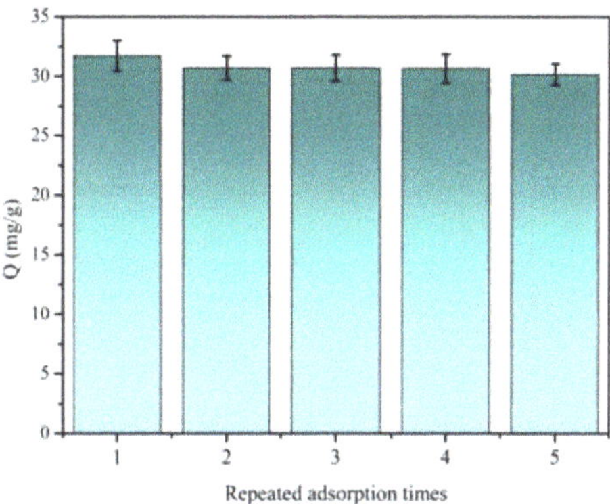

**Figure 7.** Regeneration rebinding performance of the Bi-MIPs.

## 4. Conclusions

The Bi-MIPs were prepared by precipitation polymerization using methacrylic acid and acrylamide as functional monomers and ethylene glycol dimethacrylate as a crosslinking agent. Their physicochemical properties and chemical structures were analyzed and characterized. The Bi-MIP has excellent adsorption performance for chrysin in methanol solution and has a larger specific surface area and thermodynamic properties. The adsorption of chrysin on the Bi-MIPs followed the PSO kinetic model and Langmuir isothermal model, and the adsorption process was dominated by homogeneous monolayer adsorption. Compared with the Si-MIPs, the Bi-MIPs showed excellent adsorption performance and specificity in the adsorption process of chrysin and its analogs. At the same time, the Bi-MIPs showed excellent adsorption performance in multiple cycles and were conducive to the extraction and purification of an organic solvent-soluble bioactive component from natural products. Under the synergistic action of AM, the force of MAA on the hydrogen bonds, electrostatic and $\pi$-$\pi$ stacking of chrysin increased. This synthetic method is green, efficient, eco-friendly and cost-effective. Thus, this paper provides new ideas and methods for the synthesis of high adsorption performance, green and safe molecular-imprinted polymers and build the foundation for the large-scale extraction and separation of the Bi-MIPs in chrysin.

**Author Contributions:** Methodology, L.L. (Long Li) and L.L. (Lanfu Li); formal analysis, L.L. (Long Li), X.L., G.C. (Guoning Chen) and Q.H.; investigation, L.L. (Long Li), L.L. (Lanfu Li), Q.H. and S.W.; data creation, L.L. (Lanfu Li), W.W., G.C. (Gege Cheng), Y.W., G.C. (Guoning Chen) and X.L.; writing original draft preparation, L.L. (Long Li), W.W.; writing review and editing, L.L. (Long Li), X.L. and

Polymers **2022**, 14, 2771

Q.H.; supervision, X.L.; project administration, Q.H., G.C. (Guoning Chen) and X.L. All authors have read and agreed to the published version of the manuscript.

**Funding:** The research was supported by the National Natural Science Foundation of China (31860192), the Natural Science Foundation of Guangxi (2020GXNSFBA159009), the China Postdoctoral Science Foundation (2020M683209), the Opening Project of Guangxi Key Laboratory of Clean Pulp & Paper-making and Pollution Control (2019KF28), the Scientific Research Foundation of Guangxi Minzu University (2019KJQD10), the Natural Science Funds of Guangdong Province (2021A1515012425), and the International Collaboration Programs of Guangdong Province (2020A0505100010).

**Institutional Review Board Statement:** Not applicable.

**Informed Consent Statement:** Informed consent was obtained from all subjects involved in the study.

**Conflicts of Interest:** The authors declare no conflict of interest.

## References


1. Ong, S.; Shanmugam, M.; Fan, L.; Fraser, S.; Arfuso, F.; Ahn, K.; Sethi, G.; Bishayee, A. Focus on Formononetin: Anticancer Potential and Molecular Targets. *Cancers* **2019**, *11*, 611. [CrossRef] [PubMed]
2. Zheng, X.; Wu, F.; Lin, X.; Shen, L.; Feng, Y. Developments in drug delivery of bioactive alkaloids derived from traditional Chinese medicine. *Drug Deliv.* **2018**, *25*, 398–416. [CrossRef]
3. Subramanya, S.B.; Venkataraman, B.; Meeran, M.F.N.; Goyal, S.N.; Patil, C.R.; Ojha, S. Therapeutic Potential of Plants and Plant Derived Phytochemicals against Acetaminophen-Induced Liver Injury. *Int. J. Mol. Sci.* **2018**, *19*, 3776. [CrossRef] [PubMed]
4. Ramesh, P.; Rao, V.S.; Hong, Y.-A.; Reddy, P.M. Molecular Design, Synthesis, and Biological Evaluation of 2-Hydroxy-3-Chrysino Dithiocarbamate Derivatives. *Molecules* **2019**, *24*, 3038. [CrossRef] [PubMed]
5. Al-Oudat, B.; Ramapuram, H.; Malla, S.; Audat, S.; Hussein, N.; Len, J.; Kumari, S.; Bedi, M.; Ashby, C.; Tiwari, A. Novel Chrysin-De-Allyl PAC-1 Hybrid Analogues as Anticancer Compounds: Design, Synthesis, and Biological Evaluation. *Molecules* **2020**, *25*, 3063. [CrossRef]
6. Liu, C.; Kou, X.; Wang, X. Novel chrysin derivatives as hidden multifunctional agents for anti-Alzheimer's disease: Design, synthesis and in vitro evaluation. *Eur. J. Pharm. Sci.* **2021**, *166*, 105976. [CrossRef]
7. Yu, C.-H.; Suh, B.; Shin, I.; Kim, E.-H.; Kim, D.; Shin, Y.-J.; Chang, S.-Y.; Baek, S.-H.; Kim, H.; Bae, O.-N. Inhibitory Effects of a Novel Chrysin-Derivative, CPD 6, on Acute and Chronic Skin Inflammation. *Int. J. Mol. Sci.* **2019**, *20*, 2607. [CrossRef]
8. Maruhashi, R.; Eguchi, H.; Akizuki, R.; Hamada, S.; Furuta, T.; Matsunaga, T.; Endo, S.; Ichihara, K.; Ikari, A. Chrysin enhances anticancer drug-induced toxicity mediated by the reduction of claudin-1 and 11 expression in a spheroid culture model of lung squamous cell carcinoma cells. *Sci. Rep.* **2019**, *9*, 13753. [CrossRef]
9. Xu, M.; Shi, H.; Liu, D. Chrysin protects against renal ischemia reperfusion induced tubular cell apoptosis and inflammation in mice. *Exp. Ther. Med.* **2019**, *17*, 2256–2262. [CrossRef]
10. Schindler, R.M.R. Flavonoids and vitamin E reduce the release of the angiogenic peptide vascular endothelial growth factor from human tumor cells. *J. Nutr.* **2006**, *136*, 1477–1482. [CrossRef]
11. Gilles Comte, J.B.D.; Bayet, C. C-Isoprenylation of Flavonoids Enhances Binding Affinity toward P-Glycoprotein and Modulation of Cancer Cell Chemoresistance. *J. Med. Chem.* **2001**, *44*, 763–768. [CrossRef] [PubMed]
12. Gharari, Z.; Bagheri, K.; Danafar, H.; Sharafi, A. Simultaneous determination of baicalein, chrysin and wogonin in four Iranian Scutellaria species by high performance liquid chromatography. *J. Appl. Res. Med. Aromat. Plants* **2019**, *16*, 100232. [CrossRef]
13. Gu, Y.; Chen, X.; Wang, R. Comparative two-dimensional HepG2 and L02/cell membrane chromatography/C18/time-of-flight mass spectrometry for screening selective anti-hepatoma components from Scutellariae Radix. *J. Pharm. Biomed. Anal.* **2019**, *164*, 550–556. [CrossRef] [PubMed]
14. Mohammad Reza, H.; Saman, N.; Kamyar, K. Determination of Flavonoid Markers in Honey with SPE and LC using Experimental Design. *Chromatographia* **2009**, *69*, 1291–1297.
15. Maciejewicz, W.; Daniewski, M.; Bal, K.; Markowski, W. GC-MS identification of the flavonoid aglycones isolated from propolis. *Chromatographia* **2001**, *53*, 343–346. [CrossRef]
16. Zhang, Y.; Wang, B.; Jia, Z.; Scarlett, C.J.; Sheng, Z. Adsorption/desorption characteristics and enrichment of quercetin, luteolin and apigenin from Flos populi using macroporous resin. *Rev. Bras. Farm.* **2018**, *29*, 69–76. [CrossRef]
17. Ma, Y.; Wang, H.; Guo, M. Stainless Steel Wire Mesh Supported Molecularly Imprinted Composite Membranes for Selective Separation of Ebracteolata Compound B from Euphorbia fischeriana. *Molecules* **2019**, *24*, 565. [CrossRef]
18. Kamaruzaman, S.; Nasir, N.M.; Faudzi, S.M.M.; Yahaya, N.; Hanapi, N.S.M.; Ibrahim, W.N.W. Solid-Phase Extraction of Active Compounds from Natural Products by Molecularly Imprinted Polymers: Synthesis and Extraction Parameters. *Polymers* **2021**, *13*, 3780. [CrossRef]
19. Ariani, M.D.; Zuhrotun, A.; Manesiotis, P.; Hasanah, A.N. Magnetic Molecularly Imprinted Polymers: An Update on Their Use in the Separation of Active Compounds from Natural Products. *Polymers* **2022**, *14*, 1389. [CrossRef]
20. Sun, X.; Zhang, Y.; Zhou, Y.; Lian, X.; Yan, L.; Pan, T.; Jin, T.; Xie, H.; Liang, Z.; Qiu, W.; et al. NPCDR: Natural product-based drug combination and its disease-specific molecular regulation. *Nucleic Acids Res.* **2021**, *50*, D1324–D1333. [CrossRef]

21. Wang, T.; Wang, Q.; Guo, Q.; Li, P.; Yang, H. A hydrophobic deep eutectic solvents-based integrated method for efficient and green extraction and recovery of natural products from Rosmarinus officinalis leaves, Ginkgo biloba leaves and Salvia miltiorrhiza roots. *Food Chem.* **2021**, *363*, 130282. [CrossRef] [PubMed]

22. Luliński, P.; Maciejewska, D. Examination of Imprinting Process with Molsidomine as a Template. *Molecules* **2009**, *14*, 2212–2225. [CrossRef] [PubMed]

23. Peeters, M.M.; Van Grinsven, B.; Foster, C.W.; Cleij, T.J.; Banks, C.E. Introducing Thermal Wave Transport Analysis (TWTA): A Thermal Technique for Dopamine Detection by Screen-Printed Electrodes Functionalized with Molecularly Imprinted Polymer (MIP) Particles. *Molecules* **2016**, *21*, 552. [CrossRef]

24. Zhang, W.; Li, Q.; Cong, J.; Wei, B.; Wang, S. Mechanism Analysis of Selective Adsorption and Specific Recognition by Molecularly Imprinted Polymers of Ginsenoside Re. *Polymers* **2018**, *10*, 216. [CrossRef] [PubMed]

25. Chen, J.; Bai, L.-Y.; Liu, K.-F.; Liu, R.-Q.; Zhang, Y.-P. Atrazine Molecular Imprinted Polymers: Comparative Analysis by Far-Infrared and Ultraviolet Induced Polymerization. *Int. J. Mol. Sci.* **2014**, *15*, 574–587. [CrossRef]

26. Si, Z.; Yu, P.; Dong, Y.; Lu, Y.; Tan, Z.; Yu, X.; Zhao, R.; Yan, Y. Thermo-Responsive Molecularly Imprinted Hydrogels for Selective Adsorption and Controlled Release of Phenol From Aqueous Solution. *Front. Chem.* **2019**, *6*, 674. [CrossRef]

27. Xing, R.; Ma, Y.; Wang, Y.; Wen, Y.; Liu, Z. Specific recognition of proteins and peptides *via* controllable oriented surface imprinting of boronate affinity-anchored epitopes. *Chem. Sci.* **2018**, *10*, 1831–1835. [CrossRef]

28. Piletska, E.; Yawer, H.; Canfarotta, F.; Moczko, E.; Smolinska-Kempisty, K.; Piletsky, S.S.; Guerreiro, A.; Whitcombe, M.J.; Piletsky, S.A. Biomimetic Silica Nanoparticles Prepared by a Combination of Solid-Phase Imprinting and Ostwald Ripening. *Sci. Rep.* **2017**, *7*, 11537. [CrossRef]

29. Liu, R.; Poma, A. Advances in Molecularly Imprinted Polymers as Drug Delivery Systems. *Molecules* **2021**, *26*, 3589. [CrossRef]

30. Soufi, G.J.; Iravani, S.; Varma, R.S. Molecularly imprinted polymers for the detection of viruses: Challenges and opportunities. *Analyst* **2021**, *146*, 3087–3100. [CrossRef]

31. Ramanavicius, S.; Jagminas, A.; Ramanavicius, A. Advances in Molecularly Imprinted Polymers Based Affinity Sensors (Review). *Polymers* **2021**, *13*, 974. [CrossRef] [PubMed]

32. Zhao, H.; Gao, W.-C.; Li, Q.; Khan, M.R.; Hu, G.-H.; Liu, Y.; Wu, W.; Huang, C.-X.; Li, R.K. Recent advances in superhydrophobic polyurethane: Preparations and applications. *Adv. Colloid Interface Sci.* **2022**, *303*, 102644. [CrossRef] [PubMed]

33. He, X.; Wang, Y.; Li, H. Specific recognition of protein by deep eutectic solvent-based magnetic beta-cyclodextrin molecularly imprinted polymer. *Mikrochim. Acta* **2021**, *188*, 232. [CrossRef] [PubMed]

34. Xie, L.; Xiao, N.; Li, L.; Xie, X.; Li, Y. Theoretical Insight into the Interaction between Chloramphenicol and Functional Monomer (Methacrylic Acid) in Molecularly Imprinted Polymers. *Int. J. Mol. Sci.* **2020**, *21*, 4139. [CrossRef] [PubMed]

35. Song, X.; Li, J.; Wang, J.; Chen, L. Quercetin molecularly imprinted polymers: Preparation, recognition characteristics and properties as sorbent for solid-phase extraction. *Talanta* **2009**, *80*, 694–702. [CrossRef]

36. Sobiech, M.; Luliński, P.; Wieczorek, P.P. Quantum and carbon dots conjugated molecularly imprinted polymers as advanced nanomaterials for selective recognition of analytes in environmental, food and biomedical applications. *TrAC Trends Anal. Chem.* **2021**, *142*, 116306. [CrossRef]

37. Wan, Y.; Wang, M.; Fu, Q.; Wang, L.; Wang, D.; Zhang, K.; Xia, Z.; Gao, D. Novel dual functional monomers based molecularly imprinted polymers for selective extraction of myricetin from herbal medicines. *J. Chromatogr. B* **2018**, *1097–1098*, 1–9. [CrossRef]

38. Zeng, H.; Wang, Y.; Liu, X.; Kong, J.; Nie, C. Preparation of molecular imprinted polymers using bi-functional monomer and bi-crosslinker for solid-phase extraction of rutin. *Talanta* **2012**, *93*, 172–181. [CrossRef]

39. Thach, U.D.; Thi, H.H.N.; Pham, T.D.; Mai, H.D.; Nhu-Trang, T.-T. Synergetic Effect of Dual Functional Monomers in Molecularly Imprinted Polymer Preparation for Selective Solid Phase Extraction of Ciprofloxacin. *Polymers* **2021**, *13*, 2788. [CrossRef]

40. Gao, W.-C.; Wu, W.; Chen, C.-Z.; Zhao, H.; Liu, Y.; Li, Q.; Huang, C.-X.; Hu, G.-H.; Wang, S.-F.; Shi, D.; et al. Design of a Superhydrophobic Strain Sensor with a Multilayer Structure for Human Motion Monitoring. *ACS Appl. Mater. Interfaces* **2021**, *14*, 1874–1884. [CrossRef]

41. Jiang, Y.; Wang, Z.; Zhou, L.; Jiang, S.; Liu, X.; Zhao, H.; Huang, Q.; Wang, L.; Chen, G.; Wang, S. Highly efficient and selective modification of lignin towards optically designable and multifunctional lignocellulose nanopaper for green light-management applications. *Int. J. Biol. Macromol.* **2022**, *206*, 264–276. [CrossRef] [PubMed]

42. Li, L.; Liu, X.; Li, L. Preparation of Rosin-Based Composite Membranes and Study of Their Dencichine Adsorption Properties. *Polymers* **2022**, *14*, 2161. [CrossRef] [PubMed]

43. Zahara, S.; Minhas, M.A.; Shaikh, H.; Ali, M.S.; Bhanger, M.I.; Malik, M.I. Molecular imprinting-based extraction of rosmarinic acid from Salvia hypoleuca extract. *React. Funct. Polym.* **2021**, *166*, 104984. [CrossRef]

44. Jiang, Y.; Wang, Z.; Liu, X.; Yang, Q.; Huang, Q.; Wang, L.; Dai, Y.; Qin, C.; Wang, S. Highly Transparent, UV-Shielding, and Water-Resistant Lignocellulose Nanopaper from Agro-Industrial Waste for Green Optoelectronics. *ACS Sustain. Chem. Eng.* **2020**, *8*, 17508–17519. [CrossRef]

45. Cheng, B.-X.; Gao, W.-C.; Ren, X.-M.; Ouyang, X.-Y.; Zhao, Y.; Zhao, H.; Wu, W.; Huang, C.-X.; Liu, Y.; Liu, X.-Y.; et al. A review of microphase separation of polyurethane: Characterization and applications. *Polym. Test.* **2022**, *107*, 107489. [CrossRef]

46. Azizi, S.; Shahri, M.M.; Mohamad, R. Green Synthesis of Zinc Oxide Nanoparticles for Enhanced Adsorption of Lead Ions from Aqueous Solutions: Equilibrium, Kinetic and Thermodynamic Studies. *Molecules* **2017**, *22*, 831. [CrossRef]

47. Bagbi, Y.; Sarswat, A.; Mohan, D.; Pandey, A.; Solanki, P.R. Lead and Chromium Adsorption from Water using L-Cysteine Functionalized Magnetite ($Fe_3O_4$) Nanoparticles. *Sci. Rep.* **2017**, *7*, 7672. [CrossRef]
48. Moon, G.H.; Jung, Y.; Shin, B. On-Chip Chemiresistive Sensor Array for On-Road NO x Monitoring with Quantification. *Adv. Sci.* **2020**, *7*, 2002014. [CrossRef]
49. Pham, T.; Bui, T.T.; Nguyen, V.T.; Van Bui, T.K.; Tran, T.T.; Phan, Q.C.; Hoang, T.H. Adsorption of Polyelectrolyte onto Nanosilica Synthesized from Rice Husk: Characteristics, Mechanisms, and Application for Antibiotic Removal. *Polymers* **2018**, *10*, 220. [CrossRef]
50. Liu, H.; Zhang, F.; Peng, Z. Adsorption mechanism of Cr(VI) onto GO/PAMAMs composites. *Sci. Rep.* **2019**, *9*, 3663. [CrossRef]
51. Duan, C.; Zhang, Y.; Li, J.; Kang, L.; Xie, Y.; Qiao, W.; Zhu, C.; Luo, H. Rapid Room-Temperature Preparation of Hierarchically Porous Metal–Organic Frameworks for Efficient Uranium Removal from Aqueous Solutions. *Nanomaterials* **2020**, *10*, 1539. [CrossRef] [PubMed]
52. Ni, X.; Zhao, Z.; Li, Z.; Li, Q. The adsorptive behaviour of kaolinite to sodium dodecyl benzene sulphonate and the structural variation of kaolinite. *Sci. Rep.* **2021**, *11*, 1796. [CrossRef] [PubMed]
53. Dasgupta-Schubert, N.; Tiwari, D.K.; Cendejas, L.M.V. Comment on 'Carbon and fullerene nanomaterials in plant system'. *J. Nanobiotechnol.* **2016**, *14*, 28. [CrossRef] [PubMed]
54. Raibaut, L.; Cargoët, M.; Ollivier, N.; Chang, Y.M.; Drobecq, H.; Boll, E.; Desmet, R.; Monbaliu, J.-C.M.; Melnyk, O. Accelerating chemoselective peptide bond formation using bis(2-selenylethyl)amido peptide selenoester surrogates. *Chem. Sci.* **2016**, *7*, 2657–2665. [CrossRef] [PubMed]

**Article**

# Improvement of Gas Barrier Properties for Biodegradable Poly(butylene adipate-co-terephthalate) Nanocomposites with MXene Nanosheets via Biaxial Stretching

Xionggang Wang, Xia Li, Lingna Cui, Yuejun Liu and Shuhong Fan *

Key Laboratory of Advanced Packaging Materials and Technology of Hunan Province, School of Packaging and Materials Engineering, Hunan University of Technology, Zhuzhou 412007, China; cheersun@126.com (X.W.); xiali_2019@163.com (X.L.); lncui@hut.edu.cn (L.C.); liuyuejun@hut.edu.cn (Y.L.)
* Correspondence: yangling2020@hut.edu.cn

**Abstract:** In order to ease the white pollution problem, biodegradable packaging materials are highly demanded. In this work, the biodegradable poly (butylene adipate-co-terephthalate)/MXene (PBAT/Ti$_3$C$_2$T$_X$) composite casting films were fabricated by melt mixing. Then, the obtained PBAT/Ti$_3$C$_2$T$_X$ composite casting films were biaxially stretched at different stretching ratios so as to reduce the water vapor permeability rate (WVPR) and oxygen transmission rate (OTR). It was expected that the combination of Ti$_3$C$_2$T$_X$ nanosheets and biaxial stretching could improve the water vapor and oxygen barrier performance of PBAT films. The scanning electron microscope (SEM) observation showed that the Ti$_3$C$_2$T$_X$ nanosheets had good compatibility with the PBAT matrix. The presence of Ti$_3$C$_2$T$_X$ acted as a nucleating agent to promote the crystallinity when the content was lower than 2 wt%. The mechanical tests showed that the incorporation of 1.0 wt% Ti$_3$C$_2$T$_X$ improved the tensile stress, elongation at break, and Young's modulus of the PBAT/Ti$_3$C$_2$T$_X$ nanocomposite simultaneously, as compared with those of pure PBAT. The mechanical dynamical tests showed that the presence of Ti$_3$C$_2$T$_X$ significantly improved the storage modulus of the PBAT nanocomposite in a glassy state. Compared with pure PBAT, PBAT-1.0 with 1.0 wt% Ti$_3$C$_2$T$_X$ exhibited the lowest OTR of 782 cc/m$^2$·day and 10.2 g/m$^2$·day. The enhancement in gas barrier properties can be attributed to the presence of Ti$_3$C$_2$T$_X$ nanosheets, which can increase the effective diffusion path length for gases. With the biaxial stretching, the OTR and WVPR of PBAT-1.0 were further reduced to 732 cc/m$^2$·day and 6.5 g/m$^2$·day, respectively. The PBAT composite films with enhanced water vapor and water barrier performance exhibit a potential application in green packaging.

**Keywords:** PBAT; MXene; nanocomposite; gas barrier properties; biaxial stretching

**Citation:** Wang, X.; Li, X.; Cui, L.; Liu, Y.; Fan, S. Improvement of Gas Barrier Properties for Biodegradable Poly(butylene adipate-co-terephthalate) Nanocomposites with MXene Nanosheets via Biaxial Stretching. *Polymers* **2022**, *14*, 480. https://doi.org/10.3390/polym14030480

Academic Editors: Wei Wu, Hao-Yang Mi, Chongxing Huang, Hui Zhao, Tao Liu and Alexey Iordanskii

Received: 3 December 2021
Accepted: 23 January 2022
Published: 25 January 2022

**Publisher's Note:** MDPI stays neutral with regard to jurisdictional claims in published maps and institutional affiliations.

## 1. Introduction

Green packaging materials are highly demanded in the recent years because of the ever-increasing plastic pollution problem. The traditional packaging materials, such as polyethylene (PE) and poly(vinyl chloride) (PVC), have been gradually replaced by biodegradable polymers such as poly(lactic acid) (PLA), poly(butylene adipate-co-terephthalate) (PBAT), and poly(butylene succinate) (PBS) [1–4]. Compared with other biodegradable polyesters, PBAT has adjustable properties due to the copolymerization of 1,4-butanediol, adipic acid, and terephthalic acid [5]. In addition, it has good ductility, good thermal resistance, and high impact performance, which is similar to PE. However, it was reported that the oxygen transmission rate (OTR) of PBAT under ambient conditions was around 1050 cc/m$^2$·day, whereas the water vapor permeability rate (WVPR) was 3.3 × 10$^{-11}$ g·m/m$^2$·s·Pa, which made it difficult to meet the requirements for packaging applications [6–8]. The poor oxygen and water vapor barrier performances limit the broad applications of PBAT in packaging. Therefore, it is necessary to improve the oxygen and water vapor barrier performance of PBAT so as to prolong the shelf life and maintain good quality of food.

It is widely accepted that the incorporation of nanofillers is a simple and effective method to reduce the OTR and WVPR of polymer films [9–12] because the presence of fillers can have barrier effects that increase the escape distance of oxygen and water molecules [13,14]. Li et al. reported that the well-aligned graphene nanosheets simultaneously reduced the oxygen permeability and enhanced the aging resistance of the PBAT composite film [15]. The oxygen-containing groups on graphene nanosheets enhanced the interactions between water molecules and altered the diffusion paths of water molecules. Mondal et al. found that the WVPR of PBAT could be reduced by 25% with the addition of 4 wt% organically modified montmorillonite (OMMT) [16]. This was because the impermeable OMMT in the PBAT matrix increased the tortuosity of the path for water molecules. Li et al. mixed graphene nanosheets with PBAT by the solution casting method [15]. The presence of graphene resulted in an 80% reduction in water permeation and a 99% reduction in oxygen transmission of PBAT nanocomposite films, which was ascribed to the fact that the graphene nanosheets enlarged the effective diffusion path length of water and oxygen across the films.

MXene is a novel family of (2D) transition metal carbides and/or nitrides [17–21]. The abundance of functional groups on the surface of MXene, such as oxygen (=O), hydroxyl (-OH), or fluorine (-F), endows it with good compatibility with many polar polymer matrices [22]. MXene has attracted considerable research interest for various applications, such as energy storage, sensors, electromagnetic shielding, and so on [23–26]. $Ti_3C_2T_X$ nanosheets have high stiffness and strength, which can serve as effective, reinforced fillers to improve the mechanical properties of polymer/$Ti_3C_2T_X$ nanocomposites. In addition, Wu et al. demonstrated that a small amount of $Ti_3C_2T_X$ improved the complex viscosity and storage modulus of PVDF nanocomposites significantly [22]. The ultrahigh molecular weight polyethylene (UMWPE) composites containing 0.75 wt% $Ti_3C_2T_X$ had the best creep performance [27]. With the addition of 1.9 vol% of MXene, the $Ti_3C_2T_X$/polystyrene nanocomposites exhibited a 54% higher storage modulus than that of neat polystyrene [28]. Yu et al. demonstrated that the addition of MXene improved the thermal stability and fire safety of polystyrene [29]. However, to the best of our knowledge, the PBAT/$Ti_3C_2T_X$ nanocomposites have not been reported on yet. It is expected that the impermeable $Ti_3C_2T_X$ nanosheets in the PBAT matrix via biaxial stretching can not only improve the gas barrier performance, but also enhance the thermal stability and stiffness of the nanocomposites.

Biaxial stretching processing is the process of stretching hot polymeric films in the cross-machine direction. It is also reported that the biaxial stretching process endows the polymer matrix with an ordered structure and improved gas barrier properties [30–32]. This is because the biaxial stretching can help to reduce the free volume of the amorphous region of the polymers, resulting in an enhancement in gas barrier performance [33]. In addition, the biaxial stretching can help the 2D filler in the polymer matrices form an orientation structure, which can benefit for the enhancement of the gas barrier performance [34]. Li et al. reported that the orientated OMMT in the PBAT matrix prepared by biaxial stretching significantly reduced the WVPR and caused a 99% reduction in OTR with an enhancement in elongation at break [15]. Yoksan et al. demonstrated that the stretched PLA/PBAT/thermoplastic starch composite films had stacked-layer planar morphology, which contributed to the improvement in crystallinity, impact strength, water vapor, and oxygen barrier properties [35].

## 2. Materials and Methods

### 2.1. Materials

Poly(butylene adipate-co-terephthalate) (PBAT) pellets were supplied by BASF. The melt flow index (190 °C, 2.16 kg) was 5.0 g/10 min. The MXene ($Ti_3C_2T_X$, 400 mesh, purity > 99%) nanosheets etched by HF were supplied by 11 Technology Co., Ltd. (Changchun, China). The specific surface area of the obtained $Ti_3C_2T_X$ was approximately 31.5 m$^2$/g.

## 2.2. Preparation of PBAT/Ti$_3$C$_2$T$_X$ Nanocomposite Biaxial Stretching Films

The preparation diagram of PBAT/Ti$_3$C$_2$T$_X$ nanocomposite biaxial stretching films is shown in Scheme 1. To achieve a better dispersion of Ti$_3$C$_2$T$_X$ nanosheets in the PBAT matrix, Ti$_3$C$_2$T$_X$ was first mixed with PBAT pellets by melt compounding via a twin extruder to obtain composite casting films. Then the PBAT/Ti$_3$C$_2$T$_X$ nanocomposite casting films were biaxially stretched. The effects of Ti$_3$C$_2$T$_X$ content on the morphology, thermal stability, and crystallization behavior, as well as mechanical properties of PBAT/Ti$_3$C$_2$T$_X$ nanocomposites, were comprehensively evaluated. The PBAT nanocomposite films containing the optimized 1 wt% Ti$_3$C$_2$T$_X$ were biaxially stretched under different parameters.

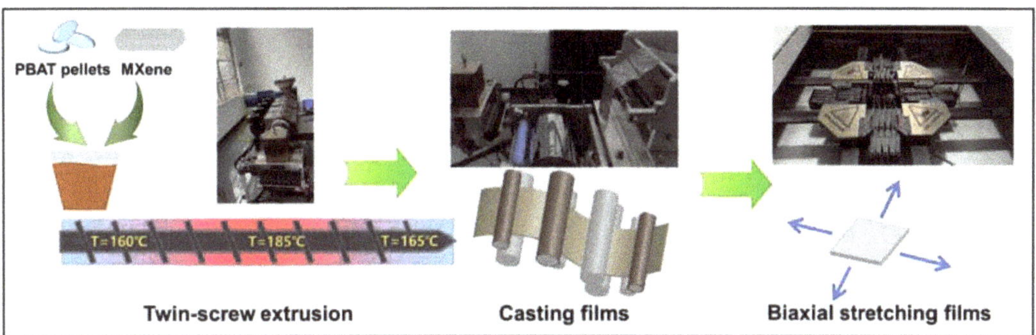


**Scheme 1.** Schematic diagram of preparation of PBAT/Ti$_3$C$_2$T$_X$ nanocomposite biaxial stretching films.

The PBAT/Ti$_3$C$_2$T$_X$ nanocomposite biaxial stretching films were prepared by two steps. Prior to the experiment, the PBAT pellets were dried in a vacuum oven at 80 °C for 12 h. Firstly, the PBAT/Ti$_3$C$_2$T$_X$ nanocomposite casting film was prepared by extrusion casting (FDHU-35, Potpo, Guangzhou, China) so that the Ti$_3$C$_2$T$_X$ could have a better dispersion in the PBAT matrix. The temperatures from the hopper to the nozzle were set at 130-150-170-170-170-150 °C. The screw speed was set at 60 rpm. The thickness of the obtained PBAT/Ti$_3$C$_2$T$_X$ nanocomposite casting films was approximately 100 μm. The code of the samples was abbreviated as PBAT-X, where X represents the weight ratio of Ti$_3$C$_2$T$_X$ in the PBAT/Ti$_3$C$_2$T$_X$ nanocomposites.

The extruded PBAT nanocomposite casting films containing 1 wt% Ti$_3$C$_2$T$_X$ nanosheets were cut into squares (80 mm side length) for biaxial stretching. The biaxially oriented films were then prepared at different stretching ratios on a KARO 5.0 (Brückner MaschinenbauGmbH & Co., Siegsdorf, Germany) equipped with mechanically driven clamps. The stretched films were thermally annealed at a temperature of 110 °C for 60 s. Finally, the biaxially oriented nanocomposite films were obtained to characterize the crystal orientation and the gas barrier properties.

## 2.3. Characterization

The fracture surfaces of the PBAT/Ti$_3$C$_2$T$_X$ nanocomposite were characterized by scanning electron microscopy (SEM, Quanta 250, FEI, Hillsboro, OR, USA). The samples were fractured in liquid nitrogen for 30 min. They were observed at an accelerating voltage of 5 kV. Prior to the observation, all the samples were coated with a layer of gold.

Thermogravimetric analysis (TGA) of all PBAT/Ti$_3$C$_2$T$_X$ nanocomposites was conducted on a TG-209F1 thermal analyzer (Netzsch, Selb, Germany) to measure the thermal stability under air atmosphere. The samples of about 10 mg were heated from room temperature to 600 °C at a heating rate of 10 °C/min.

The crystallization and melting behaviors of PBAT/Ti$_3$C$_2$T$_X$ nanocomposites were conducted on a DSC-204F1 (Netzsch, Selb, Germany). The samples of approximately 5 mg were first heated to 180 °C at a heating rate of 10 °C/min to establish the thermal history, then cooled to 20 °C at a cooling rate of 10 °C/min and followed by a second heating rate of 10 °C/min to 180 °C. The peak crystallization temperature ($T_{cp}$), the peak melting temperature ($T_{mp}$), the crystallization enthalpy ($\Delta H_c$), and the melting enthalpy ($\Delta H_m$) of these samples were summarized. The degree of crystallinity of PBAT ($\chi_c$) was calculated by the following equation:

$$\chi_c = \frac{\Delta H_m}{\Delta H_0 (1 - \varphi_c)} \tag{1}$$

where $\Delta H_0$ is the 100% melting enthalpy of PBAT (114 J/g) [36], and $\varphi_c$ presents the weight ratio of $Ti_3C_2T_X$ in the nanocomposites.

The tensile test was performed on an Instron 5566 universal electron tensile machine. The specimens were cut into rectangle shape with a dimension of 1 cm × 8 cm × 100 μm. The tensile speed was fixed at 10 mm/min. The reported results were the average values of at least five successful specimens.

The dynamical mechanical analysis was conducted on a Netzsch DMA 242E (Netzsch, Selb, Germany) analyzer. Tensile measurements were taken on specimens with dimensions of 30 mm at a fixed frequency of 1 Hz and from 90 °C to 70 °C at a ramping rate of 3 °C/min.

The 2D, wide-angle, X-ray scattering (2D-WAXS) measurements were carried out on an X-ray diffractometer (Rigaku UltimaIV, The Woodlands, TX, USA). The data were collected in the scanning range of 10–60° with a step of 0.02°.

The oxygen transmission rate (OTR) of the oriented films was measured with a MO-CON OX-TRAN (Ametek Mocon, Brooklyn Park, MN, USA) 2/21 at 23 °C, 1 atm, and 85% RH. The water vapor transmission rate (WVTR) was determined according to ASTM E96-80 at MOCON PERMATRAN-W 3/33 (Ametek Mocon, Brooklyn Park, MN, USA). The reported values were the average results of three successful samples.

## 3. Results and Discussions

### 3.1. Morphology

Figure 1 shows the SEM images of the cross-section fracture surfaces of PBAT/$Ti_3C_2T_X$ nanocomposite films. In Figure 1a,b, PBAT-0 exhibits a relatively smooth surface due to its brittle fracture after immersion in liquid nitrogen [37]. With the addition of $Ti_3C_2T_X$ nanosheets, the surfaces of the PBAT/$Ti_3C_2T_X$ nanocomposites become gradually rough in Figure 1c–f when the $Ti_3C_2T_X$ content is lower than 2.0 wt%. This is due to the presence of $Ti_3C_2T_X$, which served as a rigid filler to transfer the stress during facture. In addition, it can be seen that an agglomeration phenomenon existed on the surface of PBAT-2.0 (Figure 1g,h). In Figure 1, no pores and holes are observable between the exposed $Ti_3C_2T_X$ nanosheets on the surfaces (Figure 1f,h) and the PBAT matrix, indicating that $Ti_3C_2T_X$ nanosheets have good compatibility with the PBAT matrix. It is attributed to the large number of polar groups of $Ti_3C_2T_X$ that can react with the polyester groups of PBAT.

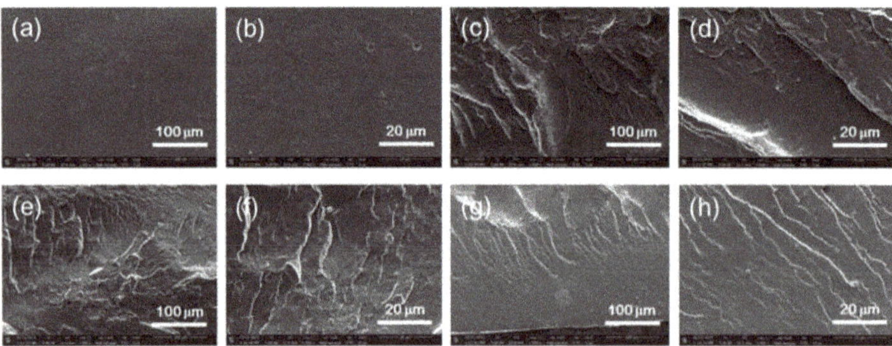



**Figure 1.** SEM images of the cross-section PBAT/$Ti_3C_2T_X$ nanocomposite films. (**a,b**) PBAT-0, (**c,d**) PBAT-0.5, (**e,f**) PBAT-1.0, and (**g,h**) PBAT-2.0.

### 3.2. Thermal Stability

The thermal stability of PBAT/Ti$_3$C$_2$T$_X$ nanocomposites is shown in Figure 2, and the corresponding data, including the temperatures at 10% weight loss ($T_{10}$), the temperatures at the maximum weight loss rate ($T_{max}$), and the char yields at 600 °C, are listed in Table 1. In Figure 2a, two thermal decomposition stages are observable for all the samples. The first stage between 300 °C and 420 °C can be attributed to the random, main-chain scission and thermo-oxidative reactions of PBAT [38]. The second stage, in the range of 420–550 °C, corresponds to cis-elimination and thermo-oxidative reactions [32]. In Table 1, it can be seen that the values of $T_{10}$ showed an increasing trend with the increase of Ti$_3$C$_2$T$_X$ content. When compared to PBAT-0, the $T_{10}$ of PBAT-2.0 dropped from 373.5 °C to 379.2 °C. In addition, the $T_p$ of PBAT-2.0 gradually increased to 412.5 °C with the addition of 2 wt% Ti$_3$C$_2$T$_X$. That is because the presence of Ti$_3$C$_2$T$_X$ rapidly catalyzed the formation of a char layer that served as a thermal barrier to protect the underlying polymer matrices [39]. The improvement of char yield benefited from the isolating of volatile gases and oxygen; therefore, improving the thermal stability of PBAT. Furthermore, the char yield at 600 °C for PBAT-0, PBAT-0.5, PBAT-1.0, and PBAT-2.0 was 0.7%, 1.0%, 1.4%, and 1.7%, respectively. It was mainly due to the introduction of Ti$_3$C$_2$T$_X$, which promoted charring, and partial polymers could not be completely thermally decomposed, resulting in enhanced char residues [40].

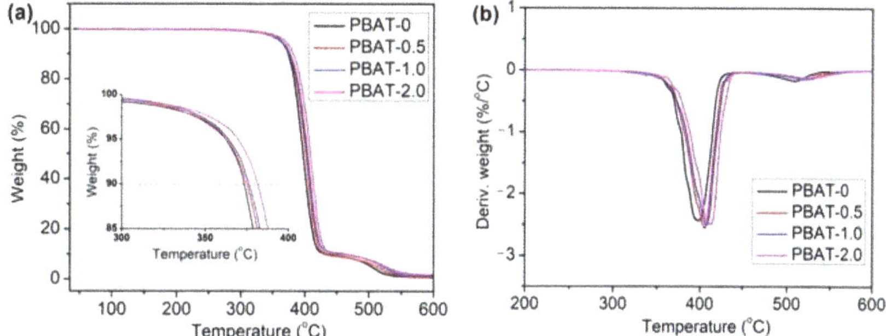


**Figure 2.** TGA curves of PBAT/Ti$_3$C$_2$T$_X$ nanocomposites at air atmospheres. (**a**) TGA and (**b**) DTG.

**Table 1.** Thermal stability of PBAT/Ti$_3$C$_2$T$_X$ nanocomposite films.

| Samples | $T_{10}$ (°C) | $T_p$ (°C) | Char Yield at 600 °C (wt%) |
|---|---|---|---|
| PBAT-0 | 373.5 | 398.3 | 0.7 |
| PBAT-0.5 | 375.1 | 404.7 | 1.0 |
| PBAT-1.0 | 376.4 | 408.5 | 1.4 |
| PBAT-2.0 | 379.2 | 412.5 | 1.7 |

### 3.3. Crystallization and Melting Behavior

The DSC curves of the PBAT/Ti$_3$C$_2$T$_X$ nanocomposites are shown in Figure 3. In Figure 3a, the onset crystallization temperatures of PBAT nanocomposites exhibit an increasing trend with the increase of Ti$_3$C$_2$T$_X$ content. In addition, the values of $T_{cp}$ for PBAT nanocomposites in Table 2 are 72.1, 73.1, 73.7, and 75.2 °C, respectively. The increase in $T_{cp}$ indicated that the presence of Ti$_3$C$_2$T$_X$ had a heterogeneous nucleation effect, accelerating the formation of crystallites in the PBAT matrix during cooling [41]. It was noted that the values of $\Delta H_m$ were lower than $\Delta H_c$ for the PBAT/Ti$_3$C$_2$T$_X$ composites, which can be ascribed to the fast cooling rate. In Figure 3b, the values of $T_{mp}$ for PBAT nanocomposites are 119.3, 121.0, 121.6, and 121.0 °C, respectively. The increase in $T_{mp}$ suggests that the filling Ti$_3$C$_2$T$_X$ contributes to the formation of perfect crystallinity of PBAT during the cooling procedure. Moreover, the crystallization degree of PBAT-1.0 had the highest value of 13.4% with the addition of 1.0 wt% Ti$_3$C$_2$T$_X$. When the content of Ti$_3$C$_2$T$_X$ was further increased up to 2.0 wt%, the crystallization degree of PBAT-2.0 showed a slight decrease.

This may be due to the excessive addition of $Ti_3C_2T_X$, which led to agglomeration, to some extent. On the other hand, the inhibition effects of the excessive $Ti_3C_2T_X$ nanosheets were more profound than the nucleating effect that led to smaller crystallization and decreased crystallinity [42].

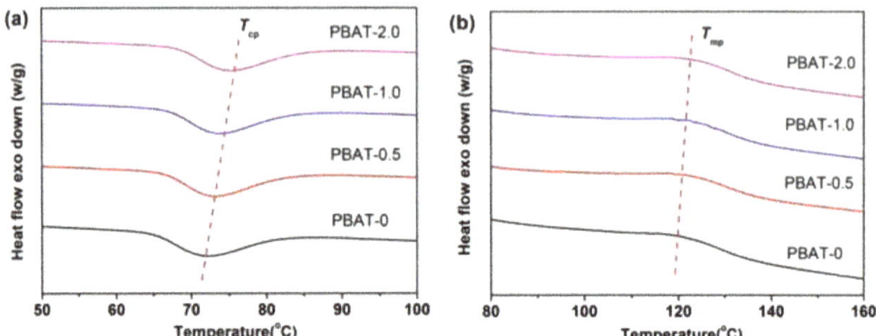


**Figure 3.** Differential scanning calorimetry (DSC) thermograms of PBAT/$Ti_3C_2T_X$ nanocomposites. (**a**) first cooling, (**b**) second heating.

**Table 2.** DSC thermograms of PBAT/$Ti_3C_2T_X$ nanocomposites.

| Samples | $T_{cp}$ (°C) | $\Delta H_c$ (J/g) | $T_{mp}$ (°C) | $\Delta H_m$ (J/g) | $\chi$ (%) |
|---|---|---|---|---|---|
| PBAT-0 | 72.1 | 16.8 | 119.3 | 14.6 | 12.8 |
| PBAT-0.5 | 73.1 | 17.0 | 121.0 | 14.7 | 13.0 |
| PBAT-1.0 | 73.7 | 16.8 | 121.6 | 15.1 | 13.4 |
| PBAT-2.0 | 75.2 | 16.9 | 121.0 | 14.2 | 12.7 |

*3.4. Mechanical Properties of Casting Films*

Figure 4 shows the typical stress–strain curves for pure PBAT and PBAT/$Ti_3C_2T_X$ nanocomposite casting films, and the corresponding data are summarized in Table 3. It was observed that PBAT-0 exhibited a high ductility (elongation at break ~ 1442%) but low tensile strength (~22.6 MPa), which is consistent with previous report [14]. With the addition of 0.5 wt% $Ti_3C_2T_X$, the tensile strength of the PBAT/$Ti_3C_2T_X$ nanocomposite increased by 19.8% with a slight increase in the elongation at break. As depicted in Figure 1, the $Ti_3C_2T_X$ had good interfacial interaction with the PBAT matrix; therefore, the addition of $Ti_3C_2T_X$ nanosheets can transform the stress during the tensile process. When the $Ti_3C_2T_X$ content was increased to 1.0%, the PBAT/$Ti_3C_2T_X$ nanocomposite had the maximum tensile strength (31.6 MPa). This enhancement can be ascribed to the reinforcement effects of the nanofillers and the interaction between the stress concentration zones around the $Ti_3C_2T_X$ nanosheets [43,44]. It is worth noting that PBAT-2.0 showed a decreasing tendency in both tensile stress and elongation at break as compared with PBAT-1.0. This may be due to the aggregation of $Ti_3C_2T_X$ nanosheets in the PBAT matrix.

**Table 3.** Tensile properties of PBAT/$Ti_3C_2T_X$ nanocomposite casting films.

| Samples | Tensile Stress (MPa) | Young's Modulus (MPa) | Elongation at Break (%) |
|---|---|---|---|
| PBAT-0 | 22.6 ± 3.2 | 24.5 ± 4.1 | 1442.3 ± 104.5 |
| PBAT-0.5 | 27.1 ± 3.6 | 28.6 ± 2.0 | 1524.8 ± 98.7 |
| PBAT-1.0 | 31.6 ± 4.7 | 31.4 ± 2.9 | 1483.2 ± 132.4 |
| PBAT-2.0 | 24.3 ± 5.1 | 32.1 ± 3.3 | 1350.3 ± 329.7 |

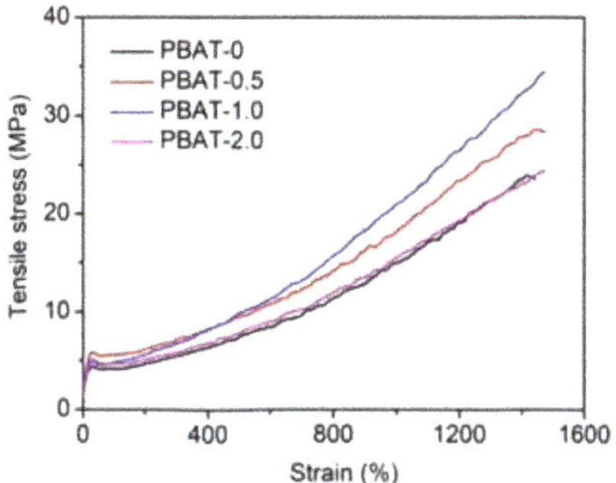

**Figure 4.** Tensile stress versus strain of PBAT/Ti$_3$C$_2$T$_X$ nanocomposite casting films.

### 3.5. Dynamic Mechanical Analysis

The storage modulus as a function of temperature for PBAT/Ti$_3$C$_2$T$_X$ nanocomposite casting films is shown in Figure 5a. It is clear that the storage modulus of the PBAT nanocomposite reinforced with Ti$_3$C$_2$T$_X$ was higher than that of PBAT-0 in the glassy states. In addition, the reinforcement effect was more obvious with the increase of Ti$_3$C$_2$T$_X$ content. When compared to PBAT-0, the storage modulus of PBAT-2.0 at 80 °C increased from 1220 MPa to 2342 MPa. This is due to the stiffening effect of rigid Ti$_3$C$_2$T$_X$ nanosheets. Aside from this, the polar groups on the surface of Ti$_3$C$_2$T$_X$ may have had intramolecular interaction with the PBAT. matrix, which may have also improved the storage modulus of the PBAT/Ti$_3$C$_2$T$_X$ nanocomposites [43]. The loss factor peak (tan$\delta$) is usually defined as the glass transition temperature ($T_g$). It is observable from Figure 5b that the $T_g$ shifted to a lower temperature when the PBAT matrix was incorporated with Ti$_3$C$_2$T$_X$. With the addition of 2 wt% Ti$_3$C$_2$T$_X$, the PBAT-2.0 shifted from −11.9 °C to −15.0 °C, as compared with that of PBAT-0. It can be attributed to the incorporation of Ti$_3$C$_2$T$_X$, which can improve the chain mobility of the amorphous regions of PBAT due to the liberation effect of Ti$_3$C$_2$T$_X$. In addition, the height of tan$\delta$ also showed a slight increase, indicating that an increase in Ti$_3$C$_2$T$_X$ content will result in higher dissipative energy [45].

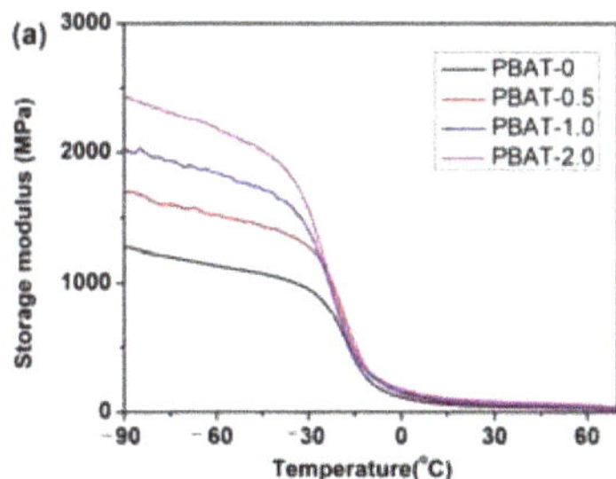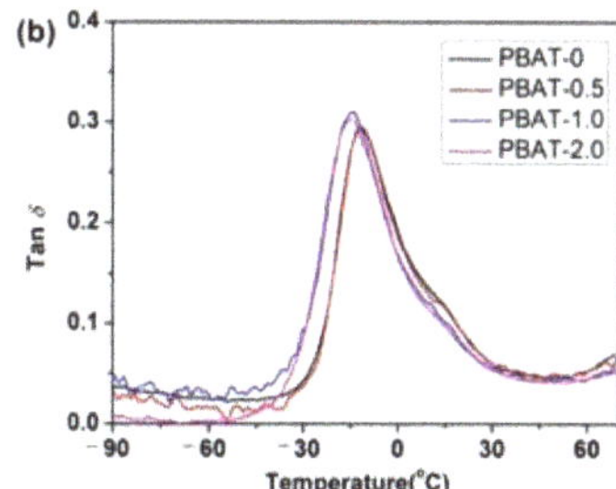

**Figure 5.** (**a**) Storage modulus and (**b**) loss factor of PBAT/Ti$_3$C$_2$T$_X$ nanocomposite casting films.

### 3.6. 2D-WAXS Patterns of Biaxial Stretching Films

Figure 6 shows the 2D-WAXS images of the PBAT-1.0 casting films under different biaxial stretching ratios. It is observable that there are four crystal planes (111), (100), (110), and (010) in the PBAT film (1 × 1) in Figure 6a, and these crystal planes belong to the PBAT phase [10]. With the increase of the stretching ratio in the machine direction (MD), the crystal planes (111), (100), (110), and (010) in the PBAT composite films (Figure 6b,c) had more obvious orientation. In addition, the larger the stretching ratio, the more obvious the orientation effect, which indicates that uniaxial stretching can promote the orientation of the PBAT/Ti$_3$C$_2$T$_X$ biaxial stretching films' crystal form along the MD. In Figure 6d,e, there is no obvious crystal orientation in the 2D-WAXS diffraction pattern in the biaxially stretched PBAT/Ti$_3$C$_2$T$_X$ films, indicating that the biaxial stretching will not cause the film to have an obvious crystal orientation in a certain direction. The crystal orientation of PBAT/Ti$_3$C$_2$T$_X$ composite films further confirms that the biaxially oriented PBAT/Ti$_3$C$_2$T$_X$ film has excellent isotropy.

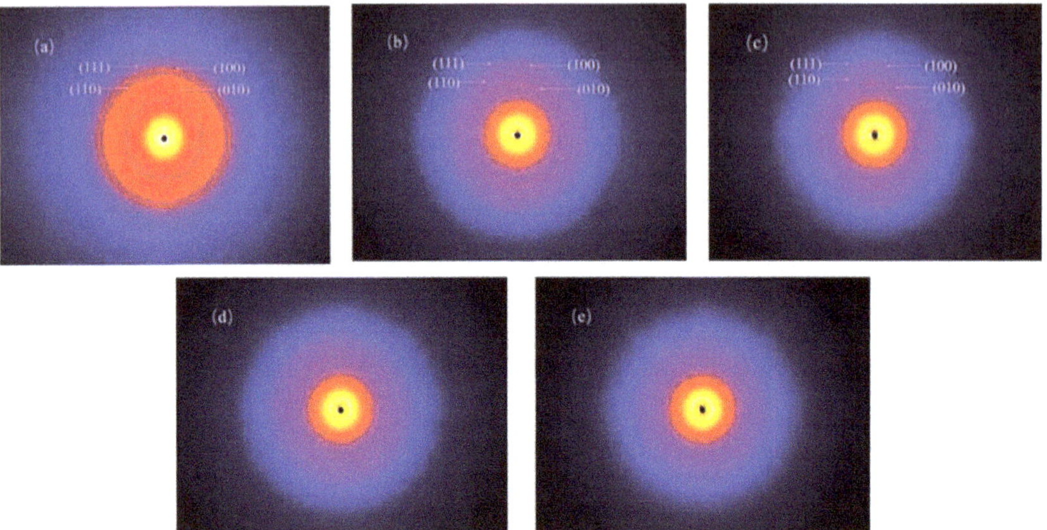


**Figure 6.** The films of PBAT-1.0 under different biaxial stretching ratio (transverse direction × machine direction). (**a**) 1 × 1, (**b**) 1 × 2, (**c**) 1 × 3, (**d**) 1.5 × 1.5, and (**e**) 2 × 2.

*3.7. Gas Barrier Properties of Biaxial Stretching Films*

The gas barrier properties of PBAT/Ti$_3$C$_2$T$_X$ nanocomposite casting films are shown in Figure 7. In Figure 7a, the OTR of PBAT nanocomposite casting films shows a decreasing trend with the increase of Ti$_3$C$_2$T$_X$ content. The lowest OTR was achieved for PBAT-1.0, which decreased from 1030 to 782 cc/m$^2$·day. Similarly, the water vapor transmission rate (WVTR) of PBAT/Ti$_3$C$_2$T$_X$ nanocomposite casting films decreased as the Ti$_3$C$_2$T$_X$ content increased in the PBAT matrix. In Figure 7b, the WVTR for PBAT-0, PBAT-0.5, PBAT-1.0, and PBAT-2.0 is determined to be 14.3, 12.7, 10.2, and 11.7 g/m$^2$·day, respectively. It is speculated that the addition of Ti$_3$C$_2$T$_X$ nanosheets can serve as a barrier to form a tortuous path, increasing the effective diffusion path length. Furthermore, the abundant hydroxyl groups on the surface of Ti$_3$C$_2$T$_X$ will contribute to the interactions with water molecules, delaying the diffusion to some extent. However, the aggregation of Ti$_3$C$_2$T$_X$ will result in a deterioration of the gas barrier performance when the content of Ti$_3$C$_2$T$_X$ is increased by up to 2.0 wt%.

To investigate the effects of the stretching ratio on the gas barrier performance of PBAT/Ti$_3$C$_2$T$_X$ nanocomposite stretching films, the OTR and WVTR data of PBAT-1.0 stretching film under different stretching ratios are shown in Figure 8. In Figure 8a, it is clear that the OTR of PBAT-1.0 stretching film decreased from 782 to 732 cc/m$^2$·day with the stretching ratio increasing to 3 under uniaxial stretching. This can be attributed to the enhanced orientation of PBAT crystallites formed during the uniaxial stretching process, which is demonstrated in 2D-WAXS patterns (Figure 6). Meanwhile, the WVTR for PBAT-1.0 stretching film achieved the lowest value of 6.5 g/m$^2$·day under 2 × 2 biaxial stretching condition, shown in Figure 8b. This is because the biaxial stretching process can contribute to the formation of an amorphous phase of PBAT and the exfoliation of Ti$_3$C$_2$T$_X$ sheets. However, the barrier effect of Ti$_3$C$_2$T$_X$ sheets is more profound than the effect of the PBAT amorphous phase, resulting in a further decrease in OTR and WVTR. The combination of two-dimensional, inorganic nanofillers with the biaxial stretching process paves the way for the preparation of a biodegradable polymer with enhanced gas barrier performance.

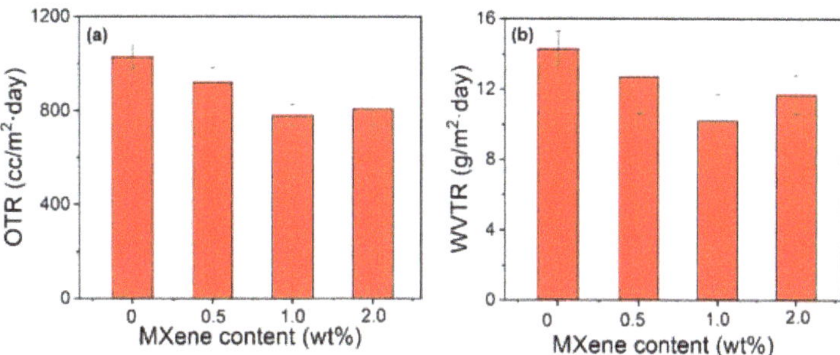

**Figure 7.** The gas barrier properties of PBAT/Ti$_3$C$_2$T$_X$ nanocomposite casting films. (**a**) OTR, (**b**) WVTR.

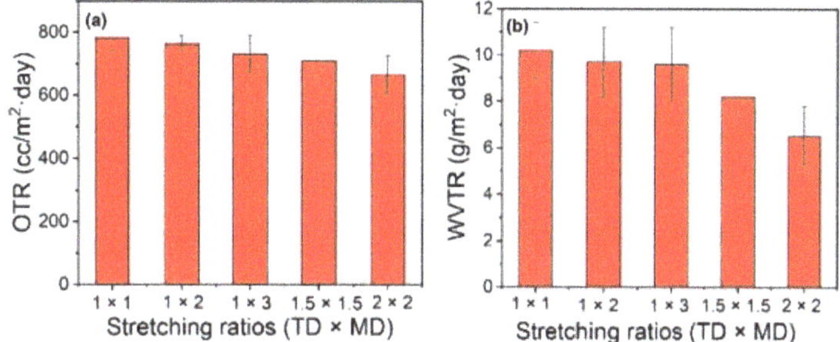

**Figure 8.** The gas barrier properties of PBAT-1.0 films under different biaxial stretching ratio. (**a**) OTR, (**b**) WVTR.

## 4. Conclusions

In this work, two-dimensional MXene (Ti$_3$C$_2$T$_X$) nanosheets were mixed with PBAT by melt compounding. The effects of Ti$_3$C$_2$T$_X$ content on the morphology, thermal stability, crystallization behavior, and gas barrier performance of PBAT were investigated. Furthermore, the effects of the biaxial stretching ratio on the gas barrier properties were further discussed. The TGA results showed that the addition of Ti$_3$C$_2$T$_X$ improved the thermal stability of the PBAT nanocomposite. In addition, the tensile tests showed that the addition of 1.0 wt% Ti$_3$C$_2$T$_X$ improved the maximum tensile stress without losing ductility. The storage modulus of PBAT was significantly improved in the glassy state with the addition of Ti$_3$C$_2$T$_X$. After biaxial stretching, the PBAT-1.0 film (1 × 3) exhibited an oxygen transmission rate of 732 cc/m$^2$·day, which was 28.9% lower than that of pure PBAT casting film. When the stretching ratio was 2 × 2, the WVTR of PBAT-1.0 biaxial stretching film was 6.5 g/m$^2$·day, which was 36.3% lower than that of 1 × 1 PBAT-1.0 film. The enhancement in gas barrier properties can be attributed to the presence of Ti$_3$C$_2$T$_X$ nanosheets, which can increase the effective diffusion path length for gases. The results of this work indicate the need for further studies on the influence of the orientation and surface functionalization of Ti$_3$C$_2$T$_X$ nanosheets, as well as the incorporation of compatbilizers in the PBAT composite films for packaging applications.

**Author Contributions:** Investigation, X.W., X.L., L.C. and S.F.; writing—original draft preparation, X.W. and X.L.; methodology, X.W. and X.L.; writing—review and editing, Y.L. and S.F.; supervision, Y.L. and S.F. All authors have read and agreed to the published version of the manuscript.

**Funding:** This research was funded by the Natural Science Foundation of Hunan Province (no. 2019JJ50132) and the Innovation Platform Open Fund of Hunan Province (no. 18K079).

**Institutional Review Board Statement:** This study did not involve humans or animals.

**Informed Consent Statement:** This study did not involve patients.

**Data Availability Statement:** The raw/processed data required to reproduce these findings cannot be shared at this time as the data also form part of an ongoing study.

**Conflicts of Interest:** The authors declare no conflict of interest.

# References


1. Ellingford, C.; Samantaray, P.K.; Farris, S.; McNally, T.; Tan, B.W.; Sun, Z.Y.; Huang, W.J.; Ji, Y.; Wan, C.Y. Reactive extrusion of biodegradable PGA/PBAT blends to enhance flexibility and gas barrier properties. *J. Appl. Polym. Sci.* **2021**, *139*, e51617. [CrossRef]
2. Cao, X.W.; Chi, X.N.; Deng, X.Q.; Sun, Q.J.; Gong, X.J.; Yu, B.; Yuen, A.C.Y.; Wu, W.; Li, R.K.Y. Facile synthesis of phosphorus and cobalt co-doped graphitic carbon nitride for fire and smoke suppressions of polylactide composite. *Polymers* **2020**, *12*, 1106. [CrossRef]
3. Karkhanis, S.S.; Stark, N.M.; Sabo, R.C.; Matuana, L.M. Potential of extrusion-blown poly(lactic acid)/cellulose nanocrystals nanocomposite films for improving the shelf-life of a dry food product. *Food Packag. Shelf Life* **2021**, *29*, 100689. [CrossRef]
4. Cao, X.W.; Huang, J.S.; He, Y.; Hu, C.Y.; Zhang, Q.C.; Yin, X.M.; Wu, W.; Li, R.K.Y. Biodegradable and renewable UV-shielding polylactide composites containing hierarchical structured POSS functionalized lignin. *Int. J. Biol. Macromol.* **2021**, *188*, 323–332. [CrossRef]
5. Huang, F.F.; Wu, L.B.; Li, B.G. Sulfonated biodegradable PBAT copolyesters with improved gas barrier properties and excellent water dispersibility: From synthesis to structure-property. *Polym. Degrad. Stab.* **2020**, *182*, 109391. [CrossRef]
6. Venkatesan, R.; Rajeswari, N. ZnO/PBAT nanocomposite films: Investigation on the mechanical and biological activity for food packaging. *Polym. Adv. Technol.* **2017**, *28*, 20–27. [CrossRef]
7. Li, J.X.; Lai, L.; Wu, L.B.; Severtson, S.J.; Wang, W.J. Enhancement of water vapor barrier properties of biodegradable poly(butylene adipate-co-terephthalate) films with highly oriented organomontmorillonite. *ACS Sustain. Chem. Eng.* **2018**, *6*, 6654–6662. [CrossRef]
8. Bumbudsanpharoke, N.; Wongphan, P.; Promhuad, K.; Leelaphiwat, P.; Harnkarnsujarit, N. Morphology and permeability of bio-based poly (butylene adipate-co-terephthalate)(PBAT), poly (butylene succinate)(PBS) and linear low-density polyethylene (LLDPE) blend films control shelf-life of packaged bread. *Food Control* **2021**, *132*, 108541. [CrossRef]
9. Botta, L.; Titone, V.; Mistretta, M.C.; La Mantia, F.P.; Modica, A.; Bruno, M.; Sottile, F.; Lopresti, F. PBAT based composites reinforced with microcrystalline cellulose obtained from softwood almond shells. *Polymers* **2021**, *13*, 2643. [CrossRef]
10. Ren, P.G.; Liu, X.H.; Ren, F.; Zhong, G.J.; Ji, X.; Xu, L. Biodegradable graphene oxide nanosheets/poly-(butylene adipate-coterephthalate) nanocomposite film with enhanced gas and water vapor barrier properties. *Polym. Test.* **2017**, *58*, 173–180. [CrossRef]
11. Yao, Q.R.; Song, Z.Y.; Li, J.; Zhang, L. Micromorphology, mechanical, crystallization and permeability properties analysis of HA/PBAT/PLA (HA, hydroxyapatite; PBAT, poly(butylene adipate-co-butylene terephthalate); PLA, polylactide) degradability packaging films. *Polym. Int.* **2020**, *69*, 301–307. [CrossRef]
12. Sangroniz, A.; Sangroniz, L.; Gonzalez, A.; Santamaria, A.; del Rio, J.; Iriarte, M.; Etxeberria, A. Improving the barrier properties of a biodegradable polyester for packaging applications. *Eur. Polym. J.* **2019**, *115*, 76–85. [CrossRef]
13. Wadaugsorn, K.; Panrong, T.; Wongphan, P.; Harnkarnsujarit, N. Plasticized hydroxypropyl cassava starch blended PBAT for improved clarity blown films: Morphology and properties. *Ind. Crops Prod.* **2022**, *176*, 114311. [CrossRef]
14. Qin, P.K.; Wu, L.B.; Li, B.G.; Li, N.X.; Pan, X.H.; Dai, J.M. Superior gas barrier properties of biodegradable PBST vs. PBAT copolyesters: A comparative study. *Polymers* **2021**, *13*, 3449. [CrossRef]
15. Li, J.X.; Wang, S.L.; Lai, L.; Liu, P.W.; Wu, H.Q.; Xu, J.L.; Severtson, S.J.; Wang, W.J. Synergistic enhancement of gas barrier and aging resistance for biodegradable films with aligned graphene nanosheets. *Carbon* **2021**, *172*, 31–40. [CrossRef]
16. Mondal, D.; Bhowmick, B.; Mollick, M.M.R.; Maity, D.; Saha, N.R.; Rangarajan, V.; Rana, D.; Sen, R.; Chattopadhyay, D. Antimicrobial activity and biodegradation behavior of poly(butylene adipate-co-terephthalate)/clay nanocomposites. *J. Appl. Polym. Sci.* **2014**, *131*, 40079. [CrossRef]
17. Gao, L.F.; Li, C.; Huang, W.C.; Mei, S.; Lin, H.; Ou, Q.; Zhang, Y.; Guo, J.; Zhang, F.; Xu, S.X.; et al. MXene/polymer membranes: Synthesis, properties, and emerging applications. *Chem. Mater.* **2020**, *32*, 1703–1747. [CrossRef]
18. Anasori, B.; Lukatskaya, M.R.; Gogotsi, Y. 2D metal carbides and nitrides (MXenes) for energy storage. *Nat. Rev. Mater.* **2017**, *2*, 16098. [CrossRef]
19. Sun, Z.M. Progress in research and development on MAX phases: A family of layered ternary compounds. *Int. Mater. Rev.* **2011**, *56*, 143–166. [CrossRef]
20. Ihsanullah, I. MXenes (two-dimensional metal carbides) as emerging nanomaterials for water purification: Progress, challenges and prospects. *Chem. Eng. J.* **2020**, *388*, 124340. [CrossRef]

21. Ihsanullah, I. Potential of MXenes in water desalination: Current status and perspectives. *Nano-Micro Lett.* **2020**, *12*, 1–20. [CrossRef]

22. Wu, W.; Zhao, W.J.; Sun, Q.J.; Yu, B.; Yin, X.M.; Cao, X.W.; Feng, Y.H.; Li, R.K.Y.; Qu, J.P. Surface treatment of two dimensional MXene for poly(vinylidene fluoride) nanocomposites with tunable dielectric permittivity. *Compos. Commun.* **2021**, *23*, 100562. [CrossRef]

23. Nan, J.X.; Guo, X.; Xiao, J.; Li, X.; Chen, W.H.; Wu, W.J.; Liu, H.; Wang, Y.; Wu, M.H.; Wang, G.X. Nanoengineering of 2D MXene-based materials for energy storage applications. *Small* **2021**, *17*, 1902085. [CrossRef]

24. Liu, C.; Wu, W.; Shi, Y.Q.; Yang, F.Q.; Liu, M.H.; Chen, Z.X.; Yu, B.; Feng, Y.Z. Creating MXene/reduced graphene oxide hybrid towards highly fire safe thermoplastic polyurethane nanocomposites. *Compos. Part B-Eng.* **2020**, *203*, 108486. [CrossRef]

25. Lan, C.T.; Jia, H.; Qiu, M.H.; Fu, S.H. Ultrathin MXene/polymer coatings with an alternating structure on fabrics for enhanced electromagnetic interference shielding and fire-resistant protective performances. *ACS Appl. Mater. Interfaces* **2021**, *13*, 38761–38772. [CrossRef]

26. Wang, Z.X.; Han, X.S.; Zhou, Z.J.; Meng, W.Y.; Han, X.W.; Wang, S.J.; Pu, J.W. Lightweight and elastic wood-derived composites for pressure sensing and electromagnetic interference shielding. *Compos. Sci. Technol.* **2021**, *213*, 108931. [CrossRef]

27. Zhang, H.; Wang, L.B.; Chen, Q.; Li, P.; Zhou, A.G.; Cao, X.X.; Hu, Q.K. Preparation, mechanical and anti-friction performance of MXene/polymer composites. *Mater. Des.* **2016**, *92*, 682–689. [CrossRef]

28. Sun, R.H.; Zhang, H.B.; Liu, J.; Xie, X.; Yang, R.; Li, Y.; Hong, S.; Yu, Z.Z. Highly conductive transition metal carbide/carbonitride(MXene)@polystyrene nanocomposites fabricated by electrostatic assembly for highly efficient electromagnetic interference shielding. *Adv. Funct. Mater.* **2017**, *27*, 1702807. [CrossRef]

29. Yu, B.; Yuen, A.C.Y.; Xu, X.D.; Zhang, Z.C.; Yang, W.; Lu, H.D.; Fei, B.; Yeoh, G.H.; Song, P.A.; Wang, H. Engineering MXene surface with POSS for reducing fire hazards of polystyrene with enhanced thermal stability. *J. Hazard. Mater.* **2021**, *401*, 123342. [CrossRef]

30. Gao, H.W.; Cao, W.K.; He, J.M.; Bai, Y.P. Highly transparent biaxially oriented poly(ester amide) film with improved gas barrier properties and good mechanical strength. *Eur. Polym. J.* **2021**, *156*, 110620. [CrossRef]

31. Kanai, T.; Okuyama, Y.; Takashige, M. Dynamics and structure development for biaxial stretching PA6/MXD6 blend packaging films. *Adv. Polym. Technol.* **2018**, *37*, 2828–2837. [CrossRef]

32. Jung, B.N.; Jung, H.W.; Kang, D.H.; Kim, G.H.; Shim, J.K. A study on the oxygen permeability behavior of nanoclay in a polypropylene/nanoclay nanocomposite by biaxial stretching. *Polymers* **2021**, *13*, 2760. [CrossRef] [PubMed]

33. Kim, D.Y.; Lee, J.B.; Lee, D.Y.; Seo, K.H. Plasticization effect of poly(lactic acid) in the poly(butylene adipate-co-terephthalate) blown film for Tear resistance improvement. *Polymers* **2020**, *12*, 1904. [CrossRef] [PubMed]

34. Wu, F.; Misra, M.; Mohanty, A.K. Challenges and new opportunities on barrier performance of biodegradable polymers for sustainable packaging. *Prog. Polym. Sci.* **2021**, *117*, 101395. [CrossRef]

35. Yoksan, R.; Dang, K.M.; Boontanimitr, A.; Chirachanchai, S. Relationship between microstructure and performances of simultaneous biaxially stretched films based on thermoplastic starch and biodegradable polyesters. *Int. J. Biol. Macromol.* **2021**, *190*, 141–150. [CrossRef]

36. Wang, X.G.; Cui, L.N.; Fan, S.H.; Li, X.; Liu, Y.J. Biodegradable poly(butylene adipate-co-terephthalate) antibacterial nanocomposites reinforced with MgO nanoparticles. *Polymers* **2021**, *13*, 507. [CrossRef] [PubMed]

37. Wu, W.; Cao, X.W.; Luo, J.; He, G.J.; Zhang, Y.J. Morphology, thermal, and mechanical properties of poly(butylene succinate) reinforced with halloysite nanotube. *Polym. Compos.* **2014**, *35*, 847–855. [CrossRef]

38. Al-Itry, R.; Lamnawar, K.; Maazouz, A. Improvement of thermal stability, rheological and mechanical properties of PLA, PBAT and their blends by reactive extrusion with functionalized epoxy. *Polym. Degrad. Stab.* **2012**, *97*, 1898–1914. [CrossRef]

39. Wu, W.; Zhao, W.J.; Gong, X.J.; Sun, Q.J.; Cao, X.W.; Su, Y.J.; Yu, B.; Li, R.K.Y.; Vellaisamy, R.A.L. Surface decoration of halloysite nanotubes with POSS for fire-safe thermoplastic polyurethane nanocomposites. *J. Mater. Sci. Technol.* **2022**, *101*, 107–117. [CrossRef]

40. Sheng, X.X.; Li, S.H.; Zhao, Y.F.; Zhai, D.S.; Zhang, L.; Lu, X. Synergistic effects of two-dimensional MXene and ammonium polyphosphate on enhancing the fire safety of polyvinyl alcohol composite aerogels. *Polymers* **2019**, *11*, 1964. [CrossRef]

41. Xiong, S.J.; Bo, P.; Zhou, S.J.; Li, M.K.; Yang, S.; Wang, Y.Y.; Shi, Q.T.; Wang, S.F.; Yuan, T.Q.; Sun, R.C. Economically competitive biodegradable PBAT/lignin composites: Effect of lignin methylation and compatibilizer. *ACS Sustain. Chem. Eng.* **2020**, *8*, 5338–5346. [CrossRef]

42. Yan, D.S.; Wang, Z.Y.; Guo, Z.Y.; Ma, Y.M.; Wang, C.Y.; Tan, H.Y.; Zhang, Y.H. Study on the properties of PLA/PBAT composite modified by nanohydroxyapatite. *J. Mater. Res. Technol.* **2020**, *9*, 11895–11904. [CrossRef]

43. Kargarzadeh, H.; Galeski, A.; Pawlak, A. PBAT green composites: Effects of kraft lignin particles on the morphological, thermal, crystalline, macro and micromechanical properties. *Polymer* **2020**, *203*, 122748. [CrossRef]

44. Blyakhman, F.A.; Makarova, E.B.; Fadeyev, F.A.; Lugovets, D.V.; Safronov, A.P.; Shabadrov, P.A.; Shklyar, T.F.; Melnikov, G.Y.; Orue, I.; Kurlyandskaya, G.V. The contribution of magnetic nanoparticles to ferrogel biophysical properties. *Nanomaterials* **2019**, *9*, 232. [CrossRef]

45. Jaszkiewicz, A.; Bledzki, A.K.; Meljon, A. Dynamic mechanical thermal analysis of biocomposites based on PLA and PHBV-A comparative study to PP counterparts. *J. Appl. Polym. Sci.* **2013**, *130*, 3175–3183.

*Article*

# Preparation and Properties of Self-Healing Waterborne Polyurethane Based on Dynamic Disulfide Bond

Gongbo Ye and Tao Jiang *

Hubei Collaborative Innovation Center for Advanced Organic Chemical Materials, Ministry of Education Key Laboratory for the Green Preparation and Application of Functional Materials, School of Materials Science and Engineering, Hubei University, Wuhan 430062, China; 2015011113000048@stu.hubu.edu.cn
* Correspondence: jiangtao@hubu.edu.cn

**Abstract:** A self-healing waterborne polyurethane (WPU) materials containing dynamic disulfide (SS) bond was prepared by introducing SS bond into polymer materials. The zeta potential revealed that all the synthesized WPU emulsions displayed excellent stability, and the particle size of them was about 100 nm. The characteristic peaks of N-H and S-S in urethane were verified by FTIR, and the chemical environment of all elements were confirmed by the XPS test. Furthermore, the tensile strength, self-healing process and self-healing efficiency of the materials were quantitatively evaluated by tensile measurements. The results showed that the self-healing efficiency could reach 96.14% when the sample was heat treated at 70 °C for 4 h. In addition, the material also showed a good reprocessing performance, and the tensile strength of the reprocessed film was 3.39 MPa.

**Keywords:** waterborne polyurethane; self-healing; dynamic disulfide bond

**Citation:** Ye, G.; Jiang, T. Preparation and Properties of Self-Healing Waterborne Polyurethane Based on Dynamic Disulfide Bond. *Polymers* **2021**, *13*, 2936. https://doi.org/10.3390/polym13172936

Academic Editors: Wei Wu, Hao-Yang Mi, Chongxing Huang, Hui Zhao and Tao Liu

Received: 11 August 2021
Accepted: 26 August 2021
Published: 31 August 2021

**Publisher's Note:** MDPI stays neutral with regard to jurisdictional claims in published maps and institutional affiliations.



## 1. Introduction

To date, polymer materials have been widely used in various fields because of their excellent corrosion resistance and mechanical and barrier properties. However, it is accessible to produce micro-slaps on the surface and inside of the traditional polymer materials due to the influence of mechanical, light and chemical substances in the manufactural processing and practical applications. These cracks are difficult to detect with the naked eye and may cause further damage to the materials, thus reducing the mechanical properties and safety of the materials and shortening their service time. Therefore, some researchers introduced the biological self-healing characteristics and mechanisms into polymer materials and formed functional self-healing materials through bionic design. These materials can recover their original properties spontaneously after damage or under external stimulation, resulting in new intelligent/smart materials with longer lives and more reliable performances, such as shaped memory polymer [1–3] and self-healing polymer [4,5] intelligent polymer materials.

From the perspective of bionics, self-healing materials can be divided into extrinsic and intrinsic types according to whether they contain additional repair reagents or not. Extrinsic self-healing can preinstall some special structural components (such as microcapsules or micro-vessels) in the polymer composites by implantation technology. When the material is damaged, the repair reagent at the damaged part is released rapidly, which promotes the polymerization reaction and repairs the material structure in time [6], whereas intrinsic self-healing materials can achieve multiple internal repairs through the breaking and recombination of reversible chemical bonds within or between molecules. Therefore, it is not necessary to add additional repair reagents in advance [7,8].

As the earliest discovered dynamic polymerization reaction, disulfide exchange reaction can be carried out at a low temperature and has great advantages in preparing dynamic polymers [9,10] For instance, aromatic disulfide compounds can form a rapid self-healing system [11]. In 2007, Nitschke et al. found that disulfide bonds in aromatic disulfides were easier to reach the translocation exchange equilibrium than that in aliphatic disulfides [12]. Ibon Odriozola et al. successfully prepared a catalyst-free disulfide bond polymer elastomer that can be repaired at room temperature by the condensation reaction of 4, 4′-dithiodianiline with isocyanate [13]. Kim et al. developed a transparent and easy-to-process polyurethane elastomer (IP-SS), which can be rapidly self-healed at room temperature and possess a maximum tensile strength of 6.8 MPa [14]. Zhang et al. prepared a self-healing polyurethane material based on solar light-induced disulfide bond metathesis reaction [15]. In addition to the electron-donating groups connected with disulfide bonds, it helps to reduce their bond energy and form hydrogen bonds between macromolecular chains, the amorphous structures of aliphatic monomers and soft and hard segments also play a key role in the self-healing of hydrogen sulfide. However, there are still some problems in the development of the disulfide exchange reaction. Aromatic disulfide monomers are usually expensive and not conducive to large-scale industrial production. At the same time, the molecular design usually leads to the yellowish appearance and low transparency of self-healing materials.

Herein, the low-cost aliphatic bis (2-alkylethyl) disulfide (HEDS) with flexible segments was used to prepare the self-healing waterborne polyurethane materials, which was synthesized by introducing HEDS-containing −OH groups into the WPU and reacted with −NCO groups. The successful introduction of SS bond in the polymer chain was verified by FTIR. Then, the influence of SS bond content on the self-healing efficiency of WPU film was studied by the tensile test, the dispersion and stability of WPU were assessed by the particle size and zeta potential analysis and the thermal properties of the polymer were tested by DMA and TGA. In addition, the reprocessing performance of the film was investigated by a plate vulcanizing machine.

## 2. Materials and Methods

### 2.1. Materials

Poly-tetrahydrofuran (PTMG, Mn = 1000), Triethylamine (TEA), 2,2-Dimthylolpropionic acid (DMPA), Dibutyltin dilaurate (DBTDL) and bis (2-alkylethyl) disulfide (HEDS) were obtained from Shanghai Aladdin Company, Shanghai, China. Hexamethylene diisocyanate (HDI) was purchased from Bayer Technology, Germany. PTMG and DMPA should be dehumidified under vacuum at 100 °C for 2 h before the experiment.

### 2.2. Preparation of WPU-SS Emulsion

WPU-SS was synthesized by three steps, in which acetone was replaced by DMAc to reduce the viscosity of polyurethane prepolymer. Table 1 lists the experimental formulation of different WPU samples. The molar ratio of the isocyanate group to hydroxyl group was kept at 1.4, and the ratio of PTMG to HEDS was 1/2 (WPU1), 1/1(WPU2) and 2/1(WPU3).

**Table 1.** Molar ratio formulations of different WPU samples.

|        | HDI | PTMG | DMPA | HEDS | TEA  |
|--------|-----|------|------|------|------|
| WPU1   | 3.5 | 0.55 | 0.85 | 1.10 | 0.85 |
| WPU2   | 3.5 | 0.83 | 0.85 | 0.83 | 0.85 |
| WPU3   | 3.5 | 1.10 | 0.85 | 0.55 | 0.85 |

According to the experimental formulation listed in Table 1, the synthesis process of WPU2 was illustrated. As shown in Figure 1, firstly, PTMG and DMPA were added in a 500-mL four-necked round-bottomed flask with a mechanical agitator, nitrogen inlet and drying condenser tube. DMAc was used as a solvent, and DBTDL was used as a catalyst. IPDI was slowly added after thoroughly stirring. Under mechanical stirring, the reaction temperature was 80 °C, when the content of −NCO reached the theoretical value, the polyurethane prepolymer terminated with −NCO was generated. Then, the temperature was reduced to 60 °C, HEDS containing −OH was added and reacted with isocyanate under nitrogen protection until the content of $-NCO$ in the system reached the theoretical value again.

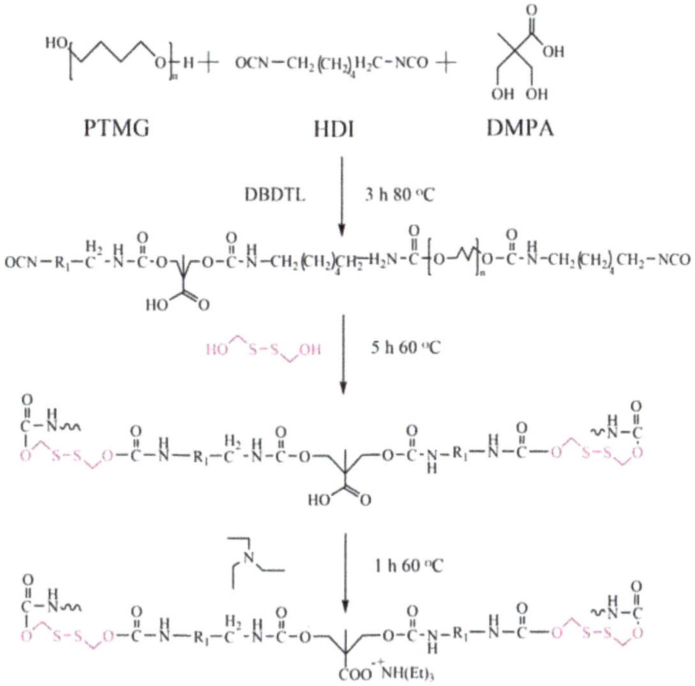

**Figure 1.** The synthesis process of WPU2 (R1 is the repeating chain segment).

TEA with DMPA and other substances was added to neutralize $-COOH$ in DMPA, and the system temperature was further reduced to 6~10 °C. The synthesized mixture was added to deionized water for high-speed dispersion for 1 h to obtain the final WPU2 emulsion. In the emulsification process, the residual $-NCO$ of the prepolymer reacted with deionized water to form urea and biuret derivatives. The final WPU2 emulsion was light blue and the solid content was about 30 wt%. The WPU samples with PTMG and HEDS ratios of 1/2 and 2/1 were synthesized by the same method.

### 2.3. Preparation of WPU Films

The predetermined amount of WPU-SS emulsion was poured into the horizontally placed PTFE mold with the size of 10 cm $\times$ 10 cm $\times$ 1.5 cm. After drying at room temperature for 48 h, most of the water evaporated. Then, the film was stripped from the mold, and the water and solvent were completely removed within 24 h in a vacuum drying oven at 60 °C to obtain a constant weight WPU-SS film. The prepared WPU-SS films were stored in the dryer containing silica gel before further characterization.

### 2.4. Measurement of Isocyanate Content by Dibutyl-Amine Method

(1) Reagent preparation:

    a. Preparation of 0.1 mol/L hydrochloric acid standard solution and calibration with anhydrous sodium carbonate.

    b. Configure 0.1-mol/L di-n-butylamine-acetone solution: 12.9-g di-n-butylamine was placed in a 1000-mL volumetric flask, diluted with acetone and shaken.

c.  Bromocresol green indicator: 0.1-g bromocresol green was dissolved in 100-mL volumetric flasks with a 1.5-mL concentration of 0.1-mol/L sodium hydroxide solution and diluted with distilled water to scale.

(2)  Operation process: Accurate weighing 1~3-g prepolymer to 150-mL conical flask, adding 20-mL dibutyl-amine-acetone solution, fully reaction 20 min after adding 5 drops of bromocresol green indicator. Use the configured 0.1-mol/L hydrochloric acid standard solution to titrate the prepolymer. When the color changes from blue to yellow, it will be the end of the reaction. Read the volume and do a blank control experiment.

(3)  Isocyanate content was determined by using the following Equation (1):

$$-\text{NCO} \% = [(V_0 - V) \times c \times 4.202]/m \times 100\% \tag{1}$$

Thereinto, c is HCL concentration (mol/L), $V$ is the sample consumed HCL volume (mL), $V_0$ is blank consumed HCL volume (mL) and m is Sample mass (g).

*2.5. Characterization*

2.5.1. Determination of Dispersion Stability

The average particle size and polydispersity index (PDI) of the dispersions were measured by dynamic light scattering (DLS) at room temperature using Zetasizer Nano ZS90 of Malvern Instrument and Equipment Company, Malvern, UK. In order to test the storage stability, all dispersions were stored in closed bottles and stored at room temperature.

2.5.2. Fourier-Transform Infrared Spectroscopy (FTIR)

The synthesized polymer was analyzed by TENSORII Fourier-transform infrared spectrometer produced by BRUKER company in Germany. The attenuated total reflection (ATR) mode was used for infrared testing of WPU-SS samples. The spectra obtained were recorded in the range of 4000–500 cm$^{-1}$.

2.5.3. Dynamic Thermomechanical Analysis (DMA)

The dynamic mechanical properties of the rectangular WPU-SS spline (30 mm $\times$ 10 mm) were tested using the tensile mode of the DMA 850 dynamic thermomechanical analyzer of the United States TA company. The test temperature range is $-100$~120 °C, and the experimental frequency is 1 Hz. The heating rate is 5 °C/min.

2.5.4. Thermogravimetric Analysis (TGA)

A certain amount of WPU-SS was investigated by the TGA55 thermogravimetric analyzer of TA company. It was carried out in a nitrogen atmosphere with an airflow velocity of 20 mL/min, a heating rate of 10 °C/min and a temperature range of 30~800 °C.

2.5.5. X-ray Photoelectron Spectroscopy (XPS)

The surface chemical compositions of WPU-SS samples were analyzed by XPS ESCALAB 250A of Thermo Electron Corporation company with an Al K$\alpha$ excitation radiation. The containment C 1s hydrocarbon peak at 284.8 eV was applied to calibrate the binding energies.

2.5.6. Determination of Self-Healing Performance

The samples were cut into dumbbell-shaped samples, and five points were randomly selected on the dumbbell-shaped samples to measure the thickness using a desktop thickness gauge, and the average value was taken. The USA Instron 3369 universal tensile testing machine was used to test the tensile strength and elongation at break at 100 mm/min, and the average value of each sample was measured after 5 times.

The dumbbell-shaped standard sample was cut from the middle position to part of the adhesion, and then the incision part was docked together. The samples were repaired under different conditions. The tensile properties of original and repaired samples were tested, and the self-healing efficiency was calculated by using the following Equation (2):

$$\eta\% = \sigma/\sigma_0 \times 100\% \tag{2}$$

where $\eta$ is self-healing efficiency, $\sigma$ is fracture strength of repaired specimen and $\sigma_0$ is fracture strength of the original sample.

## 3. Results and Discussion

### 3.1. Dispersion and Stability Analysis

The synthesized WPU-SS emulsion was analyzed by a nano-particle size analyzer to evaluate its storage stability. The particle size and distribution of suspension waterborne polyurethane particles are shown in Figure 2a. The average particle sizes of WPU1, WPU2 and WPU3 were 71.27, 77.12 and 131.6 nm, respectively, and the PDI were 0.130, 0.084 and 0.285, respectively. The particle sizes of WPU1 and WPU2 dispersion systems are similar, and the polydispersity index shows that the distribution of WPU2 dispersion is more uniform than that of WPU1 dispersion. With the increase of soft segment content, the proportion of hydrophilic groups in WPU3 emulsion decreased, the polydispersity index increased, and the dispersion became worse.

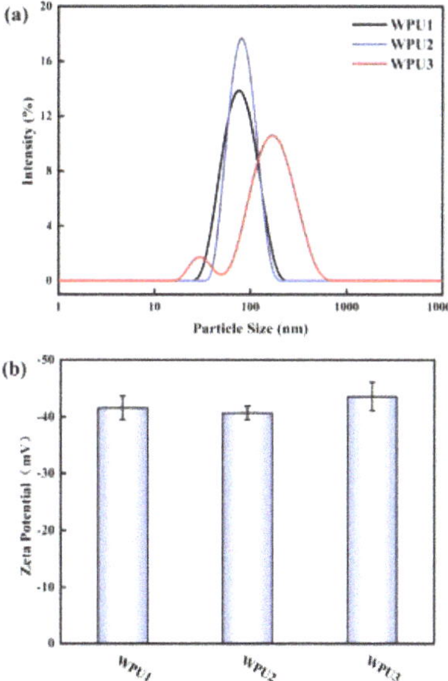


**Figure 2.** Particle size and distribution of WPU (**a**) and their zeta potential (**b**).

Zeta potential is another important indicator for evaluating emulsion stability. As shown in Figure 2b, the Zeta potential of WPU1, WPU2 and WPU3 were −41.6, −40.7 and −43.6 mV, respectively. The high absolute value of zeta potential indicates that there is abundant electrostatic repulsion on the surface of water particles. Besides, introducing SS bonds in the WPU-SS chain had little effect on the Zeta potential value, manifesting that the addition of SS bonds had no significant effect on the stability of the emulsion. In addition, there is no deposit at the bottom of WPU-SS dispersions stored in closed glass bottles at room temperature for 2 months. The WPU emulsions were diluted into 1 wt% water dispersion and looked light blue. In summary, stable WPU emulsions can be obtained by introducing SS bond into the molecule chain.

### 3.2. Structure Analysis

Figure 3 are FTIR spectra and XPS survey spectra of the linear WPU-SS samples. In Figure 3a, all curves have no absorption peak at 2270 cm$^{-1}$, which indicates that $-$NCO groups in the system are all involved in the reaction [16,17]. Compared with the three curves, the distribution of all characteristic peaks was essentially the same, indicating that $-$OH and $-$NCO on HEDS reacted to form carbamate, and no new chemical bond was formed. The three curves are in line with the characteristic peaks of waterborne polyurethane. The band located at 3321 cm$^{-1}$ corresponds to the N$-$H stretching vibration and the band at 1531 cm$^{-1}$ is assigned to the N–H in-plane bending vibration [18,19]. The absorption bands at 1460 and 1360 cm$^{-1}$ are assigned to the –CH$_2$– bending vibrations. The bands at 2940 and 2856 cm$^{-1}$ are associated with the C–H asymmetry and symmetric stretching vibration of methylene in polyurethane molecular chain [20,21]. The band at 1110 cm$^{-1}$ corresponds to bending vibration of C$-$O (aliphatic ether) in polyether polyol [22]. The characteristic band about 1707 and 1631 cm$^{-1}$ is related with nonhydrogen bond C=O hydrogen bond C=O in urea, respectively [23–25]. HEDS in polymer chain has a characteristic absorption peak of SS bond at 637 [26]. With the increase of disulfide bond content, the absorption peak of the corresponding SS bond at 637 cm$^{-1}$ became more obvious, which proved that SS bond was successfully introduced into waterborne polyurethane to form WPU-SS.

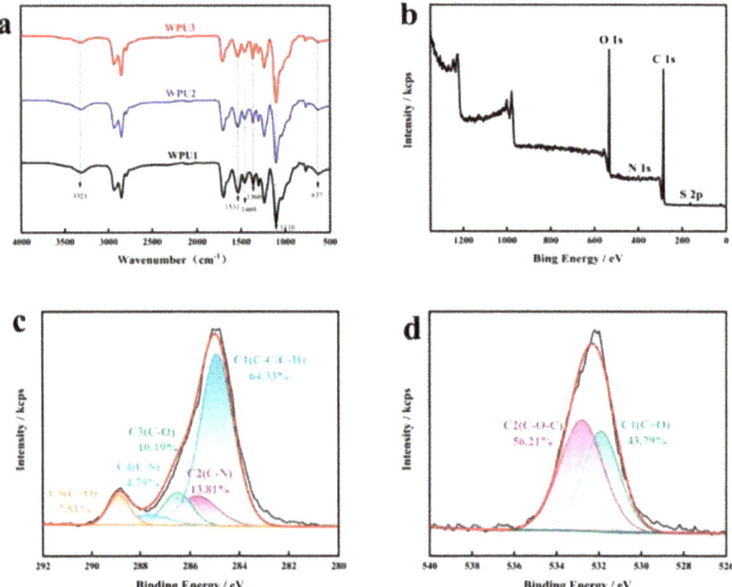

**Figure 3.** (**a**) FTIR spectra of WPU-SS samples, (**b**) XPS survey spectra of WPU2, (**c**) C 1s of WPU2, and (**d**) O 1s of WPU2.

The XPS measurement was used to further confirm the construction by analyzing the chemical compositions in the WPU-SS system. The peaks in XPS survey spectrum indicated the present of S element (163.5 eV) in the samples due to disulfide bond should be successfully chemically linked to the matrix of waterborne polyurethane. In addition, the C 1s spectrum was further measured to explore the types of carbon bonds to analyze the existence of interfacial interactions, as shown in Figure 3c. The C 1s peak was curved-fitted into four main components including C-C/C-H, C-N, C-O, C-S and C=O bonds corresponding to the peaks at 284.8, 285.7, 286.5, 287.7 and 288.8 eV, respectively. The O 1s peak was curved-fitted into four main components including C-O-C and C=O bonds.

The peak of C-S bond appeared implying that disulfide bond was successfully grafted onto polyurethane chains, as supported by FTIR results.

### 3.3. Dynamic Thermodynamic Analysis

The DMA results are shown in Figure 4. During the heating process of WPU-SS at −100~120 °C, the polymer with microphase separation structure showed two glass transition temperatures. The storage modulus and loss modulus of WPU-SS films are shown in Figure 4a,b, respectively. It could be found that the storage modulus (E′) decreased significantly at about −70 to −60 °C, while the loss modulus (E″) increased significantly at the same temperature range, which corresponding to the glass transition of the hard segments, moreover, with the increase of SS bonds, the E′ did not change much while the peak value of E″ changed obviously, which meant the move of the polymer chains took more energy. The decrease of disulfide bond content leads to the increase of soft segment content, which reduced the E′ and E″ of the films at room temperature. This is mainly due to the less restriction of the slip of the soft segment molecular chain, which leads to the easier entanglement of the molecular chain and the easier interaction with the main chain [27]. At the same time, the degree of microphase separation is reduced but the damping performance is improved.

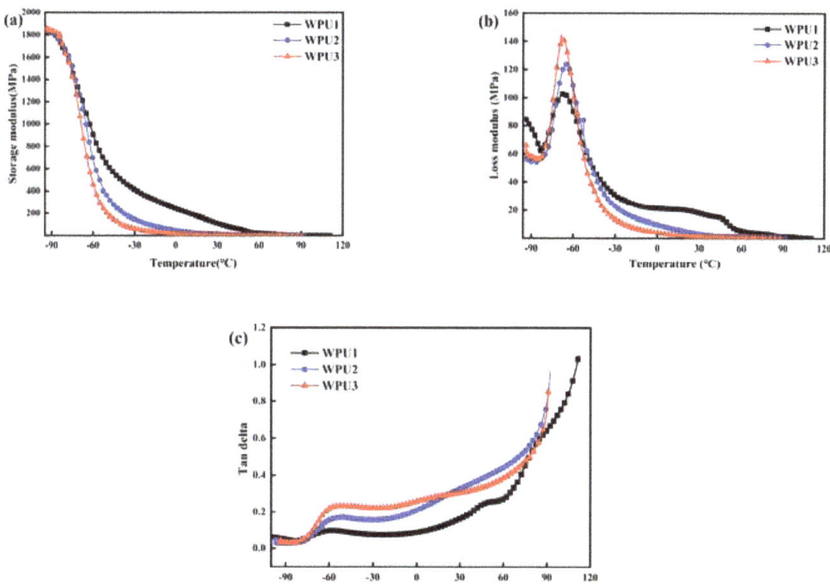

**Figure 4.** (**a**) Storage modulus, (**b**) loss modulus and (**c**) tan delta of WPU-SS samples with different contents of SS as a function of temperature.

The link between tan delta and temperature is determined in Figure 4c. With the loss of SS bond loading, the soft segment rises, and the microphase separation diminishes gradually. When the microphase separation is cut down to a certain extent, the glass transition temperature of the hard segment cannot be shown in the experiment. Since the glass transition temperature of the hard segment is much lower than room temperature, the film is in a highly elastic state and exhibits the properties of an elastomer at room temperature.

### 3.4. Thermal Stability Analysis

To investigate the thermal stability of WPU-SS films, thermogravimetric analysis (TGA) was carried out at 30~800 °C in a nitrogen atmosphere. Figure 5 shows the typical TGA and the first derivative curve DTG, respectively. The characteristic thermal degradation data display in Table 2. From the TGA curve, the thermal stability of WPU-SS films was improved with the decrease of short chain disulfide bond content and the enhancement of long chain PTMG content. In particular, the 50% weight loss temperature ($T_d$50%) increased from 361 °C to 382 °C. In the region of 230~440 °C, there are two obvious degradation stages for WPU-SS films. The partial weight loss at 230~350 °C can be attributed to the rupture of the amino ester bond in the hard segment isocyanate, and the thermal stability of the hard segment is poor, which can be decomposed into primary amines (or secondary amines), alkenes and carbon dioxide [28–30]. In contrast, polyether has better thermal stability, 350~440 °C weight loss is mainly the decomposition of the soft segment [31–33].

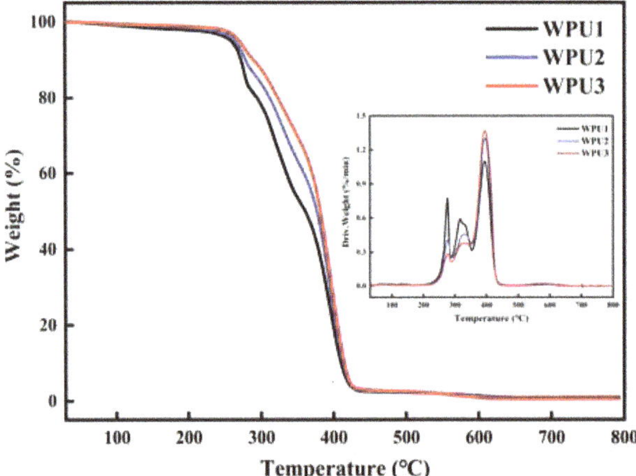


**Figure 5.** TGA curves and DTG curves of different WPU-SS films.

**Table 2.** Particular thermal degradation dates of diifferent WPU samples.

|  | $T_d$5% (°C) | $T_d$10% (°C) | $T_d$50% (°C) |
| --- | --- | --- | --- |
| WPU1 | 256 | 273 | 361 |
| WPU2 | 264 | 278 | 377 |
| WPU3 | 270 | 290 | 382 |

### 3.5. Tensile Properties and Self-Healing Efficiency of WPU-SS Films

In addition, the self-healing process of WPU films was evaluated more quantitatively through tensile measurement, and the self-healing efficiency was calculated according to the final tensile stress ratio of the repaired sample to the original sample. Figure 6 reveals the stress-strain curves of the original samples of WPU1, WPU2 and WPU3. The film of WPU1 exhibits a high tensile strength since the hard segment content is relatively high, and the content of carbamate in WPU1 may be more than that in WPU2 and WPU3. On the contrary, WPU2 and WPU3 possesses high soft segment content, and strong molecular weight mobility, so its self-healing performance is also relatively improved.

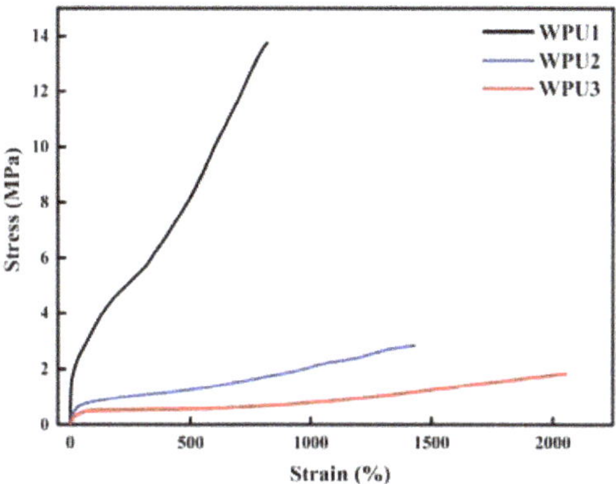

**Figure 6.** Stress-strain curve of WPU1, WPU2, WPU3.

Figure 7 shows the self-healing efficiency of WPU1 WPU2 and WPU3 under different conditions. The self-healing efficiency of WPU1 is only 29.02% after 4 h at 70 °C, while the healing efficiency of the WPU2 and WPU3 films reached 96.14% and 97.28% under the same conditions, respectively. Meanwhile, WPU2 and WPU3 films also had a self-healing efficiency of up to 84.21% and 85.86% after being placed at 25 °C for 24 h. The table of repair efficiency changing over time is shown in Table 3. There is no doubt that introducing SS bond endowed the film with excellent self-healing performance. Of course, the content of the soft segment also exerts a certain degree of influence on the repair effect of the film [34,35].

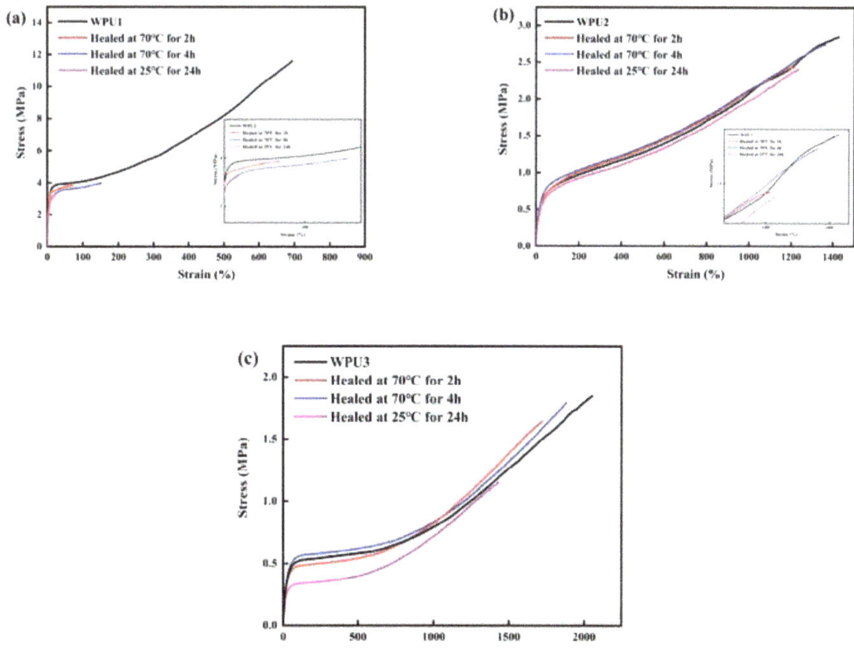

**Figure 7.** Stress-strain curves of WPU1 (**a**), WPU2 (**b**) and WPU3 (**c**) films repaired under different conditions.

**Table 3.** Details of Tensile Results and the Healing Efficiency of WPU-SS Samples under different conditions.

| Samples | Heat Treatment Time (h) | Tensile Strength (MPa) | Elongation at Break (%) | Healing Efficiency (%) |
|---|---|---|---|---|
| WPU1 | / | 13.75 | 820.01 | / |
|  | 2 | 3.85 | 71.67 | 28.00 |
|  | 4 | 3.99 | 151.12 | 29.02 |
|  | 24(25 °C) | 3.26 | 23.9 | 23.70 |
| WPU2 | / | 2.85 | 1427.78 | / |
|  | 2 | 2.43 | 1213.34 | 85.26 |
|  | 4 | 2.74 | 1365.01 | 96.14 |
|  | 24(25 °C) | 2.40 | 1235.01 | 84.21 |
| WPU3 | / | 1.84 | 2053.08 | / |
|  | 2 | 1.64 | 1719.64 | 89.13 |
|  | 4 | 1.79 | 1880.01 | 97.28 |
|  | 24(25 °C) | 1.58 | 1671.88 | 85.86 |

*3.6. Reprocessing Performance*

After being placed in a constant temperature and humidity box for 24 h, the cut dumbbell-shaped spline was used for the tensile test. Figure 8 provides the stress-strain curve of WPU2 film and WPU2 film after reprocessing. The tensile strength of WPU2 film after reprocessing was 3.39 MPa, compared with the original spline, its tensile strength increased and the elongation at break reduced, which indicated that the film had a good reprocessing performance.

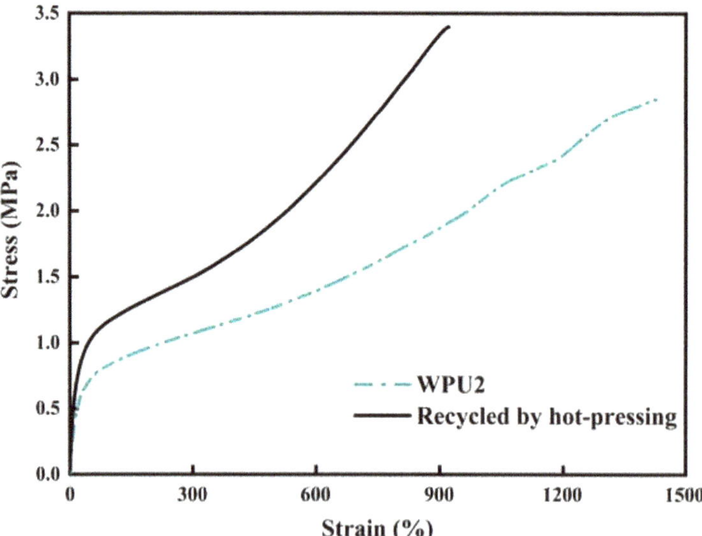


**Figure 8.** Tensile properties of the reprocessed WPU2 film.

## 4. Conclusions

In this paper, based on the principle of introducing reversible covalent bond containing SS into polymer, functional self-healing waterborne polyurethane was synthesized with PTMG as soft segment, IPDI as hard segment and DMPA containing hydrophilic group as auxiliary. The zeta potential revealed that all the synthesized WPU emulsions displayed excellent stability, and the particle sizes of WPU1 and WPU2 emulsions were relatively small. The characteristic peaks of N-H and S-S in urethane were verified by FTIR. Furthermore, the tensile strength, self-healing process and repair effect of WPU-SS film were quantitatively evaluated by tensile measurement. After the cut spline WPU2 sample was heat treated at 70 °C for 4 h, it could maintain the tensile strength and the self-healing efficiency could reach 96.14%, whilst the repair efficiency of the sample was 84.21% after 24-h self-repair at 25 °C. In addition, to realize the reprocessing of the self-healing film, the cut WPU2 film fragments were hot pressed at 130 °C for 20 min to reshape into a square film by a flat vulcanizing machine. The results indicated that the tensile strength of the reprocessed WPU2 film was 3.39 MPa, and it possessed good reprocessing performance.

**Author Contributions:** G.Y. conceived the project and wrote the paper; T.J. was in charge of all the experiments, all the authors contributed to the scientific discussion. Both authors have read and agreed to the published version of the manuscript.

**Funding:** This research received no external funding.

**Institutional Review Board Statement:** Not applicable.

**Informed Consent Statement:** Not applicable.

**Data Availability Statement:** Not applicable.

**Acknowledgments:** The authors also would like to thank Yunyun Zou from the Shiyanjia lab for the XPS analysis (www.Shiyanjia.com, accessed on 19 August 2021).

**Conflicts of Interest:** The authors declare no conflict of interest.

# References


1. Ur Rehman, H.; Chen, Y.; Hedenqvist, M.S.; Li, H.; Xue, W.; Guo, Y.; Guo, Y.; Duan, H.; Liu, H. Self-Healing Shape Memory PUPCL Copolymer with High Cycle Life. *Adv. Funct. Mater.* **2018**, *28*, 1704109. [CrossRef]
2. Zhou, L.; Liu, Q.; Lv, X.; Gao, L.; Fang, S.; Yu, H. Photoinduced triple shape memory polyurethane enabled by doping with azobenzene and GO. *J. Mater. Chem. C* **2016**, *4*, 9993–9997. [CrossRef]
3. Heo, Y.; Sodano, H.A. Self-Healing Polyurethanes with Shape Recovery. *Adv. Funct. Mater.* **2014**, *24*, 5261–5268. [CrossRef]
4. Kim, J.; Hong, P.H.; Choi, K.; Moon, G.; Kang, J.; Lee, S.; Lee, S.; Jung, H.W.; Ko, M.J.; Hong, S.W. A Heterocyclic Polyurethane with Enhanced Self-Healing Efficiency and Outstanding Recovery of Mechanical Properties. *Polymers* **2020**, *12*, 968. [CrossRef] [PubMed]
5. Yang, Y.; Lu, X.; Wang, W. A tough polyurethane elastomer with self-healing ability. *Mater. Des.* **2017**, *127*, 30–36. [CrossRef]
6. Hillewaere, X.K.D.; Du Prez, F.E. Fifteen chemistries for autonomous external self-healing polymers and composites. *Prog. Polym. Sci.* **2015**, *49–50*, 121–153. [CrossRef]
7. Campanella, A.; Dohler, D.; Binder, W.H. Self-Healing in Supramolecular Polymers. *Macromol. Rapid Commun.* **2018**, *39*, 1700739. [CrossRef] [PubMed]
8. Fan, L.F.; Rong, M.Z.; Zhang, M.Q.; Chen, X.D. Dynamic reversible bonds enable external stress-free two-way shape memory effect of a polymer network and the interrelated intrinsic self-healability of wider crack and recyclability. *J. Mater. Chem. A* **2018**, *6*, 16053–16063. [CrossRef]
9. Zhou, L.; Chen, M.; Zhao, X. Rapid degradation of disulfide-based thermosets through thiol-disulfide exchange reaction. *Polymers* **2017**, *120*, 1–8. [CrossRef]
10. Cordier, P.; Tournilhac, F.; Soulié-Ziakovic, C.; Leibler, L. Self-healing and thermoreversible rubber from supramolecular assembly. *Nature* **2008**, *451*, 977–980. [CrossRef]
11. de Luzuriaga, A.R.; Martin, R.; Markaide, N.; Rekondo, A.; Cabañero, G.; Rodríguez, J.; Odriozola, I. Correction: Epoxy resin with exchangeable disulfide crosslinks to obtain reprocessable, repairable and recyclable fiber-reinforced thermoset composites. *Mater. Horiz.* **2020**, *7*, 2460–2461. [CrossRef]
12. Sarma, R.J.; Otto, S.; Nitschke, J.R. Disulfides, imines, and metal coordination within a single system: Interplay between three dynamic equilibria. *Chemistry* **2007**, *13*, 9542–9546. [CrossRef] [PubMed]
13. Rekondo, A.; Martin, R.; de Luzuriaga, A.R.; Cabañero, G.; Grande, H.J.; Odriozola, I. Catalyst-free room-temperature self-healing elastomers based on aromatic disulfide metathesis. *Mater. Horiz.* **2014**, *1*, 237–240. [CrossRef]
14. Kim, S.M.; Jeon, H.; Shin, S.H.; Park, S.A.; Jegal, J.; Hwang, S.Y.; Oh, D.X.; Park, J. Superior Toughness and Fast Self-Healing at Room Temperature Engineered by Transparent Elastomers. *Adv. Mater.* **2018**, *30*, 1705145. [CrossRef]
15. Xu, W.M.; Rong, M.Z.; Zhang, M.Q. Sunlight driven self-healing, reshaping and recycling of a robust, transparent and yellowing-resistant polymer. *J. Mater. Chem. A* **2016**, *4*, 10683–10690. [CrossRef]
16. Zhang, P.; Lu, Y.; Fan, M.; Jiang, P.; Dong, Y. Modified cellulose nanocrystals enhancement to mechanical properties and water resistance of vegetable oil-based waterborne polyurethane. *J. Appl. Polym. Sci.* **2019**, *136*, 48228. [CrossRef]
17. Kim, M.S.; Ryu, K.M.; Lee, S.H.; Choi, Y.C.; Rho, S.; Jeong, Y.G. Chitin Nanofiber-Reinforced Waterborne Polyurethane Nanocomposite Films with Enhanced Thermal and Mechanical Performance. *Carbohydr. Polym.* **2021**, *258*, 117728. [CrossRef] [PubMed]

18. Hadjadj, A.; Jbara, O.; Tara, A.; Gilliot, M.; Malek, F.; Maafi, E.M.; Tighzert, L. Effects of cellulose fiber content on physical properties of polyurethane based composites. *Compos. Struct.* **2016**, *135*, 217–223. [CrossRef]

19. Lin, C.K.; Kuo, J.F.; Chen, C.Y.; Fang, J.J. Investigation of bifurcated hydrogen bonds within the thermotropic liquid crystalline polyurethanes. *Polymers* **2012**, *53*, 254–258. [CrossRef]

20. Wu, W.; Zhao, W.; Gong, X.; Sun, Q.; Cao, X.; Su, Y.; Yu, B.; Li, R.K.; Vellaisamy, R.A. Surface decoration of halloysite nanotubes with POSS for fire-safe thermoplastic polyurethane nanocomposites. *J. Mater. Sci. Technol.* **2021**. [CrossRef]

21. Zhao, H.; Zhao, S.Q.; Hu, G.H.; Zhang, Q.C.; Liu, Y.; Huang, C.X.; Li, W.; Jiang, T.; Wang, S.F. Synthesis and characterization of waterborne polyurethane/polyhedral oligomeric silsesquioxane composites with low dielectric constants. *Polym. Adv. Technol.* **2019**, *30*, 2313–2320. [CrossRef]

22. Lu, W.C.; Chuang, F.S.; Venkatesan, M.; Cho, C.J.; Chen, P.Y.; Tzeng, Y.R.; Yu, Y.Y.; Rwei, S.P.; Kuo, C.C. Synthesis of Water Resistance and Moisture-Permeable Nanofiber Using Sodium Alginate-Functionalized Waterborne Polyurethane. *Polymers* **2020**, *12*, 2882. [CrossRef] [PubMed]

23. Gubanska, I.; Kucinska-Lipka, J.; Janik, H. The influence of amorphous macrodiol, diisocyanate type and l-ascorbic acid modifier on chemical structure, morphology and degradation behavior of polyurethanes for tissue scaffolds fabrication. *Polym. Degrad. Stab.* **2019**, *163*, 52–67. [CrossRef]

24. Gao, Z.; Peng, J.; Zhong, T.; Sun, J.; Wang, X.; Yue, C. Biocompatible elastomer of waterborne polyurethane based on castor oil and polyethylene glycol with cellulose nanocrystals. *Carbohydr. Polym.* **2012**, *87*, 2068–2075. [CrossRef]

25. Zhao, H.; Zhao, S.Q.; Li, Q.; Khan, M.R.; Liu, Y.; Lu, P.; Huang, C.X.; Huang, L.J.; Jiang, T. Fabrication and properties of waterborne thermoplastic polyurethane nanocomposite enhanced by the POSS with low dielectric constants. *Polymers* **2020**, *209*, 122992. [CrossRef]

26. Zhao, D.; Liu, S.; Wu, Y.; Guan, T.; Sun, N.; Ren, B. Self-healing UV light-curable resins containing disulfide group: Synthesis and application in UV coatings. *Prog. Org. Coat.* **2019**, *133*, 289–298. [CrossRef]

27. Zhao, H.; Hao, T.H.; Hu, G.H.; Shi, D.; Huang, D.; Jiang, T.; Zhang, Q.C. Preparation and Characterization of Polyurethanes with Cross-Linked Siloxane in the Side Chain by Sol-Gel Reactions. *Materials* **2017**, *10*, 247. [CrossRef] [PubMed]

28. Zhao, H.; Huang, D.; Hao, T.H.; Hu, G.H.; Ye, G.B.; Jiang, T.; Zhang, Q.C. Synthesis and Investigation of Well-Defined Silane Terminated and Segmented Waterborne Hybrid Polyurethanes. *New J. Chem.* **2017**, *41*, 9268–9275. [CrossRef]

29. Chattopadhyay, D.K.; Webster, D.C. Thermal stability and flame retardancy of polyurethanes. *Prog. Polym. Sci.* **2009**, *34*, 1068–1133. [CrossRef]

30. Floros, M.; Hojabri, L.; Abraham, E.; Jose, J.; Thomas, S.; Pothan, L.; Leao, A.L.; Narine, S. Enhancement of thermal stability, strength and extensibility of lipid-based polyurethanes with cellulose-based nanofibers. *Polym. Degrad. Stab.* **2012**, *97*, 1970–1978. [CrossRef]

31. Fu, H.; Wang, Y.; Li, X.; Chen, W. Synthesis of vegetable oil-based waterborne polyurethane/silver-halloysite antibacterial nanocomposites. *Compos. Sci. Technol.* **2016**, *126*, 86–93. [CrossRef]

32. Liu, N.; Zhao, Y.; Kang, M.; Wang, J.; Wang, X.; Feng, Y.; Yin, N.; Li, Q. The effects of the molecular weight and structure of polycarbonatediols on the properties of waterborne polyurethanes. *Prog. Org. Coat.* **2015**, *82*, 46–56. [CrossRef]

33. Trovati, G.; Sanches, E.A.; Neto, S.C.; Mascarenhas, Y.P.; Chierice, G.O. Characterization of polyurethane resins by FTIR, TGA, and XRD. *J. Appl. Polym. Sci.* **2010**, *115*, 263–268. [CrossRef]

34. Zhao, H.; She, W.; Shi, D.; Wu, W.; Zhang, Q.C.; Li, R.K. Polyurethane/POSS nanocomposites for superior hydrophobicity and high ductility. *Compos. Part B Eng.* **2019**, *177*, 107441. [CrossRef]

35. Lai, Y.; Kuang, X.; Zhu, P.; Huang, M.; Dong, X.; Wang, D. Colorless, Transparent, Robust, and Fast Scratch-Self-Healing Elastomers via a Phase-Locked Dynamic Bonds Design. *Adv. Mater.* **2018**, *30*, 1802556. [CrossRef] [PubMed]

*Communication*

# Composited Film of Poly(3,4-ethylenedioxythiophene) and Graphene Oxide as Hole Transport Layer in Perovskite Solar Cells

Tian Yuan, Jin Li * and Shimin Wang *

Hubei Collaborative Innovation Center for Advanced Organic Chemical Materials, Ministry of Education Key Laboratory for the Green Preparation and Application of Functional Materials, School of Materials Science and Engineering, Hubei University, Wuhan 430062, China; tianyuan@stu.hubu.edu.cn
* Correspondence: jinli@hubu.edu.cn (J.L.); shiminwang@126.com (S.W.)

**Abstract:** It is important to lower the cost and stability of the organic–inorganic hybrid perovskite solar cells (PSCs) for industrial application. The commonly used hole transport materials (HTMs) such as Spiro-OMeTAD, poly[bis(4-phenyl)(2,4,6-trimethylphenyl)amine] (PTAA) and poly(3-hexylthiophene-2,5-diyl) (P3HT) are very expensive. Here, 3,4-ethylenedioxythiophene (EDOT) monomers are in-situ polymerized on the surface of graphene oxide (GO) as PEDOT-GO film. Compared to frequently used polystyrene sulfonic acid (PSS), GO avoids the corrosion of the perovskite and the use of $H_2O$ solvent. The composite PEDOT-GO film is between carbon pair electrode and perovskite layer as hole transport layer (HTL). The highest power conversion efficiency (PCE) is 14.09%.

**Keywords:** perovskite solar cell; hole transport layer; carbon materials; polymeric composites; solar energy materials

**Citation:** Yuan, T.; Li, J.; Wang, S. Composited Film of Poly(3,4-ethylenedioxythiophene) and Graphene Oxide as Hole Transport Layer in Perovskite Solar Cells. *Polymers* **2021**, *13*, 3895. https://doi.org/10.3390/polym13223895

Academic Editors: Wei Wu, Hao-Yang Mi, Chongxing Huang, Hui Zhao and Tao Liu

Received: 23 September 2021
Accepted: 30 October 2021
Published: 11 November 2021

**Publisher's Note:** MDPI stays neutral with regard to jurisdictional claims in published maps and institutional affiliations.



## 1. Introduction

Renewable clean energy devices are urgently demanded for the sustainable development of society. Among them, organic–inorganic metal hybrid perovskite solar cells (PSCs) have attracted ever-increasing attention owing to their excellent photovoltaic performance, simple preparation process, and relatively low cost [1]. PSCs are generally composed of FTO glass, an electron transport layer, a light absorption layer, a hole transport layer, and a counter electrode [2]. In the multi-layer structure of PSCs, the hole transport layer (HTL) is designed to promote the separation of electrons and holes, which is key to the performance and stability of the cell. However, certain problems of HTL hinder the development and application of the PSCs technology. Currently, the HTL of PSCs are based on materials such as Spiro-OMeTAD, PTAA [3] and P3HT [4]. The costs of these materials are all prohibitively high for large-scale applications [5]. What is more, the dopants in Spiro-OMeTAD show strong water absorbency, which seriously threatens the service life of PSCs [6]. Therefore, it is necessary to explore a low-cost and stable hole transport materials (HTMs) for the practical stage of PSCs.

PEDOT, usually combined with PSS, is widely used in inverted PSCs [7–9], whose price is much cheaper than the materials mentioned above. However, sulfonic acid groups contained in PSS are extremely harmful to the device life. To avoid the usage of PSS, Jiang et al. [10] synthesized 2,5-dibromo-3,4-ethylenedioxythiophene (DBEDOT) monomer, which was spin-coated on a perovskite layer and in-situ polymerized as PEDOT; a photoelectric conversion efficiency (PCE) of PSCs of about 17% was achieved. Wei et al. [11] used sulfonated acetone-formaldehyde (SAF), instead of PSS, to composite with PEDOT in inverted PSCs, which effectively increased the life of PSCs. Meanwhile, graphene oxide (GO) was also selected as the HTM of PSCs. Wu et al. [12] fabricated 2 nm thickness GO film as HTL, and the PCE of the inverted PSCs reached 12.40%.

In this work, a harmless HTL was obtained by PEDOT composited with GO. The PEDOT interacted with GO sheet via π–π stacking and hydrogen-bonding interactions, thus a conjugated system can be formed [13]. Moreover, GO functions as an excellent carrier to enable the dispersion of PEDOT in isopropanol solution, which is also harmless to the perovskite layer. Using PEDOT-GO film as HTL, a PSC with a PCE of up to 14.09% with good stability can be realized based on carbon counter electrode.

## 2. Materials and Methods

### 2.1. Material Preparation

The perovskite ($PbI_2$, MAI), FTO glasses ($3 \times 3$ cm$^2$) and the hole transport layer (HTL) ((Spiro-MeOTAD, lithium-bis (tri-fluoromethanesulfonyl) imide (Li-TFSI), and 4-tert-butylpyridine (tBP)) solution were purchased from Xi'an Polymer Light Tech-nology Co., Ltd. (Xian,China) 3,4-ethylenedioxythiophene (EDOT), graphite, acidic $(NH_4)_2S_2O_8$ (APS), $TiCl_4$, DMSO (99.9%) and 4-hydroxybutyric acid lactone (DMF) (99.9%) were purchased from Aladdin. Acetone, ethanol, and isopropyl alcohol were purchased from Sinopharm Chemical Reagent Co., Ltd. (Sinopharm Chemical Reagent Co., Ltd. ) And all the materials were used as received without further purification.

### 2.2. Fabrication of Device

FTO glasses were ultrasonically cleaned with detergent, acetone, ethanol, and iso-propyl alcohol sequentially. After that, they were dried under hot air and treated in ultraviolet-ozone for 15 min. The clean FTO substrate was then soaked in dilute 0.2 M aqueous $TiCl_4$ solution at 70 °C for 1 h, and washed with deionized water, then annealed at 200 °C for 60 min. In the Glove box, dropping the perovskite ($CH_3NH_3I$: 159 mg, $PbI_2$: 461 mg, DMF: 600 mg, and DMSO: 78 mg) onto the substrate and deposited it by spin-coating at 4000 rpm for 30 s. Solvent treatment was conducted at late 15 s, where 150 µL chlorobenzene was dropped on the spinning substrate followed by annealing at 100 °C for 10 min. After cooling to room temperature, the hole transport materials prepared before (3 mg/mL in dimethylcarbinol) were spin-coated onto the perovskite layer at 2000 rpm for 30 s and followed by 10 min of thermal annealing at 90 °C. Besides, Spiro-OMeTAD hole transport material was used in the comparative experiment: 20 µL spiro-OMeTAD solution, containing 36.1 mg spiro-OMeTAD, 14.4 µL t-BP and 9 µL Li-TFSI solution (520 mg in acetonitrile), was spin-coated on the perovskite layer at 4000 rpm for 30 s. Lastly, a carbon black counter electrode was coated on the top of the device by blade coating and sintered at 80 °C for 30 min.

### 2.3. Device Characterization

The X-ray diffraction (XRD) patterns of the samples and perovskite films were measured with a Bruker-AXS D8 Advance (Malvern Panalytical, Malvern, UK). $MAPbI_3$ perovskite films morphology was measured with scanning electron microscope (SEM, sigma 500, Krefeld, Germany). The GO nanosheets were characterized via transmission electron microscopy (TEM, TecnaiG2 F20, FEI Company, Hillsboro, OR, USA). The photocurrent-voltage (J-V) characteristics of PSCs were analyzed under simulated AM 1.5 G radiation (100 mW/cm$^2$ irradiance) by using a solar simulator (Oriel, model 91192-1000) and a source meter (Keithley 2400, USA). Electrochemical impedance spectroscopy (EIS) was measured with an electrochemical workstation (Zennium, IM6, Kronach, Germany) over the frequency range of 100 mHz$^{-2}$ MHz with 10 mV AC amplitude at $-1$ V bias under simulated AM 1.5 G radiation (100 mW/cm$^2$ irradiance). The steady-state photoluminescence (PL) measurements were acquired using an Edinburgh Instruments FLS920 fluorescence spectrometer (Oxford In-struments, Abingdon, UK). Raman spectroscopy analysis was performed by a micro-Raman instrument (XperRam 200, Nanobase, Seoul, South Korea), using 542 nm excitation with an incident power of 5 mW. The devices were measured under ambient conditions (15% < relative humidity (RH) < 60%) every time (winter, summer). After the measurements, the devices were stored in a humidity-controlled dry room (20% < RH < 40%).

## 3. Results and Discussion

Figure 1A–C respectively correspond to XRD, FT-IR, and Raman analyses to explore the structural properties of the as-prepared materials [14–16]. As shown in Figure 1A, an intense and sharp peak centered at 10.65° in the (GO) curve, which corresponds to the (001) crystal surface of the GO nanoflakes. The pattern of pure PEDOT depicts a broad peak in the region of 25.82°, which corresponds to the polymer chain structure of PEDOT. However, neither of the two peaks appeared in the composite samples. This is due to the influence of conjugation and the coating effect of PEDOT on GO sheets, indicating that the in-situ polymerization changes the growth state of the polymer chain [14,17].

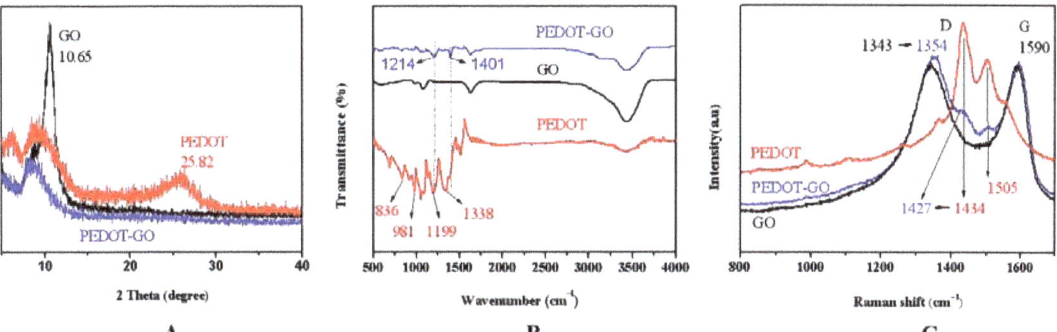

**Figure 1.** (**A**) XRD of GO, PEDOT and PEDOT-GO (**B**) FTIR spectra of GO, PEDOT and PEDOT-GO (**B**) Raman spectra of GO, PEDOT and PEDOT-GO.

In Figure 1B, the PEDOT curve shows two peaks at 981 $cm^{-1}$ and 836 $cm^{-1}$ are duo to C–S–C bond stretching of thiophene ring. The tensile vibration of the C–O–C bond at 1199 $cm^{-1}$ was detected. The peak at 1338 $cm^{-1}$ is due to C=C and C–C in the thiophene ring indicating that PEDOT was successfully synthesized. Combining three curves, the characteristic peaks of GO and PEDOT are all reflected in PEDOT-GO. Furthermore, two peaks at 1199 $cm^{-1}$ and 1338 $cm^{-1}$ on PEDOT skewed to 1214 $cm^{-1}$ and 1401 $cm^{-1}$ on the curve PEDOT-GO. This redshift phenomenon was due to the $\pi$–$\pi$ stacking interaction between GO and PEDOT [13].

In Raman spectra, the three characteristic peaks of 441 $cm^{-1}$, 1434 $cm^{-1}$, and 1505 $cm^{-1}$ in the red curve indicate the successful synthesis of PEDOT [18]. Meanwhile, the characteristic peaks of 1343 $cm^{-1}$ and 1590 $cm^{-1}$ (black curve) correspond to the respiratory vibration peaks of SP2 hybrid carbon atoms and the symmetric stretching motion peaks of SP2 hybrid atoms in the carbon ring, respectively, which are the characteristic peaks (D and G) of GO. Meanwhile, the characteristic peaks at 1434 $cm^{-1}$ (red curve) are assigned to $C_\alpha = C_\beta$ symmetric stretching vibration in PEDOT, which moves to 1427 $cm^{-1}$ in the Raman spectra of the PEDOT-GO sample (Figure 1C). This redshift phenomenon demonstrates that the PEDOT polymer changed to the quinoid form, and thus enabled the increase of conductivity [19]. Moreover, the characteristic peaks of GO and PEDOT are reflected in the curve of PEDOT-GO, confirming the in-situ polymerization of PEDOT-GO nanocomposites.

Figure 2a depicts an SEM cross-sectional view of the device. The cell structure is clearly displayed, and the thickness of the composite film is about 50 nm. Figure 2b presents the top view of the device [20–22]. The entire composite film is thin and evenly covered on the perovskite layer. Figure 2c,d are the TEM images of the GO and PEDOT-GO composite material, respectively. It can be seen that the surface of the graphene oxide sheet is smooth and transparent (Figure 2c). By comparison, the surface of the graphene oxide sheet is coated with a large amount of PEDOT nanoparticles in the PEDOT-GO film (Figure 2d), which is mutually confirmed with the previous analysis. This tight combination comes from the presence of conjugated heterocyclic structures and electronegative oxygen atoms. Simultaneously, PEDOT rich in free electrons and GO rich in carboxyl group form a good conjugated structure.

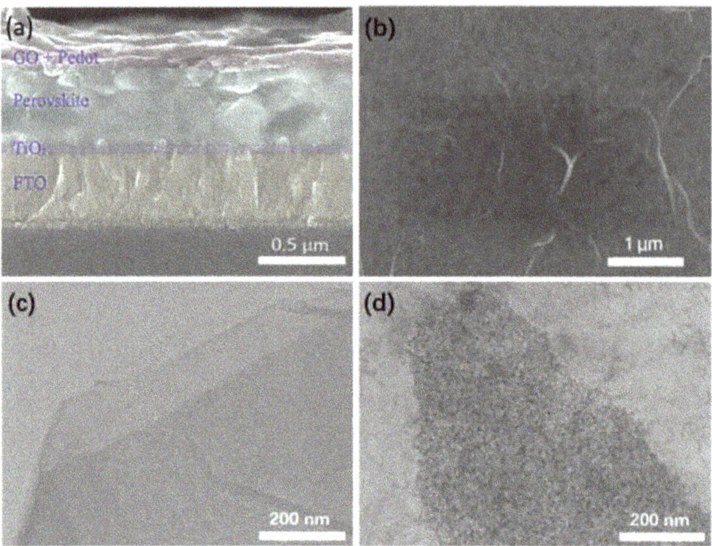

**Figure 2.** (**a**,**b**) SEM images of the device (**c**) TEM images of GO (**d**) TEM images of PEDOT-GO.

In order to investigate the optimum component, samples with PEDOT/GO ratio as 0, 0.5, 0.75, 1 were prepared and tested, respectively. Figure 3a is the J–V curves of the mesoporous PSCs with different mass ratio PEDOT-GO composite films. The FTO/cp-TiO$_2$/MAPbI$_3$/C structure of PSC was fabricated as a control group compared with FTO/cp-TiO$_2$/MAPbI$_3$/PEDOT-GO (or spiro-OMeTAD)/C structure. Figure 3b shows the Nyquist plots and the equivalent circuit model of the PSCs [23]. The high frequency arc is reflected to the hole transport and extraction between the PEDOT-GO and the carbon cathode; the low frequency arc shows charge recombination of PSCs [24].

The corresponding photovoltaic parameters are shown in Table 1. The device exhibits the highest performance when the PEDOT-GO mass ratio is 0.75, the PCE reached the 14.09% with the voltage ($V_{oc}$) as 1.10 V, the short-circuit current ($J_{sc}$), as 20.36 mA/cm$^2$, and the fill factor (FF) as 0.63. The PCE of the modified sample was increased by 26.6% compared to the HTL-free one. It is worth noting that, the FTO/TiO$_2$/MAPbI$_3$/spiro-OMeTAD/C sample with the PCE of 13.49%, Voc of 1.10 V, Jsc of 20.70 mA/cm$^2$, and FF of 0.59 shows similar performance to the PEDOT-GO (0.75) sample. Meanwhile, the lowest Rtr value of the PEDOT-GO (0.75) sample as 23.9 Ω indicates the excellent charge transfer performance, and the highest Rrec value as 187.4 Ω indicates its best anti-recombination property among all cells [25]. When the PEDOT/GO mass ratio is lower than 0.75, the $J_{sc}$ increases with the PEDOT content, but the $V_{oc}$ remains unchanged, indicating that the hole transport performance of the composite material is effectively optimized and enhanced. However, when the PEDOT/GO mass ratio gets higher than 0.75, the dispersion of the composite material in the solvent becomes worse, suggesting the insufficient addition of GO, which leads to the deterioration of the film quality and negatively affects both the $J_{sc}$ and $V_{oc}$. Finally, the mass ratio of PEDOT / GO is determined to be 0.75, the film quality and hole transport performance of the composites reach a balance, and the highest PCE is obtained. In addition, the hysteresis in the J-V curve of PEDOT-GO based devices is significantly reduced compared with the HTL-free devices (Figure 3c) [26]. PEDOT-GO composite material effectively optimized the hole extraction and transfer ability of the device, and reduces the built-in electric field at the interface of perovskite and HTL. Figure 3d shows the corresponding incident photon-to-electron conversion efficiency (IPCE) curves: the integrated current value for PEDOT-GO (0.75) sample was 19.69 mA cm$^{-2}$, which is consistent with the $\mathbf{J_{sc}}$ values extracted from the J-V curves.

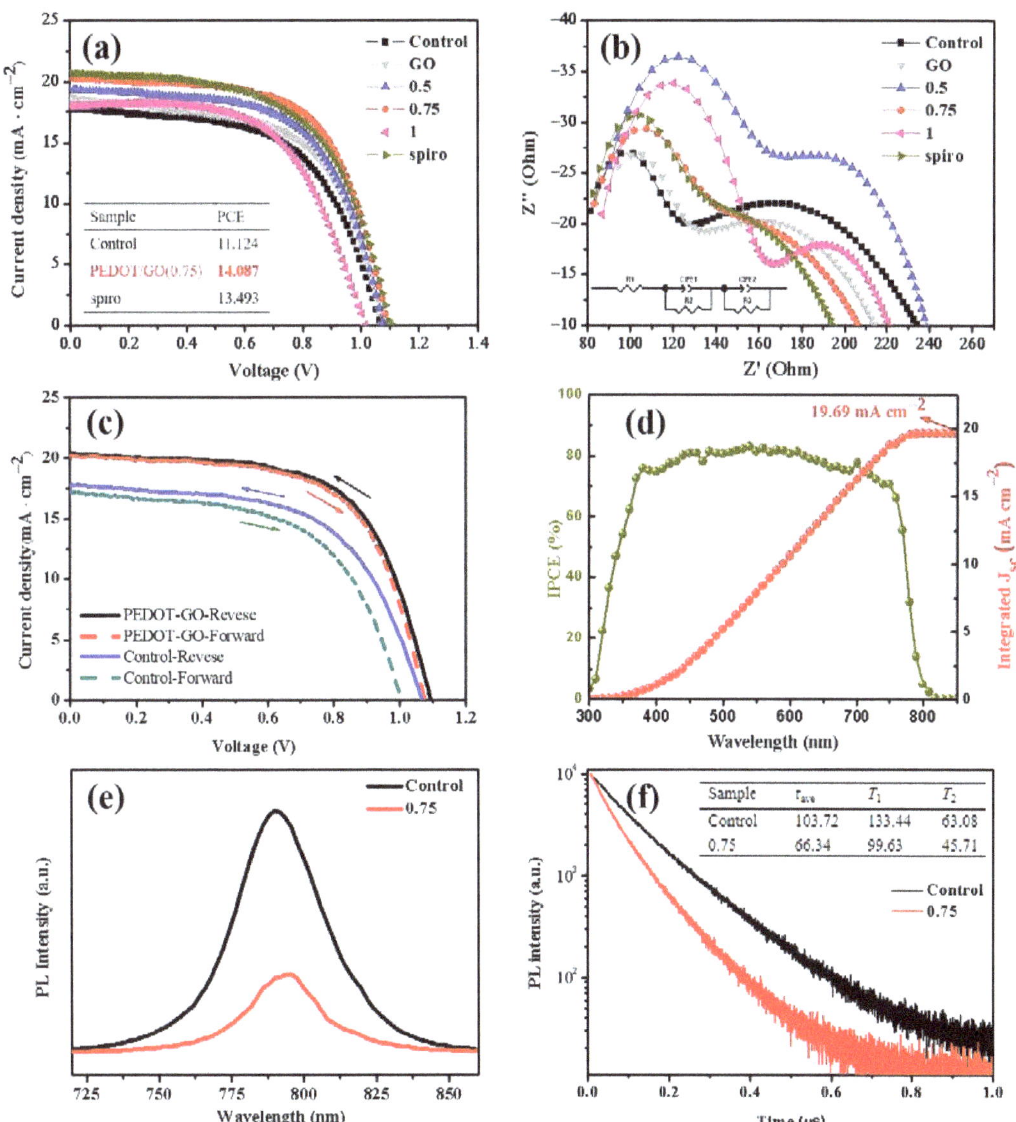

**Figure 3.** (**a**) J-V curves of control (HTL-free) sample and mass ratio PEDOT/GO: 0/1, 0.5/1, 0.75/1, 1/1, spiro-OMeTAD samples (**b**) Nyquist plots of resistance for the above samples (**c**) current-voltage characteristics with forward and reverse scans of PEDOT-GO and control sample (**d**) IPCE spectra of the PEDOT-GO (0.75) devices (**e**) steady-state PL spectra of PSCs of HTL-free sample and PSCs with PEDOT-GO (0.75) (**f**) time-resolved PL spectra of HTL-free sample and PSCs with PEDOT-GO (0.75).

The steady-state photoluminescence (PL) and the time-resolved photoluminescence (TRPL) tests were also conducted to evaluate the hole extraction ability of the HTL [27]. In Figure 3e, with the introduction of PEDOT-GO as HTL, the intense fluorescence at 790 nm is obviously suppressed. As shown in Figure 3f, for the control sample, the fast decay time ($\tau_1$) was 133.44 ns, and the slow decay time ($\tau_2$) was 63.08 ns, with an amplitude $\tau_{ave}$ ($\tau_{ave} = \Sigma A_i\tau_i^2/\Sigma A_i\tau_i$, where $A_1$ and $A_2$ are pre-exponential factors) of 103.72 ns. For the PEDOT-GO sample, $\tau_1$ was 99.63 ns, and $\tau_2$ was 45.71 ns, derived in an amplitude $\tau$ave of 66.34 ns. Obviously, the sharp decrease in the average fluorescence lifetime indicates that the PEDOT-GO film effectively inhibits the charge recombination. This is consistent with the analysis of the polymer structure obtained by Raman. The above experiments further verify the positive effect of the PEDOT-GO film, which dramatically promote the separation and directional transmission of electrons and holes, thus explaining the increased PCE in the device [28].

**Table 1.** Photovoltaic parameters of PSCs (Sweep speed of 0.25 V s$^{-1}$, voltage range 0 V–1.2 V, electrode area of 0.06 cm$^2$).

| PEDOT/GO | Voc | Jsc | FF | PCE | Highest PCE | Rs | Rtr | Rrec |
|---|---|---|---|---|---|---|---|---|
| Control | 1.05 ± 0.01 | 16.58 ± 1.15 | 0.61 ± 0.02 | 10.66 ± 0.46 | 11.12 | 72.05 | 74.12 | 95.13 |
| 0/1 | 1.05 ± 0.02 | 17.62 ± 1.11 | 0.61 ± 0.01 | 11.44 ± 0.57 | 12.01 | 51.06 | 23.23 | 175.1 |
| 0.5/1 | 1.06 ± 0.01 | 18.31 ± 1.09 | 0.62 ± 0.01 | 12.19 ± 0.6 | 12.79 | 53.4 | 30.02 | 140.7 |
| 0.75/1 | 1.09 ± 0.01 | 19.65 ± 0.72 | 0.63 ± 0.01 | 13.60 ± 49 | 14.09 | 52.3 | 23.9 | 187.4 |
| 1/1 | 0.99 ± 0.02 | 17.1 ± 0.95 | 0.57 ± 0.04 | 9.82 ± 1.12 | 10.94 | 75.04 | 76.82 | 80.54 |
| Spiro | 1.09 ± 0.01 | 20.20 ± 0.50 | 0.60 ± 0.01 | 13.18 ± 31 | 13.49 | 50.2 | 25.1 | 147.9 |

The stability of solar cells samples, without encapsulation, was further evaluated and compared in the air with the humidity of ~35% [29]. As shown in Figure 4, after ten-days placement, the PCE of spiro-OMeTAD-based solar cells decreased to 75% of the initial value, while the PCE of PEDOT-GO-based solar cells still maintained 90% of the initial value. As an approximation, the time at which the efficiency has degraded to 80% of the initial value was denoted as $T_{s80}$ [30]. We can observe that the $T_{s80}$ of spiro-OMeTAD-based solar cells was about 5 days, while it needed more than 10 days for that of PEDOT-GO-based solar cells. The device with PEDOT-GO HTL takes twice as long to fall to the same level of spiro-OMeTAD-based solar cells, which suggests that better durability can be realized by the introduction of PEDOT-GO composite films.

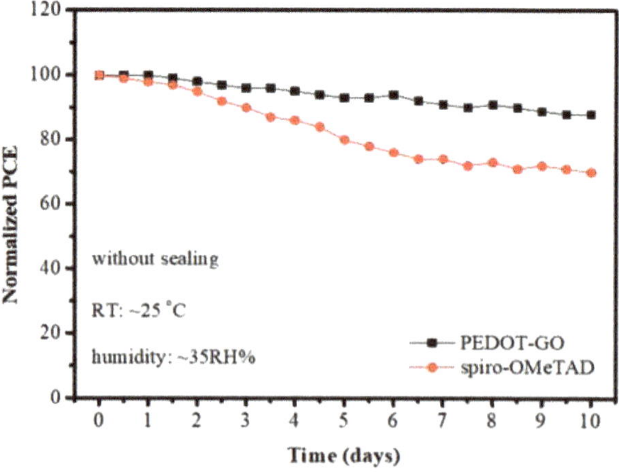


**Figure 4.** The durability test of spiro-OMeTAD-based solar cells and PEDOT-GO-based solar cells.

## 4. Conclusions

In this paper, the PEDOT-GO composite film was successfully prepared as a hole transport layer for the PSCs. The functional thin film significantly inhibited the recombination of holes and electrons, improved the current density, and finally enhanced the PCE of the PSCs. With the mass ratio of 0.75, the highest PCE reaches 14.09%, which is 26% higher than that of the HTL-free sample, and similar results were obtained from Spiro-OMeTAD devices. Compared with Spiro-OMeTAD, PTAA, P3HT, and other traditional hole transport materials, it is much cheaper and more suitable for large-scale applications.

**Author Contributions:** Conceptualization, J.L. and S.W.; Data curation, T.Y.; Formal analysis, T.Y.; Investigation, J.L. and T.Y.; Methodology, S.W.; Resources, T.Y. and S.W.; Writing—original draft, T.Y.; Writing—review & editing, J.L. and S.W. All authors have read and agreed to the published version of the manuscript.

**Funding:** Key Program for Intergovernmental S&T Innovation Cooperation Projects of National Key R&D Program of China (No. 2019YFE0107100).

**Institutional Review Board Statement:** Not applicable.

**Informed Consent Statement:** Not applicable.

**Data Availability Statement:** Not applicable.

**Acknowledgments:** The authors would like to thank Xinfang Cui from the shiyanjia lab for the SEM analysis (www.Shiyanjia.com).

**Conflicts of Interest:** The authors declare no conflict of interest.

## References

1. Song, Z.N.; McElvany, C.L.; Phillips, A.B.; Celik, I.; Krantz, P.W.; Watthage, S.C.; Liyanage, G.K.; Apul, D.; Heben, M.J. A technoeconomic analysis of perovskite solar module manufacturing with low-cost materials and techniques. *Energy Environ. Sci.* **2017**, *10*, 1297–1305. [CrossRef]
2. Chen, X.; Xu, G.; Zeng, G.; Gu, H.W.; Chen, H.Y.; Xu, H.T.; Yao, H.F.; Li, Y.W.; Hou, J.H.; Li, Y.F. Realizing Ultrahigh Mechanical Flexibility and >15% Efficiency of Flexible Organic Solar Cells via a Welding Flexible Transparent Electrode. *Adv. Mater.* **2020**, *32*, 1908478. [CrossRef] [PubMed]
3. Correa-Baena, J.P.; Tress, W.; Domanski, K.; Anaraki, E.H.; Turren-Cruz, S.H.; Roose, B.; Boix, P.P.; Grätzel, M.; Saliba, M.; Abate, A.; et al. Identifying and suppressing interfacial recombination to achieve high open-circuit voltage in perovskite solar cells. *Energy Environ. Sci.* **2017**, *10*, 1207–1212. [CrossRef]
4. Jung, E.H.; Jeon, N.J.; Park, E.Y.; Moon, C.S.; Shin, T.J.; Yang, T.Y.; Noh, J.H.; Seo, J.W. Efficient, stable and scalable perovskite solar cells using poly(3-hexylthiophene). *Nature* **2019**, *567*, 511–515. [CrossRef] [PubMed]
5. Arora, N.; Dar, M.I.; Hinderhofer, A.; Pellet, N.; Schreiber, F.; Zakeeruddin, S.M.; Graetzel, M. Perovskite solar cells with cuscn hole extraction layers yield stabilized efficiencies greater than 20%. *Science* **2017**, *358*, 768–771. [CrossRef]
6. Zhao, Q.; Wu, R.; Zhang, Z.; Xiong, J.; He, Z.; Fan, B.; Dai, Z.; Yang, B.; Xue, X.; Cai, P.; et al. Achieving efficient inverted planar perovskite solar cells with nondoped ptaa as a hole transport layer. *Org. Electron.* **2019**, *71*, 106–112. [CrossRef]
7. Han, W.; Ren, G.; Liu, J.; Li, Z.; Guo, W.; Bao, H.; Liu, C.; Guo, W. Recent progress of inverted perovskite solar cells with a modified PEDOT:PSS hole transport layer. *ACS Appl. Mater. Interfaces* **2020**, *12*, 49297–49322. [CrossRef]
8. Lian, X.; Chen, J.; Zhang, Y.; Tian, S.; Qin, M.; Li, J.; Andersen, T.R.; Wu, G.; Lu, X.; Chen, H. Two-dimensional inverted planar perovskite solar cells with efficiency over 15% via solvent and interface engineering. *J. Mater. Chem. A* **2019**, *7*, 18980–18986. [CrossRef]
9. Lin, Y.J.; Ni, W.S.; Lee, J.Y. Effect of incorporation of ethylene glycol into PEDOT:PSS on electron phonon coupling and conductivity. *J. Appl. Phys.* **2015**, *117*, 215501–215504. [CrossRef]
10. Jiang, X.; Yu, Z.; Zhang, Y.; Lai, J.B.; Li, J.J.; Gurzadyan, G.G.; Yang, X.C.; Sun, L.C. High-Performance Regular Perovskite Solar Cells Employing Low-Cost Poly(ethylenedioxythiophene) as a Hole-Transporting Material. *Sci. Rep.* **2017**, *7*, 42564. [CrossRef]
11. Yu, W.; Wang, K.; Guo, B.; Qiu, X.; Hao, Y.; Chang, J.J.; Li, Y. Effect of ultraviolet absorptivity and waterproofness of poly(3,4-ethylenedioxythiophene) with extremely weak acidity, high conductivity on enhanced stability of perovskite solar cells. *J. Power Sources* **2017**, *358*, 29–38. [CrossRef]
12. Wu, Z.; Bai, S.; Xiang, J.; Yuan, Z.; Yang, Y.; Cui, W.; Gao, X.Y.; Liu, Z.; Jin, Y.Z.; Sun, B.Q. Efficient planar heterojunction perovskite solar cells employing graphene oxide as hole conductor. *Nanoscale* **2014**, *6*, 10505–10510. [CrossRef]
13. Wang, M.; Duan, X.; Xu, Y.; Duan, X. Functional three-dimensional graphene/polymer composites. *ACS Nano* **2016**, *10*, 7231–7247. [CrossRef]

14. Oger, N.; Lin, Y.T.F.; Labrugere, C.; le Grognec, E.; Rataboul, F.; Felpin, F.X. Practical and scalable synthesis of sulfonated graphene. *Carbon* **2016**, *69*, 342–350. [CrossRef]
15. Yu, J.C.; Hong, J.A.; Jung, E.D.; Kim, D.B.; Baek, S.M.; Lee, S.; Cho, S.; Park, S.S.; Choi, K.J.; Song, M.H. Highly efficient and stable inverted perovskite solar cell employing PEDOT:GO composite layer as a hole transport layer. *Sci. Rep.* **2018**, *8*, 1070. [CrossRef]
16. Zhao, H.; Zhao, S.Q.; Li, Q.; Khan, M.R.; Liu, Y.; Lu, P.; Huang, C.X.; Huang, L.J.; Jiang, T. Fabrication and properties of waterborne thermoplastic polyurethane nanocomposite enhanced by the POSS with low dielectric constants. *Polymer* **2020**, *209*, 122992. [CrossRef]
17. Zhao, H.; She, W.; Shi, D.; Wu, W.; Zhang, Q.C.; Li, R.K.Y. Polyurethane/POSS nanocomposites for superior hydrophobicity and high ductility. *Compos. Part B Eng.* **2019**, *177*, 107441. [CrossRef]
18. Xue, T.Y.; Chen, G.S.; Hu, X.T.; Su, M.; Huang, Z.Q.; Meng, X.C.; Jin, Z.; Ma, J.; Zhang, Y.Q.; Song, Y.L. Song.Mechanically Robust and Flexible Perovskite Solar Cells via a Printable and Gelatinous Interface. *ACS Appl. Mater. Interfaces* **2021**, *13*, 19959–19969. [CrossRef]
19. Xu, B.; Gopalan, S.A.; Gopalan, A.I.; Muthuchamy, N.; Lee, K.P.; Lee, J.S.; Jiang, Y.; Lee, S.W.; Kim, S.W.; Kim, J.S.; et al. Functional solid additive modified PEDOT:PSS as an anode buffer layer for enhanced photovoltaic performance and stability in polymer solar cells. *Sci. Rep.* **2021**, *7*, 45079. [CrossRef]
20. Zhao, H.; Zhao, S.Q.; Hu, G.H.; Zhang, Q.C.; Liu, Y.; Huang, C.X.; Li, W.; Jiang, T.; Wang, S.F. Synthesis and characterization of waterborne polyurethane/polyhedral oligomeric silsesquioxane composites with low dielectric constants. *Polym. Adv. Technol.* **2019**, *30*, 2313–2320. [CrossRef]
21. Wang, Y.; Hu, Y.; Han, D.; Yuan, Q.; Cao, T.; Chen, N.; Zhou, D.; Cong, H.; Feng, L. Ammonia-treated graphene oxide and PEDOT:PSS as hole transport layer for high-performance perovskite solar cells with enhanced stability. *Org. Electron.* **2019**, *70*, 63–70. [CrossRef]
22. Zhou, Y.; Mei, J.; Feng, J.J.; Sun, D.W.; Mei, F.; Xu, J.X.; Cao, X. Effects of PEDOT:PSS:GO composite hole transport layer on the luminescence of perovskite light-emitting diodes. *RSC Adv.* **2020**, *10*, 26381–26387. [CrossRef]
23. Cho, J.S.; Jang, W.; Mun, S.C.; Yi, M.J.; Park, J.H.; Wang, D.H. Tuning surface chemistry and morphology of graphene oxide by γ-ray irradiation for improved performance of perovskite photovoltaics. *Carbon* **2018**, *139*, 564–571. [CrossRef]
24. Liu, X.; Hu, L.; Wang, R.; Li, J.; Gu, H.; Liu, S.; Zhou, Y.; Tu, G. Flexible perovskite solar cells via surface-confined silver nanoparticles on transparent polyimide substrates. *Polymers* **2019**, *11*, 427. [CrossRef]
25. Wang, Y.; Wang, S.; Chen, X.; Li, Z.; Wang, J.; Li, T.; Deng, X. Largely enhanced voc and stability in perovskite solar cells with modified energy match by couple 2D interlayers. *J. Mater. Chem. A* **2018**, *6*, 4860–4867. [CrossRef]
26. Li, R.; Liu, M.N.; Matta, S.K.; Hiltunen, A.; Deng, Z.F.; Wang, C.; Dai, Z.C.; Russo, S.P.; Vivo, P.; Zhang, H.C. Sulfonated Dopant-Free Hole-Transport Material Promotes Interfacial Charge Transfer Dynamics for Highly Stable Perovskite Solar Cells. *Adv. Sustain. Syst.* **2021**, 2100244. [CrossRef]
27. Kim, M.; Yi, M.; Jang, W.; Kim, J.K.; Wang, D.H. Acidity Suppression of Hole Transport Layer via Solution Reaction of Neutral PEDOT:PSS for Stable Perovskite Photovoltaics. *Polymers* **2020**, *12*, 129. [CrossRef]
28. Zhang, H.C.; Liu, M.N.; Yang, W.J.; Judin, L.; Hukka, T.I.; Priimagi, A.; Deng, Z.F.; Vivo, P. Thionation Enhances the Performance of Polymeric Dopant-Free Hole-Transporting Materials for Perovskite Solar Cells. *Adv. Mater. Interfaces* **2019**, *6*, 1901036. [CrossRef]
29. Ghadiri, M.; Kang, A.K.; Gorji, N.E. XRD characterization of graphene-contacted perovskite solar cells: Moisture degradation and dark-resting recovery. *Superlattices Microstruct.* **2020**, *146*, 106677. [CrossRef]
30. Canil, L.; Salunke, J.; Wang, Q.; Liu, M.N.; Köbler, H.; Flatken, M.; Gregori, L.; Meggiolaro, D.; Ricciarelli, D.; Angelis, F.D.; et al. Halogen-Bonded Hole-Transport Material Suppresses Charge Recombination and Enhances Stability of Perovskite Solar Cells. *Adv. Energy Mater.* **2021**, *35*, 2101553. [CrossRef]

*Article*

# Synthesis and Application of Tackifying Dispersant Poly (Vinyl Alcohol-Acrylic Acid-Triallyl Cyanate)

Xiaoyan Chen [1], Weizhi Huang [1], Bobing He [1,*] and Yafeng Zhang [2]

[1]  College of Chemistry, Sichuan University, Chengdu 610065, China; cxyyyazj@163.com (X.C.); euihsm5029739@163.com (W.H.)

[2]  College of Materials Science and Engineering, Xihua University, Chengdu 610065, China; zyffcb@163.com

*  Correspondence: hebobing@scu.edu.cn

**Abstract:** At present, the thickener market is relatively advanced. Only by imparting thickeners with new properties can they meet the needs of the current market. In this work, a new modified tackifying dispersant poly (vinyl alcohol-acrylic acid-triallyl cyanate) (PVA-AA-t) was prepared via alcoholysis of a random copolymer composed of vinyl acetate (VAc), acrylic acid (AA), and triallyl cyanate (TAC) by a one-step high-temperature solution polymerization in methanol, which was a relatively simple method. The structure of the polymer was characterized by FTIR and TG. FTIR proved the successful synthesis of PVA-AA-t, while TG showed the thermal stability of PVA-AA-t at around 100 °C. The excellent thickening properties of the PVA-AA-t were observed using a nano particle size analyzer and a rotary viscometer. The nano particle size analyzer showed that the PVA-AA-t particles swelled in water to nearly nine times their initial size. The rotary viscometer showed that the viscosity of PVA-AA-t in water increased significantly, while PVA-AA-t was sensitive to electrolytes and pH, which changed the polymer molecular chain from stretched to curled, resulting in a decrease in viscosity. In addition, the dispersion properties of PVA-AA-t and a common thickener as graphene (Gr) dispersants were compared. The results indicate that PVA-AA-t has very good compatibility with Gr, and can effectively disperse Gr, because of the introduction of weak polar molecules (VAc) to the polymer molecules, changing their polarity, meaning that it is possible to use PVA-AA-t in the dispersion of Gr and other industrial applications (such as conductive textile materials, Gr batteries, etc.) derived from it.

**Keywords:** thickener; dispersant; graphene

**Citation:** Chen, X.; Huang, W.; He, B.; Zhang, Y. Synthesis and Application of Tackifying Dispersant Poly (Vinyl Alcohol-Acrylic Acid-Triallyl Cyanate). *Polymers* **2022**, *14*, 557. https://doi.org/10.3390/polym14030557

Academic Editors: Wei Wu, Hao-Yang Mi, Chongxing Huang, Hui Zhao and Tao Liu

Received: 13 December 2021
Accepted: 20 January 2022
Published: 29 January 2022

**Publisher's Note:** MDPI stays neutral with regard to jurisdictional claims in published maps and institutional affiliations.

## 1. Introduction

As we know, thickeners can be used to improve fluid viscosity and fluid rheology [1,2]. Adding a small amount of thickener to the required thickening system can significantly increase the apparent viscosity of the system. At present, thickeners are widely used in oil exploitation [3], agriculture [4], food processing [5], daily chemical products [6], textile printing and dyeing [7], the pharmaceutical industry, and other fields [8–11]. With the development of industry, the variety of thickeners is gradually increasing, among which cellulose [12] and polyacrylic acid thickeners [8] are the most widely used. At present, with the thickener market being well established, domestic and foreign researchers have begun to devote themselves to the research of various modified thickeners.

Valeria et al. [13] reported a study on the rheological properties of polymer chains hydrophobically modified with a small amount of N, N-dialkyl acrylamide (N, N-dihexylacrylamide (DHAM) and N, N-dioctylacrylamide (DOAM)). Chang-E Zhou et al. [14] prepared an associative thickener by compounding two polyacrylate-based copolymers—cationic starch and polyacrylic acid—mediated by polyethylene glycol and a polyacrylamide crosslinker, which were used for digital printing of nylon carpet with enhanced performance. Abdelrahman et al. [15] reported that the thickener of biodegradable poly (lactic acid-methacrylic acid)-crosslinked polymer hydrogel was formulated in situ via a one-pot reaction employing polycondensation of lactic acid and methacrylic acid followed by free radical polymerization with N, N-methylene diacrylamide crosslinker. Tao Guan et al. [16] reported a polymerizable acrylic-type hydrophobically modified ethoxylated urethane (HEUR), which was end-functionalized by the reactive methacrylate group and hydrophobic octadecyl tail to significantly improve the viscoelastic behavior of waterborne systems such as coatings and inks.

This modified thickener has a remarkable thickening effect; however, due to the variety of raw materials, high price, and complex synthesis steps, its commercial application value is greatly limited. Meanwhile, researchers are rarely involved in the application of modified thickeners [17]. For example, applications in wearable and implantable conductive textile materials have increased in recent years [18–20], one method of which is preparing the graphene (Gr) finished onto the surface of the fabric via multiple impregnation [20,21]. However, the surface of Gr lacks groups that form a good combination with textile materials, showing the poor adhesion between Gr and the matrix, along with weak physical fastness, seriously affecting the product quality [22]. Therefore, in order to improve Gr-Gr and Gr-fabric adhesion, as well as enhance the conductivity, thickener and dispersant should generally be added to graphene slurry [23], while the addition of a dispersant (such as alkyl diphenyl ether sulfonate, naphthalene sulfonic acid condensate, etc.) will affect the flexibility of the fabric. Therefore, in this paper, from the perspective of reducing the use of dispersant while improving the adhesion of Gr, a modified tackifying dispersant was synthesized, which can not only maintain the thickening performance, but also effectively disperse Gr.

## 2. Materials and Reagent

Vinyl acetate (VAc), acrylic acid (AA), azodiisobutyronitrile (AIBN), anhydrous methanol, and sodium hydroxide were purchased from Chengdu Kelong Chemical Co., Ltd. (Chengdu, China). Triallyl cyanate (TAC) was obtained from Shanghai McLean Biochemical Technology Co., Ltd. Graphene (Gr) (Shanghai, China) powder was purchased from Jiangsu Xianfeng Nano Material Technology Co., Ltd. (Jiangsu, China). Except for the VAc—which requires distillation in order to remove the polymerization inhibitor before use—and AIBN, which needs recrystallization, the other reagents were used as received.

### 2.1. Synthesis of PVAc-AA-t

Random copolymerization of poly (vinyl acetate-acrylic acid-triallyl cyanate) (PVAc-AA-t) was carried out as follows: Typically, calculated amounts of VAc, AA, and TAC were added sequentially to methanol (100% of monomer mass) in a 500 mL four-necked round-bottomed flask equipped with a mechanical stirrer, condenser, and nitrogen inlet. After bubbling nitrogen for 30 min to remove oxygen, the reaction was carried out at 65 °C for 3 h, with methanol solution of 5% AIBN (5‰ of monomer mass) slowly added during this process, followed by continuous stirring for 2 h. In this process, the viscosity of the system increased gradually, releasing a lot of heat, and finally forming transparent polymer gel particles (PVAc-AA-t).

The polymerization of polyacrylic acid (PAA) was carried out as follows: Typically, AA (100 mL), water as a solvent (250 mL), and the oxidant potassium persulfate (KPS) (5‰ of monomer molar mass) were added sequentially to a 500 mL four-necked round-bottomed flask equipped with a mechanical stirrer, condenser, and nitrogen inlet. After bubbling nitrogen for 30 min to remove oxygen, the reaction was carried out at −2 °C for 5 h, with the aqueous solution of 1% reducing agent (NaHSO$_3$-FeCl$_2$) (where n(KPS): n(NaHSO$_3$): n(FeCl$_2$) = 5:1:1) slowly added during this process. Throughout this process, the viscosity of the system increased gradually, releasing a lot of heat and, finally, forming PAA (the polymerization process of poly (acrylic acid-triallyl cyanate) (PAA-t) was the same as that of PAA, except for TAC).

*2.2. Alcoholysis of PVAc-AA-t*

The gelatin particles synthesized in the reaction were added to a methanol solution of 0.5 mol/L NaOH at room temperature, with mechanical stirring throughout the whole process. Meanwhile, a methanol solution of 0.5 mol·L$^{-1}$ NaOH was continuously added until the solution showed weak alkalinity, and could transform -OCOCH$_3$ on the surfaces of gel particles into -OH, and transform -COOH into -COONa. In the alcoholysis stage, a lot of heat was produced, producing some colored substances because of side reactions if the temperature in the system was too high. Then, the obtained poly (vinyl alcohol-acrylic acid-triallyl cyanate) (PVA-AA-t) was rinsed with anhydrous methanol several times and dried in a vacuum at 50 °C until all of the solvents were removed. Finally, the dried samples were ground to an appropriate size by a ball mill, the reaction and alcoholysis of PVA-AA-t is shown in Figure 1. The alcoholysis of PAA and PAA-t was carried out in aqueous sodium hydroxide solution (0.5 mol·L$^{-1}$). The specific operation steps were the same as for the alcoholysis of PVA-AA-t.

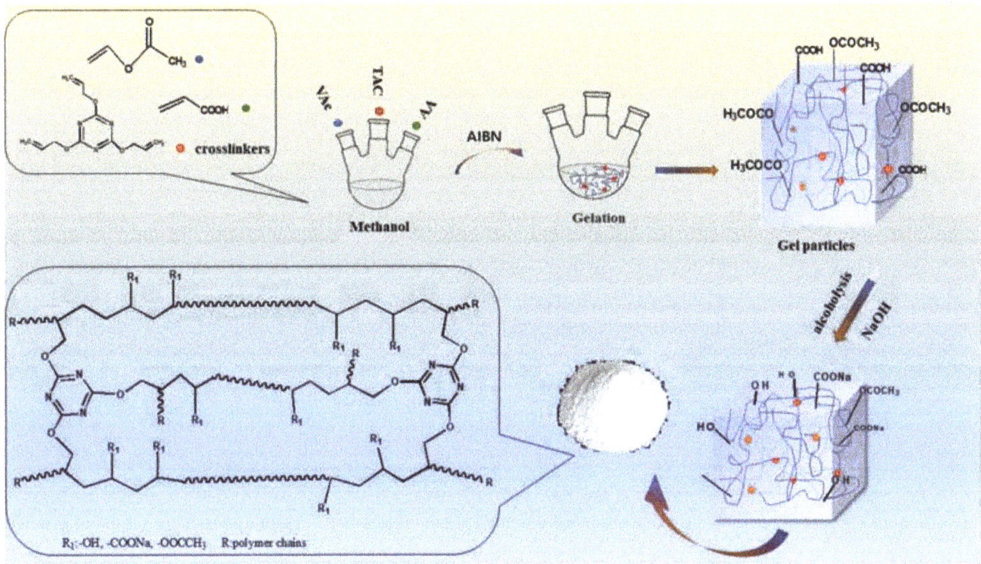


**Figure 1.** Flowchart of the reaction and alcoholysis of PVA-AA-t.

## 3. Characterizations

The viscosity of the PVA-AA-t aqueous solution (25 ± 1 °C) was measured using an NXS-11A viscometer at different shear rates; the formula is as follows:

$$\mu = \frac{T}{\Gamma} \tag{1}$$

where $\mu$ is the viscosity (Pa·s), $T$ is the shear stress (Pa), and $\Gamma$ is the shear rate (s$^{-1}$).

The FTIR spectra of the samples were recorded on a Nicolet iS50 Fourier-transform infrared spectrophotometer (Tokyo, Japan) using KBr pellets (frequency range from 4000 to 400 cm$^{-1}$; each sample was scanned 32 times). In this experiment, the size and size distribution of the polymer were measured using a HELOS KR dry wet laser particle size analyzer produced by Sympatec GmbH (Clausthal-Zellerfeld, Germany). The polymer solid powder was tested by a dry method with compressed air. The sample pool was selected for wet testing of the polymer aqueous solution. The resistivity of the Gr electrode sheet was measured using an FT-340 series double-electrometric four-probe square resistance resistivity tester (ROOKO Instruments, Tokyo, Japan). Thermogravimetric analysis (TGA) was carried out on a TGA2 (METTLER TOLEDO, Columbus, OH, USA) to characterize the thermal stability of the samples. Samples (5–10 mg) were heated from room temperature to 500 °C at a heating rate of 10 °C·min$^{-1}$.

## 4. Results and Discussion

FTIR measurement was employed to investigate the structure of PVA-AA-t, PAA-t, and PAA, as shown in Figure 2. In the three curves, the strong absorption bands at 3000–3500 cm$^{-1}$ are assigned to -OH of H$_2$O or VAc after alcoholysis (only in PVA-AA-t). The absorption bands at 2922 cm$^{-1}$ should be assigned to C-H. The strong absorption bands 1557 cm$^{-1}$ and 1406 cm$^{-1}$ can be attributed to the symmetric and antisymmetric stretching vibration peaks of COO-, respectively. However, the weak absorption at 1717 cm$^{-1}$ in the PVA-AA-t curve should also be noted, which represents the stretching vibration of carbonyl (-C=O) in the acetate group, indicating that the ester bond in the polymer molecule is incompletely alcoholized [24,25]; the C-O-C peaks of the ester bond at 1251 cm$^{-1}$ and 1024 cm$^{-1}$ also confirm this.

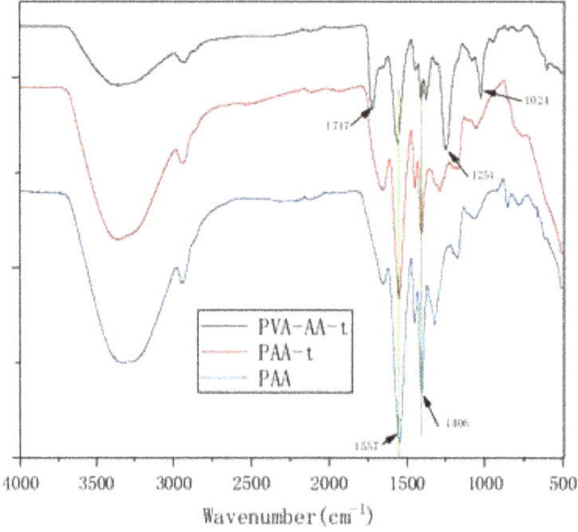


**Figure 2.** FTIR spectra of PVA-AA-t, PAA-t, and PAA (KBr).

Figure 3 shows the TG curve of PVA-AA-t. Under a nitrogen atmosphere, the mass decreases for the first time at 50–100 °C, and the weight loss rate is 13%. This is mainly because the polymer absorbs moisture at room temperature. The weight loss rate of the polymer is 10% at 180–215 °C, because the carboxyl group in the polymer molecular chain is removed. The weight loss rate at 270–340 °C is 20%, which is caused by the breaking and decomposition of the ester or hydroxyl bonds in the molecular chain. When the temperature is higher than 430 °C, the mass of the polymer decreases rapidly, due to the fracture and decomposition of C-C in the main chain of the polymer. In general, the structure of the polymer is damaged when the temperature is higher than 180 °C, so it can be used in below 150 °C without damaging the properties of the polymer.

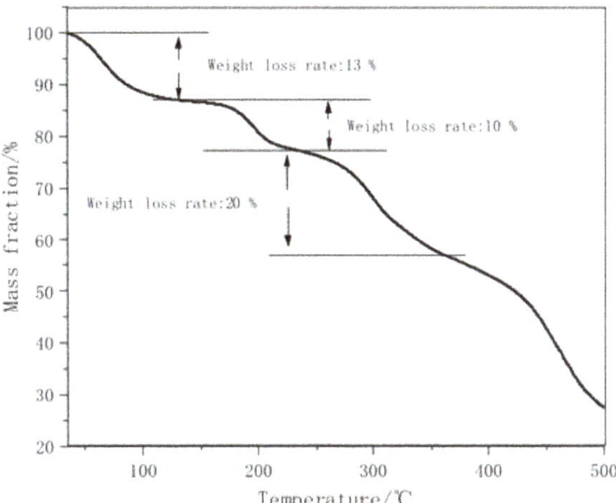

**Figure 3.** TG curve of PVA-AA-t under $N_2$.

We dissolved the PVA-AA-t in water and tested the viscosity, as shown in Figure 4. With the increase in the polymer's concentration, the viscosity of the aqueous solution increases exponentially. When the concentration is less than 0.2%, the viscosity is less than 1 Pa·s (when the solution concentration is only 0.1%, there is almost no thickening effect). When the concentration increases to 0.3%, the viscosity increases rapidly to 4 Pa·s. The viscosity of 0.5% aqueous solution can reach 11 Pa·s, showing a clear thickening effect similar to that of other thickeners. Under shear force, with the increase in shear rate, the viscosity of the solution decreases rapidly, showing the characteristics of a pseudoplastic fluid.

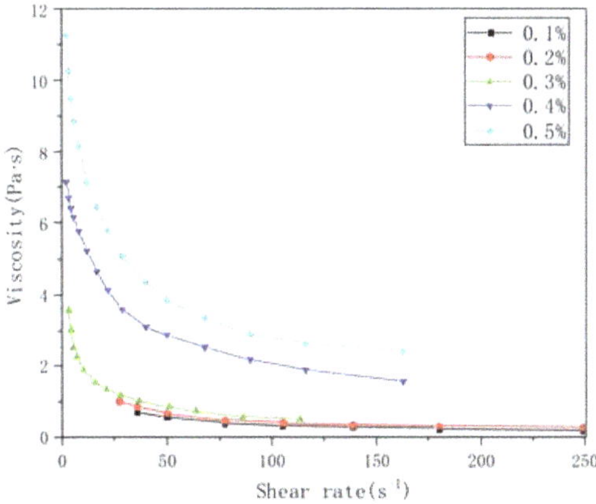

**Figure 4.** Viscosity curves of PVA-AA-t aqueous solutions with different concentrations.

Figure 5 shows the variation in the viscosity of PVA-AA-t solutions with different monomer ratios. It can be seen from the figure that the viscosity is the highest when VAc:AA = 7.5:2, reaching 3.6 Pa·s. With the increase in the AA content of the copolymer, the initial viscosity decreases gradually. When the ratio of VAc:AA is 6.5:3, the initial viscosity drops to 2.14 Pa·s. When the ratio of VAc:AA is 2:7.5, the whole system has no thickening effect, showing Newtonian fluid characteristics. As can be seen from previous literature [16,26], sodium polyacrylate has been used as a thickener for many years; however, under these experimental conditions, we found that increasing the content of acrylic acid did not further improve the thickening effect—possibly due to the simultaneous copolymerization of three monomers in the system, affecting their respective reaction rates. Moreover, at high temperatures, the reaction rate of AA is faster than that of VAc, and the formed molecular chain is shorter, which leads to a poor thickening effect. On the other hand, VAc with a relatively slow reaction rate can form long-chain molecules at this temperature (65 °C).

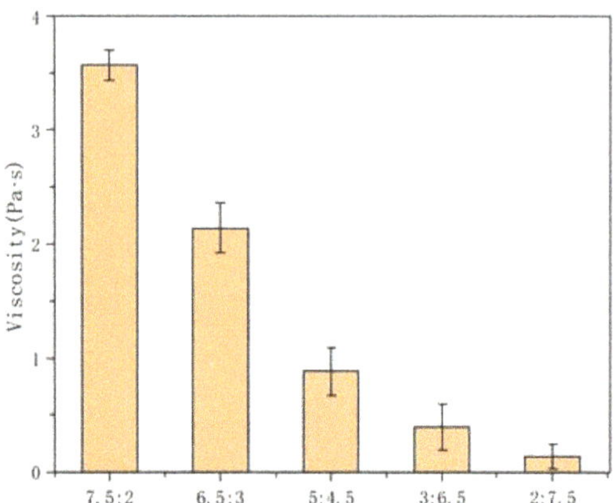

**Figure 5.** Effects of different monomer ratios on the viscosity of PVA-AA-t aqueous solution (concentration: 0.3%; TAC = 0.2 phr).

In this experiment, triallyl cyanate (TAC), which contains three ethylene groups, was used as a crosslinking agent. The crosslinking structure can be formed in the reaction process, and the more TAC is used, the higher the crosslinking density. From the Figure 6, we found that the viscosity (1.79 Pa·s) was relatively low when the dosage of TAC was 0.1 phr, which may have been due to the low dosage of the crosslinking agent, low crosslinking density of the molecular chain, and the polymer stretching and dissolving in water, resulting in the decrease in viscosity. Further increasing the amount of TAC, there was little change in viscosity, because TAC acts as a connecting point in the molecular structure [27], and the carboxyl and hydroxyl groups in the molecular chain play a thickening role. Therefore, from the perspective of application economy, the crosslinking agent is expensive; the dosage of TAC in the synthesis is 0.2 phr.

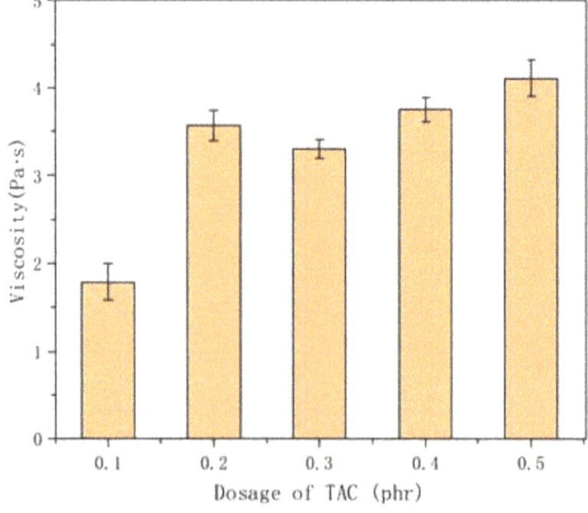

**Figure 6.** Effects of different amounts of crosslinking agent on the viscosity of polymer solution (concentration: 0.3%; VAc:AA = 7.5:2).

In this experiment, we compared the particle size distribution (PSD) of PVA-AA-t before and after swelling, as shown in Figure 7. The PSD of solid powder was in the range of 0.5–150 μm, the median diameter (D50) was 17.90 μm, and the average particle size was 24.74 μm (note: the solid polymer powder can be continuously ground by a ball mill, and the polymer powder within the fixed particle size range can be selected by standard sample separation). After swelling, the PSD of the polymer was in the range of 20–600 μm, D50 was 156.4 μm, and the average particle size was 187.2 μm. Comparing the two cases, the particle size increased nearly ninefold, showing an obvious tackifying effect. The swelling principle is the combination of non-associative thickening and associative thickening [28]; that is, -COO- in the polymer molecular chain combines with $H_2O$ to form hydrated ions, hindering the flow of molecules in the system, so as to achieve the purpose of thickening. On the other hand, the polymer molecular chain contains a small amount of acrylate chain, forming a comb-like structure. These hydrophobic short chains associate with one another to form a network structure, which can enhance the interaction between polymer particles and further increase the viscosity in water.

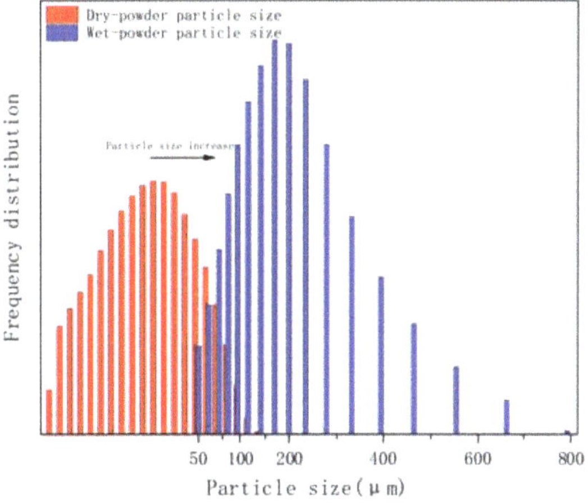


**Figure 7.** Particle size distribution of PVA-AA-t before and after swelling.

Figure 8 shows the effects of different electrolytes on the viscosity of the polymer solution. Under the same conditions, the viscosity decreases significantly (the original viscosity is nearly 4 Pa·s) after adding electrolytes. This is because the electrolyte can partially shield the carboxyl anion on the polymer molecular chain, resulting in the curling of the stretched macromolecular chain formed by the repulsion from the anions, reducing the friction between the molecular chains, and causing the viscosity to decrease rapidly with the addition of the electrolyte. The influence of different electrolytes on the viscosity follows the sequence $CaCl_2$ > KCl > NaCl > $CH_3COONa$. The greater the charge number of divalent ions at the same mole number, the stronger the shielding effect on the polymer molecules [29]. Compared with $K^+$ and $Na^+$, the larger the ion radius, the better the shielding effect on the charge of the molecular chain, making the molecular chain curl further and the viscosity decrease simultaneously. $CH_3COONa$ is a strong base and a weak acid salt. In aqueous solution, $CH_3COO-$ hydrolyzes to form $CH_3COOH$, which weakens the shielding effect on the polymer molecular chain compared with the other three electrolytes. In general, the addition of an electrolyte has a great influence on the thickening effect of PVA-AA-t.

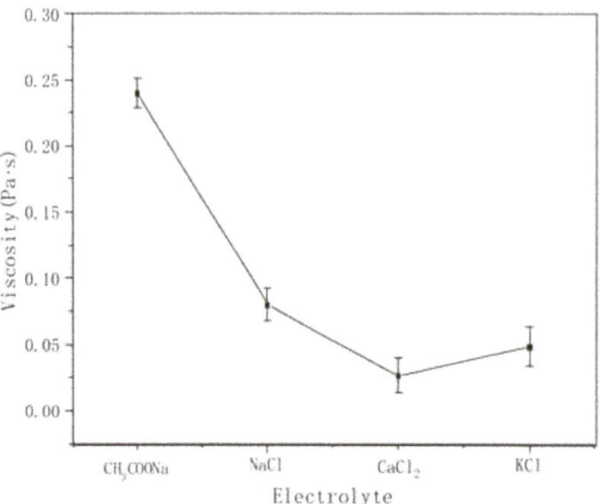

**Figure 8.** Effects of different electrolytes on the viscosity of polymer aqueous solution (polymer concentration: 0.3%, electrolyte concentration: 0.001 mol/L).

There is a carboxylic sodium salt on the polymer molecules, the pH value of which has a great influence on the viscosity of the solution. Therefore, we compared the effects of different pH values on solution viscosity. As shown in Figure 9, the viscosity of the polymer solution (about 4.0 Pa·s) is the highest in the range of pH = 5–7; with the increase in the pH value, the viscosity begins to decrease gradually. When pH = 9, the viscosity decreases to 1.2 Pa·s; further increases in the pH value cause the viscosity to further decrease. When the pH is ~13, the whole solution has no thickening effect—this is mainly because under strong alkaline conditions, the system contains a large amount of ions, which weaken the hydration of the polymer molecular chain to $H_2O$, making the molecular chain return to a curled state, reducing the intermolecular friction, and causing the decrease in viscosity.

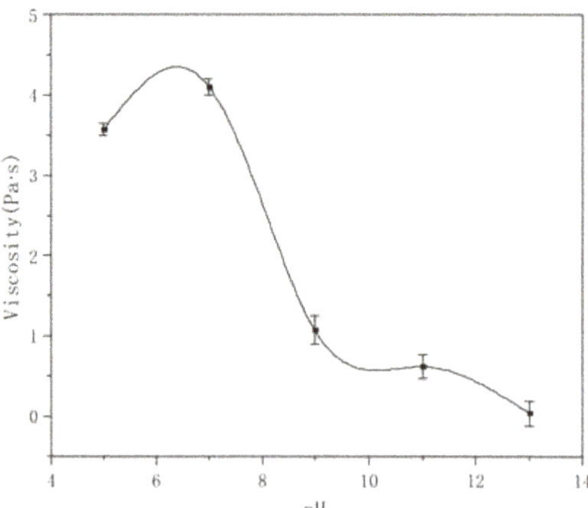

**Figure 9.** Effect of pH value on viscosity (polymer concentration: 0.3%).

On the other hand, when the pH value is reduced, white turbidity begins to appear in the solution. With the decrease in pH value, the sediment increases. We centrifuged the solution to obtain white insoluble matter, which was analyzed by FTIR, as shown in Figure 10. The peaks at 2942 cm$^{-1}$ and 2864 cm$^{-1}$ correspond to the symmetrical stretching vibration peaks of the C-H bond, while 1721 cm$^{-1}$ is the carbonyl peak of the ester bond C=O, but the peak strength is significantly higher than the infrared absorption peak before acidification. It is possible that the carboxylic acid in the molecular chain reacts with the hydroxyl group to form ester bonds during acidification. The C-O-C peaks at 1242 cm$^{-1}$ and 1164 cm$^{-1}$ demonstrate the formation of ester bonds in the molecular chain. It is also possible that the peak intensity increases due to the increase in the carbonyl peak in the acidified carboxyl group, and the infrared absorption peaks of the acidified sample at 1557 cm$^{-1}$ and 1408 cm$^{-1}$ disappear, which may indicate that Na$^+$ in the carboxylate is replaced by H$^+$. The above research shows that PVA-AA-t has excellent thickening performance, and is sensitive to electrolytes and pH.

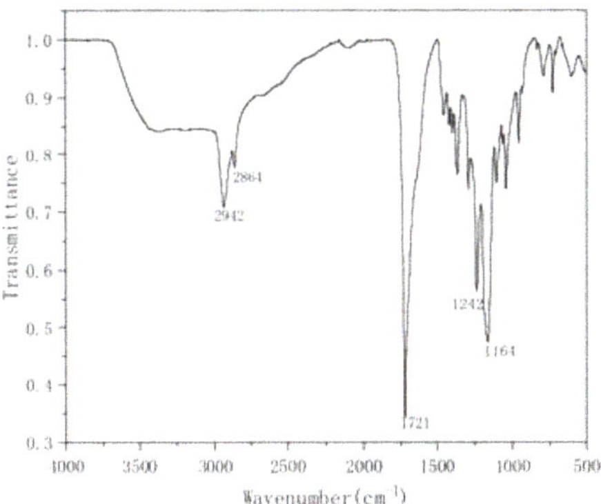

**Figure 10.** Infrared absorption of white turbidity after acidification.

In addition, we studied the dispersion of Gr by PVA-AA-t. Gr was added to the polymer solution for high-speed stirring and dispersion, and the change in the system's viscosity was measured using a portable viscometer (model: VL7-100B-d21-TS). At the same time, the dispersibility of Gr by PAA and PAA-t was compared. As shown in Figure 11, the initial viscosity of the three thickeners (PAA, PAA-t, and PVA-AA-t) was 4.3 Pa·s, 3.6 Pa·s, and 0.8 Pa·s, respectively. When t = 0 s, 0.5% Gr was added to three solutions. With the increase in dispersion time, the viscosity of the PAA solution decreased to a certain extent, and then remained stable. After long-time dispersion and grinding, there was still a large amount of agglomerated Gr in the PAA solution. This phenomenon also appeared in the PAA-t system, the viscosity of which decreased after the addition of Gr, with a large amount of agglomerated Gr. Conversely, when Gr was added to PVA-AA-t, the viscosity increased rapidly to more than 10 Pa·s under high-speed stirring, showing paste-like characteristics without fluidity. After full grinding and dispersion, the Gr slurry appeared as a dark glossy paste. Three kinds of Gr slurry were evenly coated on silicon wafers and glass (thickness: 60 μm) with an applicator, before being transferred to a vacuum-drying oven and dried at 60 °C for 5 h. The dried silicon wafers were used to measure the resistivity. The dried glass sheets were placed under a microscope (YKP-700C magnification: ×600) for observation (as shown in Figure 12a,b). As shown in Figure 12, the Gr dispersed by PVA-AA-t showed good resistivity, while the Gr dispersed by the other two thickeners (PAA, PAA-t) showed poor resistivity; this is because the Gr dispersed by PVA-AA-t was evenly distributed, without obvious agglomeration. After the Gr slurry was dried, it can be seen from Figure 12b that it showed evenly distributed pores (left by the dried polymer powder particles), and the Gr particles were overlapped with one another, showing good adhesion and good conductivity. However, the Gr dispersed by PAA and PAA-t had a lot of agglomeration, and the existence of the acrylic acid on the polymer chains likely inhibited bridging between particles, causing the force between the Gr particles to weaken, yielding obvious cracking and even falling off after drying, which affected the conductivity of the Gr electrode sheet.

16. Guan, T.; Du, Z.; Peng, J.; Zhao, D.; Sun, N.; Ren, B. Polymerizable Hydrophobically Modified Ethox-ylated Urethane Acrylate Polymer: Synthesis and Viscoelastic Behavior in Aqueous Systems. *Macromolecules* **2020**, *53*, 7420–7429. [CrossRef]

17. Umar, K.; Yaqoob, A.A.; Ibrahim, M.N.M.; Parveen, T.; Safian, M.T. Chapter Thirteen-Environmental Applications of Smart Polymer Composites. In *Smart Polymer Nanocomposites: Biomedical and Environmental Applications*; Woodhead Publishing Series in Composites Science and Engineering; Elsevier Inc.: Cambridge, MA, USA, 2021; pp. 295–312.

18. Zheng, S.; Xu, C.; Zhang, K.; Yang, X.; Li, R.; Liu, Y. Preparation and Application of Flexible Conductive Fabric Based on Silk. In *IOP Conference Series: Earth and Environmental Science*; IOP Publishing: Bristol, UK, 2020; Volume 440, p. 022036.

19. Zhang, Y.; Ren, H.; Chen, H.; Chen, Q.; Jin, L.; Peng, W.; Xin, S.; Bai, Y. Cotton Fabrics Decorated with Conductive Graphene Nanosheet Inks for Flexible Wearable Heaters and Strain Sensors. *ACS Appl. Nano Mater.* **2021**, *4*, 9709–9720. [CrossRef]

20. Woltornist, S.J.; Alamer, F.A.; McDannald, A.; Jain, M.; Sotzing, G.A.; Adamson, D.H. Preparation of conductive graphene/graphite infused fabrics using an interface trapping method. *Carbon* **2015**, *81*, 38–42. [CrossRef]

21. Ren, J.; Wang, C.; Zhang, X.; Carey, T.; Chen, K.; Yin, Y.; Torrisi, F. Environmentally-friendly conductive cotton fabric as flexible strain sensor based on hot press reduced graphene oxide. *Carbon* **2017**, *111*, 622–630. [CrossRef]

22. Yaqoob, A.A.; Ibrahim, M.N.M.; Ahmad, A.; Reddy, A.V.B. Toxicology and environmental application of carbon nanocom-posite. In *Environmental Remediation Through Carbon Based Nano Composites*; Springer: Berlin/Heidelberg, Germany, 2021; pp. 1–18.

23. Cataldi, P.; Ceseracciu, L.; Athanassiou, A.; Bayer, I.S. Healable cotton–graphene nanocomposite conductor for wearable electronics. *ACS Appl. Mater. Interfaces* **2017**, *9*, 13825–13830. [CrossRef]

24. Jiang, L.; Yang, T.; Peng, L.; Dan, Y. Acrylamide modified poly(vinyl alcohol): Crystalline and enhanced water solubility. *RSC Adv.* **2015**, *5*, 86598–86605. [CrossRef]

25. Moritani, T.; Kajitani, K. Functional modification of poly (vinyl alcohol) by copolymerization: Modification with carboxylic monomers. *Polymer* **1997**, *38*, 2933–2945. [CrossRef]

26. Wang, Y.; Hudson, N.; Pethrick, R.; Schaschke, C. Poly(acrylic acid)–poly(vinyl pyrrolidone)-thickened water/glycol de-icing fluids. *Cold Reg. Sci. Technol.* **2014**, *101*, 24–30. [CrossRef]

27. Chang, S.; Ma, A.W.K.; Lai, H. New Insight into the Preparation of Starch-Based Spherical Microgels with Tunable Volume. *Starch Stärke* **2019**, *71*, 1800288. [CrossRef]

28. Hashizaki, K.; Umeda, R.; Miura, M.; Taguchi, H.; Fujii, M. Preparation and Rheological Properties of Cross-linked Liposomes Using Hydroxypropylmethylcellulose Bearing a Hydrophobic Anchor. *Yakugaku Zasshi J. Pharm. Soc. Jpn.* **2020**, *140*, 435–441. [CrossRef] [PubMed]

29. Jong, L. Poly(acrylic acid) grafted soy carbohydrate as thickener for waterborne paints. *Mater. Today Commun.* **2020**, *23*, 100882. [CrossRef]

*Article*

# Effects of Various Polymeric Films on the Pericarp Microstructure and Storability of Longan (cv. Shixia) Fruit Treated with Propyl Disulfide Essential Oil from the Neem (*Azadirachta indica*) Plant

Muhammad Rafiullah Khan [1,*], Chongxing Huang [1,*], Rafi Ullah [2], Hakim Ullah [2], Ihsan Mabood Qazi [3], Taufiq Nawaz [3], Muhammad Adnan [4], Abdullah Khan [4], Hongxia Su [1] and Liu Ren [1]

[1] School of Light Industry and Food Engineering, Guangxi University, Nanning 530004, China; suhongxia@st.gxu.edu.cn (H.S.); liuren2020@126.com (L.R.)
[2] Department of Agriculture, University of Swabi, Swabi 25130, Pakistan; rafiullah@uoswabi.edu.pk (R.U.); hakimullah277@gmail.com (H.U.)
[3] Department of Food Science and Technology, The University of Agriculture Peshawar, Peshawar 25000, Pakistan; drqazi@aup.edu.pk (I.M.Q.); taufiqnawaz85@gmail.com (T.N.)
[4] Guangxi Key Laboratory of Sugarcane Biology, State Key Laboratory for Conservation and Utilization of Subtropical Agro-Bioresorces, Guangxi University, Nanning 530004, China; adnan.gxu786@gmail.com (M.A.); abdullahgxu@gmail.com (A.K.)
* Correspondence: khan87@gxu.edu.cn (M.R.K.); huangcx@gxu.edu.cn (C.H.)

Citation: Khan, M.R.; Huang, C.; Ullah, R.; Ullah, H.; Qazi, I.M.; Nawaz, T.; Adnan, M.; Khan, A.; Su, H.; Ren, L. Effects of Various Polymeric Films on the Pericarp Microstructure and Storability of Longan (cv. Shixia) Fruit Treated with Propyl Disulfide Essential Oil from the Neem (*Azadirachta indica*) Plant. *Polymers* **2022**, *14*, 536. https://doi.org/10.3390/polym14030536

Academic Editor: Hiroshi Ito

Received: 23 December 2021
Accepted: 24 January 2022
Published: 28 January 2022

**Publisher's Note:** MDPI stays neutral with regard to jurisdictional claims in published maps and institutional affiliations.

**Abstract:** Plant extracts represent a rich repository of metabolites with antioxidant and antimicrobial properties. Neem (*Azadirachta indica*) is a medicinal plant considered the tree of the 21st century. In this study, we investigated the antioxidant and antimicrobial effects of propyl disulfide (PD), a major volatile compound in neem seed, against the pericarp browning (BI), microbial decay incidence (DI), and water loss of longan fruit. Fresh longan cv. Shixia samples were packaged in oriented polypropylene (OPP) and polyethene (PE) packages of different thicknesses (20, 40, and 60 µm). Sterile gauze was fixed inside the packages and 500 uL of PD was placed on them to avoid the direct contact of PD with fruit samples. Packages were sealed immediately to minimize vaporization and stored at 12 ± 1 °C for 18 days. Fruit samples packaged in open net packages served as controls. The results showed that fruit treated with PD in OPP and PE packages significantly prevented losses of water, DI, and BI compared to control treatment. PD also maintained the color, TSS values, TA values, pH values, high peel firmness, high TPC content, and high TFC content, and reduced the activity levels of PPO and POD. Scanning electron microscope (SEM) analysis indicated that the exocarp, mesocarp, and endocarp of longan peel were smooth, uniform, and compact with no free space compared to control, where crakes, a damaged and loose structure, and a lot of fungal mycelia were found. The shortest shelf life of 9 days was observed in control as compared to 18 days in OPP-20 and OPP-40; 15 days in OPP-60, PE-20, and PE-40; and 12 days in PE-60 packaging films. Therefore, PD as a natural antioxidant and antimicrobial agent, in combination with OPP-20 and OPP-40 polymeric films, could successfully be applied commercially to extend the postharvest shelf life of longan.

**Keywords:** longan; fruit; polymeric films; antioxidant activity; enzymatic browning; neem; propyl disulfide; microbial decay; essential oil

## 1. Introduction

Longan (*Dimocarpus longan* L.) is an attractive subtropical fruit of the evergreen tree of the Sapindaceae family. The fruit is widely cultivated in many countries, especially China, Thailand, Vietnam, and Australia. The fruit has high nutritional value and is best when eaten fresh. However, longan is non-climacteric; the fruit is harvested at optimum maturity and does not continue to ripen once harvested. The fruit matures in high temperature and humidity, meaning it deteriorates rapidly once harvested due to pericarp browning and microbial rot. Further, due to the unique pericarp structure of longan fruit, the dehydration and microbial invasion expedite the senescence and browning, consequently shortening its postharvest life [1]. Several studies have been conducted on treatments such as chlorine dioxide [2]; adenosine triphosphate (ATP) [3]; hydrogen peroxide [4]; chitosan [5]; $SO_2$, $ClO_2$, or their combination [6]; thymol coatings and thymol fumigation [7,8]; and many other studies in order to preserve the quality and extend the shelf life. However, due to consumer awareness of the health concerns regarding the residues of the synthetic compounds and the resistance of the microbes to the existing preservatives, there is a need to develop other preservatives that are safe to humans and the environment. Therefore, the continuous use of synthetic compounds needs to be eliminated to ensure the availability of safe and fresh fruits for longer periods of time.

Recently, plant-based extracts have attracted much more interest from researchers due to their biologically active components with antioxidant and antimicrobial properties. Various essential oils have been extracted from plants and have been utilized in the food industry. The neem plant (*Azadirachta indica*) is a rich source of about 300 primary and secondary metabolites, which possess antifungal, antibacterial, and antioxidant properties [9–11]. In another study, more than 140 biologically active compounds have been isolated from different parts [9], which have anti-inflammatory, antihyperglycemic, antiulcer, antimalarial, antifungal, antibacterial, antioxidant, antimutagenic, and anticarcinogenic properties [10,12,13]. So far, most studies have been conducted on neem extracts in the pharmaceutical industry or traditional medicines. Few studies have reported on controlling plant diseases. Propyl disulfide compound is the major volatile compound in neem plant seeds, and our research group previously assessed the antifungal activity levels of propyl disulfide, which effectively inhibited the mycelial growth of fungi, which causes anthracnose [14] and stem end rot [15] in mango fruit, obtaining very promising results. Zakawa et al. [16] studied the effects of neem leaf extract on the fungi causing anthracnose in wild mango. In 2011, Suleiman [17] reported on the effects of neem leaf extract against tomato anthracnose.

Neem extracts have also been used to preserve a wide range of other food products. For example, Serrone et al. [11] preserved the quality of fresh retail meat using neem oil and reported its efficacy against a wide range of bacterial populations. In another study, neem cake oil was used to preserve the quality of fresh retail meat [18]. The antioxidant activity of neem oil was reported in regard to beef lipid oxidation reactions, which extended the shelf life of raw beef patties to 11 days at 4 °C [19].

Enzymatic browning, microbial decay, and water loss are the major concerns in longan fruit; therefore, in this study, we focused on the antioxidant and antimicrobial activity levels of propyl disulfide from neem. Different packaging films were tested to find the best storage conditions. A very simple, cost-effective, and practical method was proposed for PD fumigation. The in-depth antioxidant mechanism of propyl disulfide was assessed regarding the enzymatic browning reaction in longan fruit, which includes phenolic substrates, enzymes, and browning. Pearson's correlation coefficient analysis was carried out to find the relationships between these parameters. The effect of PD on the unique pericarp structure of longan fruit and its three components were analyzed via scanning electron microscopy and are reported in detail in this paper.

## 2. Materials and Methods

### 2.1. Fumigation and Fruit Treatment

Longan fruit cv. Shixia were harvested from a commercial farm in the Guangxi region and transported to the laboratory. Fruit samples of uniform size and color with no defects were prepared. About 500 g of fruit was packed in oriented polypropylene (Jiahang OPP Packing Bag Co., Ltd., Jinhua, Zhejiang, China) of different thicknesses (OPP-20, OPP-40, OPP-60) and polyethylene (Bailiheng PE Packaging Bag Co., Ltd., Shenzhen, China) (PE-20, PE-40, PE-60) packages. A sterile gauze was fixed inside in each package, and 500 uL of the propyl disulfide (a major volatile compound in neem (*Azadirachta indica*) seed; >99% Food grade, Aladdin Reagent Co., Ltd., Shanghai, China; Figure 1) was placed on the gauze to avoid its direct contact with the fruit. Each package was immediately sealed to reduce the evaporation of the compound from the bag. For comparison, fruit samples packaged in open net packages without propyl disulfide were considered the control treatment. All packages were stored at 12 ± 1 °C. The quality of the longan fruit was evaluated every third day using three bags (three replicates) from each treatment.

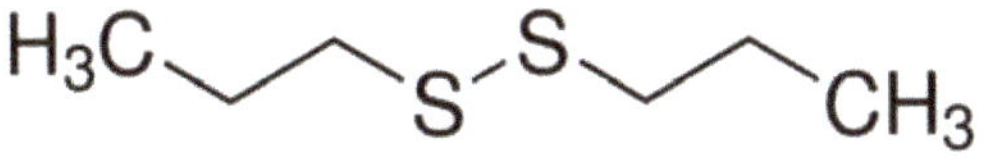

**Figure 1.** Structure of propyl disulfide.

### 2.2. Quality Evaluation Tests

### 2.2.1. Weight Loss, Decay Incidence, and Pericarp Browning Index

Weight loss was determined by weighing the fruit on day 0 and then every third day, and the results were expressed as percentages. Fruit showing visible decay symptoms on the surface were considered decayed, and all the decayed fruit were counted and the decay percent was calculated using the following formula:

$$\text{Percentage of fruit decay} = \frac{\text{Number of decayed fruit}}{\text{Total number of fruit}} \times 100 \tag{1}$$

The pericarp browning index (BI) of longan fruit was determined by observing the amount of total brown area on each fruit surface, whereby 1 = 0% (no browning), 2 = 1–10% (slight browning), 3 => 10–25% (moderate browning), 4 => 25–50% (severe browning), 5 => 50%, calculated using the below equation [20]. Fruit with a browning scale above 3 was the limit of acceptability:

$$\text{BI} = \sum \frac{\text{Browning level} \times \text{number of fruit at each browning level}}{\text{Total number of fruit in the treatment}} \tag{2}$$

### 2.2.2. Color, Firmness, Total Soluble Solids, Titratable Acidity, and pH

Pericarp color (L*, a*, b*) was measured using a Konica Minolta Spectrophotometer (CM-3600d, Konica Minolta Sensing Inc., Tokyo, Japan) using 3 fruit samples from each replicate. The firmness of the peel was measured using a texture analyzer (TA.XT Plus 10752, Godalming, UK) in compression mode. A 2-mm-diameter plunger was used to puncture the fruit to a depth of five mm at a speed of 20 mm min$^{-1}$. The maximum force needed to penetrate the fruit was recorded in newtons (N). A total of 3 fruit samples were used for each test from each replicate. For TSS, TA, and pH, the flesh of about 15 fruit samples from each replicate was ground, filtered through muslin cloth, and clear juice was obtained. TSS was determined using a digital Abbe refractometer (Way-2S, Shanghai Shenguang Instruments and Instrument Co. Ltd., Shanghai, China. Titratable acidity (TA) was carried out via the titration of juice with 0.1 N NaOH. The pH of the longan juice was determined using a pH meter (FE28, Mettler Toledo Co. Ltd., Shanghai, China). TSS, TA, and pH were conducted in triplicate.

### 2.2.3. Extraction and Determination of Phenolic and Flavonoid Contents

Phenolic and flavonoid contents were extracted from the pericarps of 15 fruit samples according to Khan et al. [20]. Reaction mixtures for TPC and TFC were prepared according to the methods of Khan et al. [20] and Dewanto et al. [21], respectively, using a UV–visible spectrophotometer (SPECORD 50 Plus, Analytik Jena, Germany). The results for TPC and TFC were expressed as milligrams of gallic acid per kg and milligrams of catechin per kg of fresh weight, respectively. The experiment was performed in triplicate.

### 2.2.4. Polyphenol Oxidase (PPO) and Peroxidase (POD) Activity Levels

The antioxidant effect of PD on the inhibition of enzymes activity levels was determined using the pericarps of 15 fruit samples. Crude enzyme extract was obtained according to the method used by Khan et al. [7] and Duan et al. [22]. The reaction mixture used to assess the PPO activity was prepared by following the protocol used by Khan et al. [7] and Jiang [23]. Absorbance was recorded at 410 nm for 5 min with a UV–visible spectrophotometer (SPECORD 50 Plus, Analytik Jena, Germany). One unit of enzyme activity was defined as the activity that caused a change of 0.001 in the absorbance per min. The reaction mixture for POD activity was prepared following the protocol used by Khan et al. [7] and Zhang et al. [24] and absorbance was measured using a UV–visible spectrophotometer (SPECORD 50 Plus, Analytik Jena, Germany). One unit of enzyme activity was defined as the amount that caused a change of 0.01 in the absorbance per min. Protein contents were determined according to the method used in [25]. Enzyme (PPO and POD) activity levels were expressed as units $min^{-1}$ $mg^{-1}$ protein. The experiments were performed in triplicate.

### 2.3. Scanning Electron Microscope and Pericarp Microstructure

Longan pericarps were prepared by following the protocols of Yao et al. [1] and Chitbanchong et al. [26], with slight modifications. Briefly, pericarps of 3 mm squares were washed twice with 0.1 M of phosphate buffer (pH 7.4). The pieces were immediately shifted to the 2.5% glutaraldehyde prepared in 0.1 M of phosphate buffer and kept overnight for 24 h at 4 °C. Samples were then rinsed in the same buffer and postfixed in 1% osmium tetroxide for 2 h and stepwise exposed to a series of ethanol–buffer mixtures of 30%, 50%, 70%, 80%, 90%, and 100% ethanol for 15 min for dehydration. The dried samples were then mounted on specimen stubs, sputter-coated with gold, and viewed under SEM (Inspect$^{TM}$, FEI company, Hillsboro, OR, USA) with an accelerating voltage of 20 kV.

### 2.4. Statistical Analysis

Data were subjected to analysis of variance (ANOVA) using Statistix. The least significant difference (LSD) tests were performed to determine the significant differences ($p \leq 0.05$) among the treatments. The correlation analysis was carried out in Microsoft Excel 2016.

## 3. Results

### 3.1. Weight Loss, Decay Incidence, and Pericarp Browning

An increase in weight loss was observed with storage time. Weight loss was significantly higher in control samples throughout the storage time, while in all other packaged samples no significant differences were observed until day 9 (Figure 2). From day 12 onward, some differences in weight loss were observed. Among all the treatments, the lowest weight loss was found in OPP-20. Statistics within the storage intervals also showed that weight loss increased significantly on each quality test day in each treatment; however, OPP-20 comparatively prevented the loss in weight more than other treatments (Figure 2). No decay symptoms were found in any treatment until day 6, except PE-60 and control, which showed decay rates of 3.91% and 6.87%, respectively. On day 9, DI increased in all treatments, being significantly ($p \leq 0.05$) higher in the control, followed by PE-20 and PE-40. Due to the high DI rate, the control treatment was discarded on day 9 (70.62%), while the storage of PE-60 and OPP-60 and PE-20 and PE-40 treatments was discontinued on day 12 and day 15, respectively (Figure 3). Among all treatments, the lowest DI rates were observed in OPP-20 and OPP-40, which extended the longan storage life to up to 18 days (Figure 3). Pericarp browning (BI) was gradually enhanced in all treatments. BI was significantly ($p \leq 0.05$) higher in control followed by PE treatments. Among the OPP films, OPP-20 and OPP-40 maintained lower BI rates than all other packaging films during the entire storage period (Figure 4).

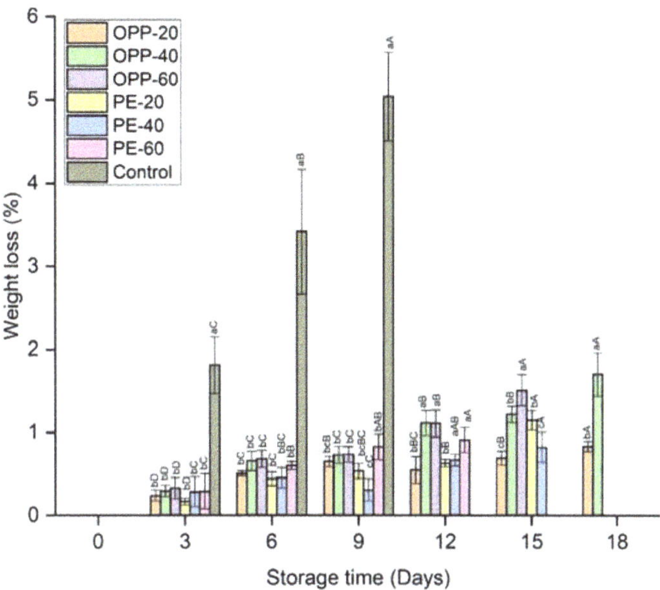

**Figure 2.** Weight loss rates (%) of longan fruit samples in different packaging films fumigated with propyl disulfide and control, stored at 12 ± 1 °C. Vertical bars represent means ± standard deviations (*n* = 3). Different small and capital letters show significant differences between the treatments within the same day and between the days in the same treatment, respectively, via LSD test at $p \leq 0.05$.

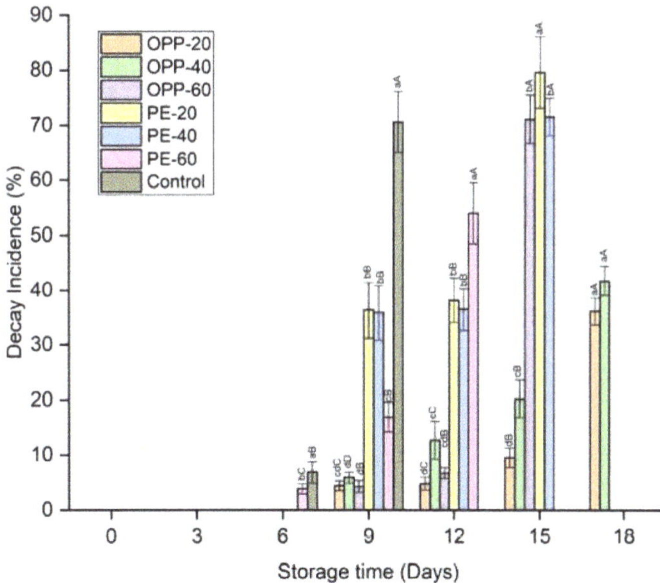

**Figure 3.** Decay incidence rates (%) of longan fruit samples in different packaging films fumigated with propyl disulfide and control, stored at 12 ± 1 °C. Vertical bars represent means ± standard deviations (*n* = 3). Different small and capital letters show significant differences between the treatments within the same day and between the days in the same treatment, respectively, via LSD test at $p \leq 0.05$.

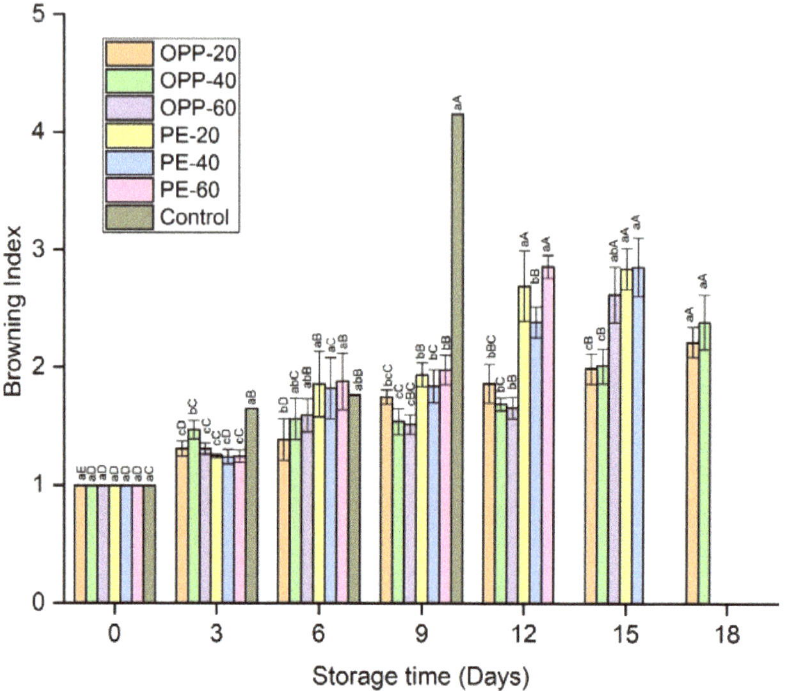



**Figure 4.** Pericarp browning rates of longan fruit samples in different packaging films fumigated with propyl disulfide and control, stored at $12 \pm 1$ °C. Vertical bars represent means $\pm$ standard deviations ($n = 3$). Different small and capital letters show significant differences ($p \leq 0.05$) between the treatments within the same day and between the days in the same treatment, respectively, via LSD test at $p \leq 0.05$.

### 3.2. Firmness of Longan Fruit Samples

The peel firmness of longan fruit samples decreased with the extension of storage time. The peel firmness significantly declined in the control treatment. Among the other treatments, the firmness fluctuated and dropped at the end of storage time in each treatment, probably associated with decay incidence. In comparison to other films, OPP-20 maintained the highest level of firmness, followed by OPP-40 (Figure 5).

### 3.3. Color (L*, a*, and b*) Values

The lightness (L*) values of the longan pericarps were reduced in all treatments with storage time, as shown in Table 1. The decrease in L* values in the control treatment was comparatively high compared to other packaging films, and storage was discontinued on day 9. Similarly, the a* values, which indicate the redness of fruit samples, increased over time, and high a* values were obtained in the control treatment. The b* values, which indicate the yellowness of the longan pericarp, decreased with storage. Statistical analysis within treatments and storage intervals revealed the greatest decrease in b* values in the control treatment (Table 1). With prolonged storage time, the changes in pericarp color became more prominent, as presented in Figure 6. The outer and inner pericarps of fruit samples changed to dark brown with storage, with more severe effects in control fruit (day 9) than other treatments.

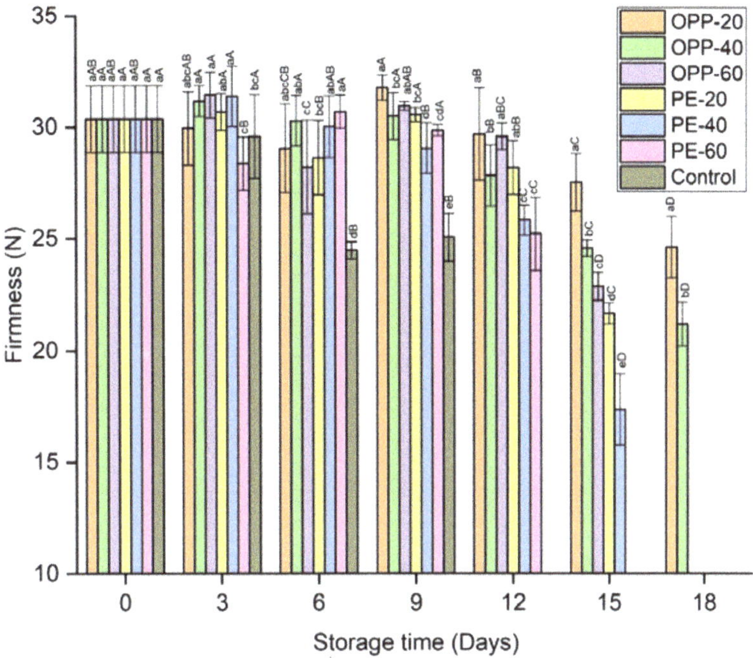

**Figure 5.** Peel firmness rates of longan fruit samples in different packaging films fumigated with propyl disulfide and control, stored at 12 ± 1 °C. Vertical bars represent means ± standard deviations (*n* = 6). Different small and capital letters show significant differences between the treatments within the same day and between the days in the same treatment, respectively, via LSD test at $p \leq 0.05$.

**Table 1.** Color values of longan pericarps fumigated with propyl disulfide in different packaging films in comparison to control, stored fruit at 12 ± 1 °C.

| | | | | Treatments | | | |
|---|---|---|---|---|---|---|---|
| **Days** | **OPP-20** | **OPP-40** | **OPP-60** | **PE-20** | **PE-40** | **PE-60** | **Control** |
| | | | | L* | | | |
| 0 | 69.44 [aAB] | 69.44 [aA] | 69.44 [aA] | 69.44 [aA] | 69.44 [aA] | 69.44 [aA] | 69.44 [aA] |
| | (2.44) | (2.44) | (2.44) | (2.44) | (2.44) | (2.44) | (2.44) |
| 3 | 70.68 [aA] | 67.47 [abAB] | 66.25 [bAB] | 65.64 [bB] | 67.13 [abA] | 64.74 [bA] | 56.59 [cB] |
| | (0.26) | (1.90) | (1.70) | (0.89) | (2.30) | (0.78) | (2.22) |
| 6 | 66.81 [aBC] | 67.00 [aAB] | 66.91 [aAB] | 66.22 [aB] | 66.61 [aA] | 65.84 [aA] | 48.44 [bC] |
| | (0.14) | (0.17) | (0.06) | (0.87) | (0.42) | (1.10) | (2.44) |
| 9 | 66.01 [aCD] | 66.47 [aB] | 64.33 [bB] | 61.88 [cC] | 62.16 [cB] | 59.39 [dB] | 35.50 [eD] |
| | (0.51) | (0.61) | (2.23) | (2.07) | (1.14) | (0.86) | (0.35) |
| 12 | 62.70 [aDE] | 64.89 [aB] | 56.91 [bC] | 53.55 [bD] | 53.27 [bC] | 52.28 [bC] | − |
| | (0.28) | (0.82) | (0.06) | (0.13) | (2.61) | (4.16) | |
| 15 | 59.37 [aEF] | 60.61 [aC] | 48.02 [bD] | 46.88 [bE] | 48.83 [bD] | − | − |
| | (1.15) | (0.73) | (1.13) | (0.93) | (2.82) | | |
| 18 | 57.94 [aE] | 56.03 [aD] | − | − | − | − | − |
| | (0.16) | (1.02) | | | | | |

**Table 1.** *Cont.*

| | | | | Treatments | | | |
|---|---|---|---|---|---|---|---|
| Days | OPP-20 | OPP-40 | OPP-60 | PE-20 | PE-40 | PE-60 | Control |
| | | | | a* | | | |
| 0 | 1.78 [aC] (0.24) | 1.78 [aE] (0.24) | 1.78 [aD] (0.24) | 1.78 [aE] (0.24) | 1.78 [aD] (0.24) | 1.78 [aD] (0.24) | 1.78 [aD] (0.24) |
| 3 | 2.36 [bC] (0.22) | 3.44 [aD] (0.30) | 2.64 [bC] (0.21) | 4.00 [aD] (0.15) | 3.38 [aC] (0.29) | 3.61 [aBC] (0.19) | 3.45 [aC] (0.13) |
| 6 | 4.02 [abcB] (0.08) | 3.65 [bcCD] (0.28) | 3.84 [abcB] (0.23) | 4.23 [abCD] (0.33) | 3.83 [abcC] (0.21) | 3.38 [cC] (0.26) | 4.51 [aB] (0.26) |
| 9 | 4.11 [bcB] (0.28) | 4.23 [bcBCD] (0.06) | 4.12 [bcB] (0.13) | 4.68 [bBC] (0.20) | 4.06 [cBC] (0.24) | 4.23 [bcB] (0.16) | 5.87 [aA] (0.35) |
| 12 | 4.43 [bB] (0.10) | 4.51 [bBC] (0.10) | 4.57 [abB] (0.16) | 4.91 [abAB] (0.18) | 4.57 [abB] (0.18) | 5.23 [aA] (0.17) | – |
| 15 | 4.68 [bAB] (0.03) | 5.07 [abAB] (0.24) | 6.02 [aA] (0.19) | 5.58 [abA] (0.43) | 5.35 [abA] (0.17) | – | – |
| 18 | 5.19 [bA] (0.19) | 5.84 [aA] (0.33) | – | – | – | – | – |
| | | | | b* | | | |
| 0 | 35.23 [aA] (0.85) | 35.23 [aA] (0.85) | 35.23 [aA] (0.85) | 35.23 [aA] (0.85) | 35.23 [aA] (0.85) | 35.23 [aA] (0.85) | 35.23 [aA] (0.85) |
| 3 | 33.96 [bA] (0.10) | 35.04 [aA] (0.35) | 30.29 [bB] (1.82) | 28.66 [bcB] (0.75) | 35.15 [aA] (0.34) | 26.14 [cdB] (1.90) | 23.82 [dB] (2.14) |
| 6 | 31.51 [aAB] (0.62) | 32.15 [aB] (0.34) | 27.74 [bBC] (0.59) | 26.66 [bcBC] (1.14) | 32.37 [aA] (0.25) | 23.92 [cdBC] (0.67) | 20.38 [dC] (1.32) |
| 9 | 28.51 [aBC] (0.93) | 30.60 [aB] (0.91) | 25.40 [bC] (0.80) | 25.33 [bC] (0.22) | 28.71 [aB] (0.60) | 20.80 [cCD] (1.29) | 18.05 [dC] (0.57) |
| 12 | 25.74 [aCD] (0.70) | 25.71 [aC] (0.96) | 20.96 [bD] (1.33) | 22.55 [abD] (1.08) | 26.15 [aB] (0.09) | 18.92 [bD] (0.51) | – |
| 15 | 22.96 [aDE] (1.59) | 22.37 [aD] (0.23) | 18.51 [bD] (0.26) | 21.33 [aD] (0.59) | 21.37 [aC] (1.05) | – | – |
| 18 | 20.96 [aE] (0.37) | 19.71 [bE] (0.21) | – | – | – | – | – |

Different small letters in each row and capital letters in each column show the significant differences between the treatments within the same day and between the days in the same treatment, respectively, via LSD test at $p \leq 0.05$. SDs are presented in parenthesis.

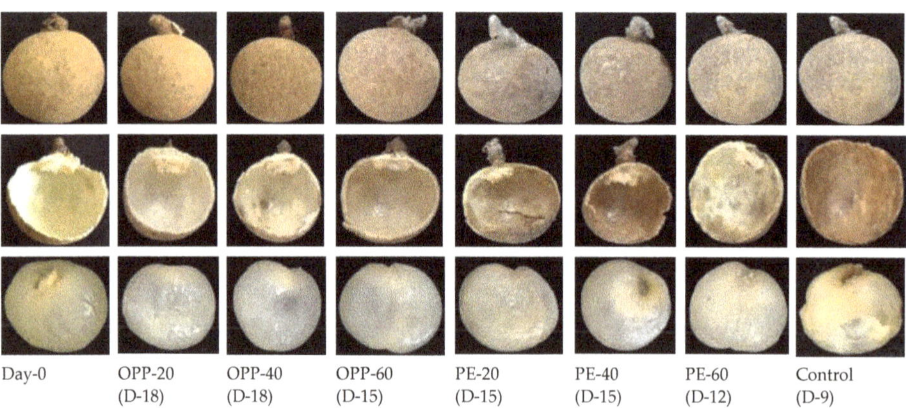



**Figure 6.** Changes in the peel and flesh of longan fruit samples during storage at $12 \pm 1$ °C treated with propyl disulfide and packaged in oriented polypropylene (OPP-20, OPP-40, and OPP-60), polyethylene (PE-20, PE-40, and PE-60), and control (open net) packages.

### 3.4. Total Phenolic (TPC) and Total Flavonoid Contents (TFC)

Figure 7A,B presents the total phenol and flavonoid contents in the longan fruit samples. A continuous decline was observed in all treatments in terms of TPC and TFC contents, being significantly ($p \leq 0.05$) higher in control samples. The contents of phenols and flavonoids varied slightly among the treatments during storage; however, OPP-20 and PE-20 showed relatively higher TPC and TFC contents than the other treatments.

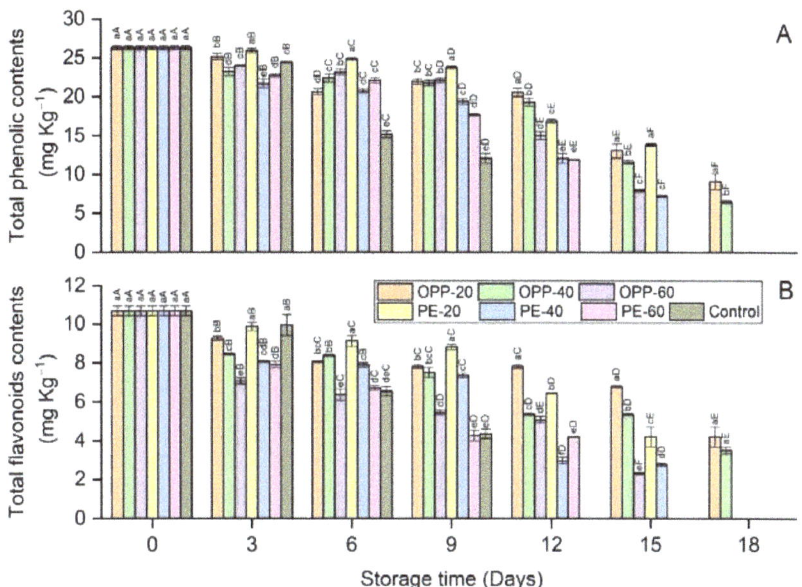


**Figure 7.** Total phenolic (**A**) and total flavonoid (**B**) contents in longan fruit samples in different packaging films fumigated with propyl disulfide and control, stored at $12 \pm 1$ °C. Vertical bars represent means $\pm$ standard deviation ($n$ = 3). Different small and capital letters show significant differences between the treatments within the same day and between the days in the same treatment, respectively, via LSD test at $p \leq 0.05$.

### 3.5. Enzymes Activity Levels

Enzyme activity levels are presented in Figure 8A,B. Polyphenol oxidase (PPO) and peroxidase (POD) activity levels significantly increased in all treatments with storage time. The highest activity levels for PPO and POD were found in control treatment at days 6 and 9. Generally, at the end of the storage period, PPO and POD activity levels were higher in OPP packages than PE packages. The longest shelf life times were obtained for OPP-20 and OPP-40 packages with no difference in PPO and POD activity levels, with few exceptions (Figure 8A,B).

### 3.6. TSS, TA, and pH

In all treatments, TSS decreased with increased storage time (Table 2). No significant differences were found between the treated and control fruit samples, with few exceptions. Similarly, packaging films did not affect the TA and pH values, as no significant differences were found in the treated and control fruit samples, with few exceptions. However, storage time had an effect, as TA and pH values increased in all treatments with increasing storage time (Table 2).

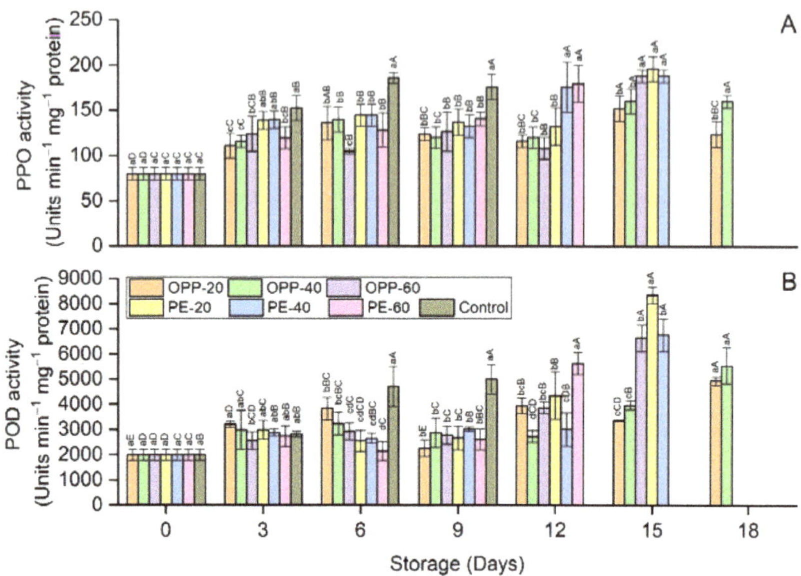

**Figure 8.** Polyphenol oxidase (**A**) and peroxidase (**B**) activity levels in longan fruit samples in different packaging films fumigated with propyl disulfide and control, stored at $12 \pm 1$ °C. Vertical bars represent means $\pm$ standard deviations ($n = 3$). Different small and capital letters show significant differences between the treatments within the same day and between the days in the same treatment, respectively, via LSD test at $p \leq 0.05$.

**Table 2.** TSS, TA, and pH values of longan fruit treated with propyl disulfide in different packaging films in comparison to control, stored samples at $12 \pm 1$ °C.

| | Treatments | | | | | | |
|---|---|---|---|---|---|---|---|
| **Days** | **OPP-20** | **OPP-40** | **OPP-60** | **PE-20** | **PE-40** | **PE-60** | **Control** |
| | | | | TSS | | | |
| 0 | 20.73 [aA] (0.12) | 20.73 [aA] (0.12) | 20.73 [aA] (0.12) | 20.73 [aA] (0.12) | 20.73 [aA] (0.12) | 20.73 [aA] (0.12) | 20.73 [aA] (0.12) |
| 3 | 18.77 [bcC] (0.15) | 18.53 [cC] (0.32) | 18.40 [cC] (0.10) | 18.80 [bcB] (0.40) | 18.60 [cD] (0.26) | 19.17 [abC] (0.29) | 19.43 [aB] (0.38) |
| 6 | 19.10 [bB] (0.10) | 18.10 [cC] (0.10) | 17.13 [eE] (0.15) | 17.73 [dC] (0.15) | 20.07 [aB] (0.06) | 18.23 [cD] (0.12) | 19.13 [bBC] (0.12) |
| 9 | 18.90 [cBC] (0.20) | 19.13 [bcB] (0.15) | 19.50 [abB] (0.17) | 17.93 [dC] (0.21) | 19.03 [cC] (0.06) | 19.90 [aB] (0.35) | 18.87 [cC] (0.35) |
| 12 | 18.97 [aBC] (0.23) | 17.13 [cdD] (0.47) | 13.63 [bcD] (0.31) | 16.97 [dD] (0.15) | 17.93 [bE] (0.35) | 17.90 [bD] (0.17) | – |
| 15 | 18.70 [aC] (0.30) | 17.50 [bD] (0.36) | 17.47 [bD] (0.15) | 16.83 [cD] (0.25) | 17.47 [bF] (0.21) | – | – |
| 18 | 17.53 [aD] (0.15) | 17.27 [aD] (0.55) | – | – | – | – | – |

**Table 2.** *Cont.*

| Days | OPP-20 | OPP-40 | OPP-60 | PE-20 | PE-40 | PE-60 | Control |
|---|---|---|---|---|---|---|---|
| | | | | Treatments | | | |
| | | | | TA | | | |
| 0 | 0.15 [aD] | 0.15 [aE] | 0.15 [aE] | 0.15 [aE] | 0.15 [aC] | 0.15 [aD] | 0.15 [aB] |
| | (0.01) | (0.01) | (0.01) | (0.01) | (0.01) | (0.01) | (0.01) |
| 3 | 0.16 [bcC] | 0.16 [bcD] | 0.17 [abD] | 0.16 [bcD] | 0.15 [cC] | 0.17 [abC] | 0.18 [aA] |
| | (0.01) | (0.01) | (0.01) | (0.01) | (0.01) | (0.01) | (0.01) |
| 6 | 0.19 [aB] | 0.17 [bD] | 0.19 [aC] | 0.16 [bD] | 0.18 [aB] | 0.19 [aB] | 0.18 [aA] |
| | (0.01) | (0.01) | (0.01) | (0.01) | (0.01) | (0.01) | (0.01) |
| 9 | 0.19 [bB] | 0.19 [bC] | 0.21 [aB] | 0.18 [bC] | 0.20 [abB] | 0.20 [abB] | 0.17 [cA] |
| | (0.01) | (0.01) | (0.01) | (0.01) | (0.01) | (0.01) | (0.01) |
| 12 | 0.20 [bB] | 0.20 [bBC] | 0.22 [aAB] | 0.20 [bB] | 0.22 [aA] | 0.23 [aA] | – |
| | (0.01) | (0.01) | (0.01) | (0.01) | (0.01) | (0.01) | |
| 15 | 0.23 [abA] | 0.21 [cAB] | 0.23 [abA] | 0.22 [bcA] | 0.24 [aA] | – | – |
| | (0.01) | (0.01) | (0.01) | (0.01) | (0.01) | | |
| 18 | 0.23 [aA] | 0.22 [aA] | – | – | – | – | – |
| | (0.01) | (0.01) | | | | | |
| | | | | pH | | | |
| 0 | 6.88 [aG] | 6.88 [aE] | 6.88 [aE] | 6.88 [aE] | 6.88 [aE] | 6.88 [aD] | 6.88 [aD] |
| | (0.02) | (0.02) | (0.02) | (0.02) | (0.02) | (0.02) | (0.02) |
| 3 | 6.94 [abcF] | 6.96 [abcD] | 6.93 [bcD] | 6.92 [cE] | 6.98 [aD] | 6.97 [abC] | 6.99 [aC] |
| | (0.02) | (0.05) | (0.03) | (0.05) | (0.02) | (0.02) | (0.01) |
| 6 | 6.99 [cE] | 6.99 [cD] | 6.97 [cC] | 6.98 [cD] | 7.11 [bC] | 7.11 [bB] | 7.20 [aB] |
| | (0.01) | (0.01) | (0.02) | (0.01) | (0.02) | (0.02) | (0.02) |
| 9 | 7.13 [cdD] | 7.18 [bC] | 7.10 [dB] | 7.14 [cC] | 7.25 [aB] | 7.10 [cdB] | 7.29 [aA] |
| | (0.04) | (0.02) | (0.01) | (0.02) | (0.02) | (0.01) | (0.03) |
| 12 | 7.20 [cC] | 7.26 [bB] | 7.33 [aA] | 7.21 [cB] | 7.25 [bcB] | 7.27 [bA] | – |
| | (0.01) | (0.03) | (0.03) | (0.02) | (0.05) | (0.02) | |
| 15 | 7.31 [bB] | 7.31 [abA] | 7.32 [abA] | 7.26 [cA] | 7.35 [aA] | – | – |
| | (0.02) | (0.03) | (0.02) | (0.03) | (0.01) | | |
| 18 | 7.36 [aA] | 7.31 [bAB] | – | – | – | – | – |
| | (0.03) | (0.02) | | | | | |

Different small letters in each row and capital letters in each column show the significant differences between the treatments within the same day and between the days in the same treatment, respectively, via LSD test at $p \leq 0.05$. SDs are presented in parentheses.

### 3.7. Scanning Electron Microscopy and Pericarp Microstructure

Longan peel consists of three major parts, namely the exocarp, mesocarp, and endocarp. In this experiment, SEM analysis of the longan fruit samples treated with propyl disulfide was conducted to observe the structural changes in these parts and to compare them with controls. The SEM micrographs of the various parts of the treated and control longan peel samples are presented in Figure 9. The exocarp is the outermost part of the longan peel. The propyl disulfide had a positive effect; as shown in Figure 9, the exocarps of the treated fruit samples were very smooth, complete, tight, and connected. This honeycomb-like structure was composed of cup-shaped cells that were uniform and more clear, while the exocarp of the control fruit was not complete, had free space, a loose structure, a non-uniform honeycomb appearance, less cup-shaped cells, and more mycelial pathogens. The mesocarp is the second layer of the longan peel and comprises about 70% of the longan pericarp. The mesocarp layer of the propyl-disulfide-treated fruit was more uniform and continuous than the control, where rough structures and many cracks and openings were observed. The rills were deeper and broader in control samples than treated samples. The endocarp is a single layer of cells. The endocarp layer of the propyl-disulfide-treated fruit showed a regular surface omamentation compared to the control fruit, where some irregular surface omamentation was observed.

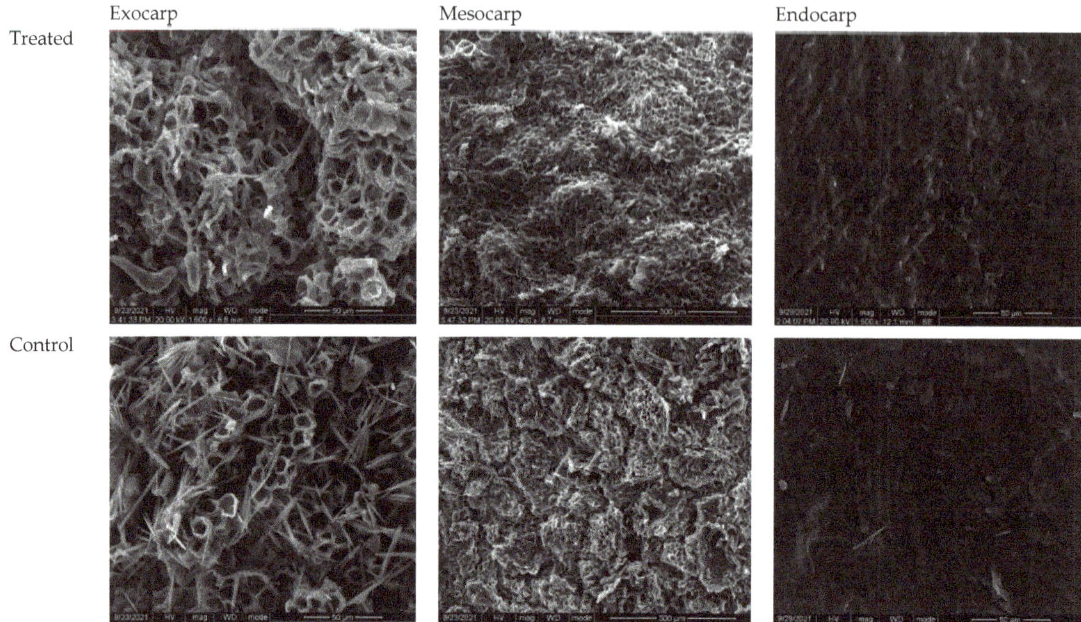

**Figure 9.** Scanning electron micrographs of treated and control longan pericarp. Exocarp samples at 1500× magnification, mesocarp samples at 400× magnification, and endocarp samples at 1500× magnification.

## 4. Discussion

Water loss is one of the main reasons for pericarp browning in longan fruit, whereby shrinkage occurs when substrates, enzymes, and other cell constituents come into contact and initiate browning. This can be seen from the strong positive correlation of water loss with BI (r = 85) and L* values (r = 87), as shown in Table 3. Weight loss was significantly higher in control fruit samples than in all the other treatments. This is one of the benefits of polymeric films, which prevent water loss from fruit samples. On the other hand, PD also might play a role in water loss prevention, probably maintaining the pericarp's integrity and reducing water loss. Additionally, it can be stated that PD worked as a barrier against water loss, as previously found in thymol-treated longan fruit samples [8] and fresh-cut beans treated with tea tree essential oil and peppermint essential oil [27].

PD effectively inhibited the growth of microbes on the longan fruit surfaces. Although DI percentages increased in all treatments at the end of storage, the DI percentage of the control longan fruit samples was comparatively higher than all other treatments (Figure 2). A possible mechanism of the decay prevention of PD could be attributed to the sulfur compound, as it is well known for its antimicrobial activity levels. Ramos et al. [28] reported that the antiadhesive mechanism of neem extract could be the hydrophobicity of the cell surfaces' and the formation of biofilm, which could affect the microbial colonization. Koul [29] stated that either the characteristic odor of sulfur compounds or some physiological mechanism of interaction make the PD an effective grain protectant against insect pests. Kumar and Kudachikar [30] reported that the antifungal mechanisms of natural plant extracts against pathogens could be attributed to the disruption of membrane integrity and cellular component leakage. As microbial growth symptoms appear on the longan fruit's surfaces, in the current research work we did not study the antimicrobial efficiency of PD on specific microorganisms, but generally evaluated the decay of longan fruit. Hence, no decay and less symptoms were found in PD-treated fruit samples than control fruit.

The term decay incidence was used to describe this process, and the data are presented in Figure 3. However, in our previous study on mango fruit, we found that PD was very efficient in inhibiting the growth of major fungi in mango fruit samples. We found that PD effectively inhibited the mycelial growth of *Colletotrichum gloeosporioides* and *Colletotrichum acutatum*, causing anthracnose, as well as *Lasiodiplodia theobromae* and *Neofusicoccum parvum*, causing stem end rots in mango [14,15]. Future research should be directed toward the specific microorganisms found in longan fruit samples.

Pericarp browning of longan is another major concern that limits its postharvest life. The combined effect of PD and polymeric films delayed the pericarp browning. Pericarp browning is a complex phenomenon that may include many interconnected factors that can be represented in terms of color values, phenol contents, activity levels of enzymes, and so on. The effect of PD on browning inhibition was obvious in maintaining the color values of longan pericarp samples (Table 1). Compared to the control treatment, fruit samples treated with PD and packaged in various polymeric films had high L* values, lowest redness (a*), and high yellowness (b*). The presence of high TPC and TFC contents in longan pericarps treated with PD further confirmed the antioxidant efficiency of neem plant extracts. Interestingly, PD also affects the enzymes involved in the browning reaction. Longan fruit samples treated with PD and packaged in polymeric films showed lower PPO and POD activity levels than those in the control treatment. Regarding the overall phenomenon of the browning reaction and the antioxidant properties of neem, the high TPC and TFC contents (Figure 7), reduced activity levels of PPO and POD (Figure 8), high color values (Table 1), and lower BI levels (Figure 4) indicate the high antioxidant effect of PD. This can be seen from the correlation coefficient values in Table 3, whereby BI is highly correlated with PPO (r = 92), POD (r = 83), phenols (r = −85), flavonoids (r = −88), L* (r = −99), a* (r = 92), and b* (r = −93) (Table 3). This good fit of the correlation coefficients in these parameters and compounds involved in the enzymatic browning reaction and their dependency on each other shows the antioxidant efficiency of PD against the enzymatic browning in longan fruit samples. The apparent color changes in the longan pericarp further confirm the antioxidant activity of PD, as can be seen in Figure 6, where the longan fruit pericarp turned brown on day 9. Changes in PD-treated fruit and packaged in the polymeric films were very slow, and the fruit samples were still acceptable on day 18 in OPP-20 and OPP-40 packages. This trend of high phenols, reduced enzymes activity levels, and consequently low BI levels was also seen in our previous study, when longan fruit cv. Daw was treated with plant essential oil (thymol) [7,8]. Valero et al. [31] fumigated table grapes with thymol, eugenol, and menthol, and reported that these essential oils maintained the phenol contents and high color values better than the control. Similar to our study on the propyl disulfide compound, other phytochemicals, such as diallyl disulfide and diallyl trisulfide obtained from garlic essential oil, showed strong antioxidant activity against nicotinamide–adenine dinucleotide phosphate (NADPH) oxidase enzymes [32].

The antioxidant activity of neem extract was also studied regarding the preservation and shelf life extension of other food products. Ouerfelli et al. [19] preserved the quality and extended the shelf life of raw beef patties to 11 days storage at 4 °C. These authors reported that neem extract prevented the loss in color, reduced the metmyoglobin formation, scavenged the DPPH free radicals, and possessed high antibacterial potential against beef patties. Serrone and Nicoletti [18] preserved fresh retail meat using neem cake oil and reported its efficacy against a wide range of bacteria. Serrone et al. [11] reported that neem oil effectively preserved the quality of fresh retail meat.

The peel firmness of the longan was higher in the treated fruit than in control. As shown in Figure 6, the peel and aril breakdown in the control on day 9 meant they were not suitable for further storage and were discarded, while the complete aril and peel in the treated fruit showed that PD treatment and storage in various polymeric films, particularly OPP-20 and OPP-40, prevented softness and aril breakdown.

**Table 3.** Pearson's correlation coefficient values for different quality parameters in longan fruit samples stored at $12 \pm 1\ ^\circ$C.

| | Wt. Loss | BI | DI | Texture | L* | a* | b* | TSS | TA | pH | TPC | TFC | PPO | POD |
|---|---|---|---|---|---|---|---|---|---|---|---|---|---|---|
| Wt. loss | 1 | | | | | | | | | | | | | |
| BI | 0.87 | 1 | | | | | | | | | | | | |
| DI | 0.68 | 0.94 | 1 | | | | | | | | | | | |
| Texture | −0.47 | −0.79 | −0.90 | 1 | | | | | | | | | | |
| L* | −0.85 | −0.99 | −0.96 | 0.79 | 1 | | | | | | | | | |
| a* | 0.70 | 0.92 | 0.95 | −0.90 | −0.92 | 1 | | | | | | | | |
| b* | −0.71 | −0.93 | −0.94 | 0.87 | 0.92 | −0.99 | 1 | | | | | | | |
| TSS | −0.48 | −0.80 | −0.91 | 0.93 | 0.80 | −0.96 | 0.94 | 1 | | | | | | |
| TA | 0.31 | 0.70 | 0.83 | −0.91 | −0.69 | 0.89 | −0.89 | −0.97 | 1 | | | | | |
| pH | 0.63 | 0.89 | 0.92 | −0.91 | −0.87 | 0.98 | −0.98 | −0.96 | 0.93 | 1 | | | | |
| TPC | −0.59 | −0.85 | −0.89 | 0.93 | 0.83 | −0.97 | 0.97 | 0.95 | −0.94 | −0.99 | 1 | | | |
| TFC | −0.61 | −0.88 | −0.94 | 0.93 | 0.88 | −0.99 | 0.98 | 0.97 | −0.94 | −0.99 | 0.99 | 1 | | |
| PPO | 0.64 | 0.92 | 0.99 | −0.91 | −0.93 | 0.95 | −0.94 | −0.92 | 0.86 | 0.92 | −0.90 | −0.95 | 1 | |
| POD | 0.51 | 0.83 | 0.96 | −0.93 | −0.86 | 0.93 | −0.91 | −0.96 | 0.91 | 0.91 | −0.89 | −0.94 | 0.97 | 1 |

Longan fruit has a unique pericarp structure, and besides the effect of PD on the physicochemical characteristics, the impact of PD was also evaluated on the pericarp structure. The scanning electron microscopy analysis indicated that PD maintained a uniform smooth surface. The SEM analysis confirmed that the PD maintained the cell integrity and compactness, prevented water loss, and minimized the chance of pathogen growth. On the other hand, the pericarp of the control treatment was cracked and damaged, showing an irregular surface and free space (Figure 9). This kind of loose structure was beneficial for the pathogens' invasion, as shown in Figure 3 (decay incidence), and enhanced the water loss, as shown in Figure 2.

Another reason for the good quality of the longan fruit being maintained could be the application of PD in the vapor phase due to the slow release time, which might maintain the quality and suppress the microbes for a longer time. Another benefit of the vapor phase application is that low quantities of PD are required, without altering the sensory properties of the products. Hence, the shelf life of the longan fruit treated with PD was extended up to 18 days compared to 9 days in control.

## 5. Conclusions

In this study, we applied a new fumigation method by placing the essential oil on the sterile gauze inside the polymeric films. Fruit samples without PD and in open net packages were considered as controls and all packages were stored at $12 \pm 1\ ^\circ$C for 18 days. In combination with plant extract (PD), polymeric films played a very effective role in the quality maintenance of longan fruit samples. Compared to control packages, all types of polymeric films in combination with PD prevented weight loss, inhibited microbial growth, and delayed the pericarp browning of longan fruit samples. PD-treated fruit samples and sealed in packaging films also exhibited high firmness and color values, better prevented the oxidation of the phenols and flavonoids, and better inhibited enzyme activity than the control treatment. Extended shelf life of 18 days was observed in OPP-20 and OPP-40 packaging films with good quality attributes. SEM analysis showed a clear uniformity in the pericarp structures, which was entirely associated with the obtained results. Overall, PD maintained the structural integrity of the pericarp, prevented pericarp browning, inhibited the growth of microorganisms, and had a very positive effect on longan fruit quality as compared to control fruit samples. Therefore, with a combination of OPP-20 and OPP-40 polymeric films, PD could be applied commercially as a potent antioxidant and antimicrobial agent in a wide range of food products to replace synthetic fumigants or preservatives.

**Author Contributions:** Conceptualization and design, C.H. and M.R.K.; methodology, M.R.K.; resources, H.S. and L.R.; software, R.U., H.U., H.S., M.A. and A.K.; writing—original draft, M.R.K.; supervision, C.H.; statistical analysis, M.R.K. and I.M.Q.; data curation, M.R.K., H.U., T.N., M.A. and A.K.; writing—review and editing, M.R.K., C.H., I.M.Q. and R.U.; funding acquisition, C.H. All authors have read and agreed to the published version of the manuscript.

**Funding:** This research was funded by the Guangxi Science and Technology Plan Project (Project No. AB18221126), the Guangxi Natural Science Foundation Program (2019JJD120012), and the Postdoctoral Project of the Guangxi University.

**Institutional Review Board Statement:** Not applicable.

**Informed Consent Statement:** Not applicable.

**Data Availability Statement:** Not applicable.

**Acknowledgments:** The authors thank Guangxi University for providing a Postdoctoral Fellowship.

**Conflicts of Interest:** The authors declare no conflict of interest.

# References

1. Yao, F.; Huang, Z.; Li, D.; Wang, H.; Xu, X.; Jiang, Y.; Qu, H. Phenolic components, antioxidant enzyme activities and anatomic structure of longan fruit pericarp following treatment with adenylatetriphosphate. *Sci. Hortic.* **2014**, *180*, 6–13. [CrossRef]
2. Chumyam, A.; Shank, L.; Uthaibutra, J.; Saengnil, K. Effects of chlorine dioxide on mitochondrial energy levels and redox status of Daw longan pericarp during storage. *Postharvest Biol. Technol.* **2016**, *116*, 26–35. [CrossRef]
3. Chen, M.; Lin, H.; Zhang, S.; Lin, Y.; Chen, Y.; Lin, Y. Effects of Adenosine Triphosphate (ATP) Treatment on Postharvest Physiology, Quality and Storage Behavior of Longan Fruit. *Food Bioprocess Technol.* **2015**, *8*, 971–982. [CrossRef]
4. Lin, Y.; Lin, H.; Lin, Y.; Zhang, S.; Chen, Y.; Jiang, X. The roles of metabolism of membrane lipids and phenolics in hydrogen peroxide-induced pericarp browning of harvested longan fruit. *Postharvest Biol. Technol.* **2016**, *111*, 53–61. [CrossRef]
5. Lin, Y.; Lin, Y.; Lin, Y.; Lin, M.; Chen, Y.; Wang, H.; Lin, H. A novel chitosan alleviates pulp breakdown of harvested longan fruit by suppressing disassembly of cell wall polysaccharides. *Carbohydr. Polym.* **2019**, *217*, 126–134. [CrossRef] [PubMed]
6. Joradol, A.; Uthaibutra, J.; Lithanatudom, P.; Saengnil, K. Induced expression of NOX and SOD by gaseous sulfur dioxide and chlorine dioxide enhances antioxidant capacity and maintains fruit quality of 'Daw' longan fruit during storage through $H_2O_2$ signaling. *Postharvest Biol. Technol.* **2019**, *156*, 110938. [CrossRef]
7. Khan, M.R.; Huang, C.; Durrani, Y.; Muhammad, A. Chemistry of enzymatic browning in longan fruit as a function of pericarp pH and dehydration and its prevention by essential oil, an alternative approach to $SO_2$ fumigation. *PeerJ* **2021**, *9*, e11539. [CrossRef]
8. Khan, M.R.; Huang, C.; Zhao, H.; Huang, H.; Ren, L.; Faiq, M.; Hashmi, M.S.; Li, B.; Zheng, D.; Xu, Y.; et al. Antioxidant activity of thymol essential oil and inhibition of polyphenol oxidase enzyme: A case study on the enzymatic browning of harvested longan fruit. *Chem. Biol. Technol. Agric.* **2021**, *8*, 61. [CrossRef]
9. Kumar, D.; Rahal, A.; Malik, K.J. Neem extract. In *Nutraceuticals. Efficacy, Safety and Toxicity*; Gupta, R.C., Ed.; Academic Press: Chennai, India, 2016; pp. 585–597. [CrossRef]
10. Hossain, M.A.; Al-Toubi, W.A.S.; Weli, A.M.; Al-Riyami, Q.A.; Al-Sabahi, J.N. Identification and characterization of chemical compounds indifferent crude extracts from leaves of Omani neem. *J. Taibah Univ. Sci.* **2013**, *7*, 181–188. [CrossRef]
11. Serrone, P.D.; Toniolo, C.; Nicoletti, M. Neem (*Azadirachta indica* A. Juss) oil: A natural preservative to control meat spoilage. *Foods* **2015**, *4*, 3–14. [CrossRef]
12. VasudhaUdupa, A.; Gowda, B.; Kumarswammy, B.E.; Shivanna, M.B. The antimicrobial and antioxidant property, GC–MS analysis of non-edible oil-seed cakes of neem, madhuca, and simarouba. *Bull. Natl. Res. Cent.* **2021**, *45*, 41. [CrossRef]
13. Pankaj, S.; Lokeshwar, T.; Mukesh, B.; Vishnu, B. Review on neem (*Aazadirachta Indica*): Thousand problems one solution. *Int. Res. J. Pharm.* **2011**, *2*, 97–102.
14. Khan, M.R.; Chonhenchob, V.; Huang, C.; Suwanamornlert, P. Antifungal Activity of Propyl Disulfide from Neem (*Azadirachta indica*) in Vapor and Agar Diffusion Assays against Anthracnose Pathogens (*Colletotrichum gloeosporioides* and *Colletotrichum acutatum*) in Mango Fruit. *Microorganisms* **2021**, *9*, 839. [CrossRef]
15. Khan, M.R.; Suwanamornlert, P.; Sangchote, S.; Chonhenchob, V. Antifungal activity of propyl disulphide from neem against *Lasiodiplodia theobromae* and *Neofusicoccum parvum* causing stem end rot in mango. *J. Appl. Microbiol.* **2020**, *129*, 1364–1373. [CrossRef] [PubMed]
16. Zakawa, N.N.; Channya, K.F.; Magga, B.; Akesa, T.M. Antifungal effect of neem (*Azadirachta indica*) leaf extracts on mango fruit postharvest rot agents in Yola, Adamawa state. *J. Pharmacogn. Phytochem.* **2018**, *7*, 23–26.
17. Suleiman, M.N. Antifungal properties of leaf extract of neem and tobacco on three fungal pathogens of tomato (*Lycopersicon Esculentum* Mill). *Adv. Appl. Sci. Res.* **2011**, *2*, 217–220.
18. Serrone, P.D.; Nicoletti, M. Antimicrobial Activity of a Neem Cake Extract in a Broth Model Meat System. *Int. J. Environ. Res. Public Health* **2013**, *10*, 3282–3295. [CrossRef]

19. Ouerfelli, M.; Villasante, J.; Kaâb, L.B.B.; Almajano, M. Effect of Neem (*Azadirachta indica* L.) on Lipid Oxidation in Raw Chilled Beef Patties. *Antioxidants* **2019**, *8*, 305. [CrossRef]
20. Khan, M.R.; Suwanamornlert, P.; Leelaphiwat, P.; Chinsirikul, W.; Chonhenchob, V. Quality and biochemical changes in longan (*Dimicarpus longan* Lour cv. Daw) under different controlled atmosphere conditions. *Int. J. Food Sci. Technol.* **2017**, *52*, 2163–2170. [CrossRef]
21. Dewanto, V.; Wu, X.; Adom, K.K.; Liu, R.H. Thermal processing enhances the nutritional value of tomatoes by increasing total antioxidant activity. *J. Agric. Food Chem.* **2002**, *50*, 3010–3014. [CrossRef]
22. Duan, X.; Su, X.; You, Y.; Qu, H.; Li, Y.; Jiang, Y. Effect of nitric oxide on pericarp browning of harvested longan fruit in relation to phenolic metabolism. *Food Chem.* **2007**, *104*, 571–576. [CrossRef]
23. Jiang, Y.-M. Purification and some properties of polyphenol oxidase of longan fruit. *Food Chem.* **1999**, *66*, 75–79. [CrossRef]
24. Zhang, Z.; Pang, X.; Xuewu, D.; Ji, Z.; Jiang, Y. Role of peroxidase in anthocyanin degradation in litchi fruit pericarp. *Food Chem.* **2005**, *90*, 47–52. [CrossRef]
25. Bradford, M.M. A rapid and sensitive method for the quantitation of microgram quantities of protein utilizing the principle of protein-dye binding. *Anal. Biochem.* **1976**, *72*, 248–254. [CrossRef]
26. Chitbanchong, W.; Sardsud, V.; Whangchai, K.; Koslanund, R.; Thobunluepop, P. Minimally of polyphenol oxidase activity and controlling of rotting and browning of longan fruits cv. Daw by sulphur dioxide (SO$_2$) treatment under cold storage conditions. *Int. J. Agric. Res.* **2009**, *4*, 349–361. [CrossRef]
27. Awad, A.H.R.; Parmar, A.; Ali, M.R.; El-Mogy, M.M.; Abdelgawad, K.F. Extending the Shelf-Life of Fresh-Cut Green Bean Pods by Ethanol, Ascorbic Acid, and Essential Oils. *Foods* **2021**, *10*, 1103. [CrossRef]
28. Ramos De Resende, A.; Ludke Falcao, L.; Salviano Barbosa, G.; Helena Marcellino, L.; Silvano Gander, E. Neem (*Azadirachta indica* A. Juss) components: Candidates for control of *Crinipellis perniciosa* and *Phytophthora* spp. *Microbiol. Res.* **2007**, *162*, 238–243. [CrossRef]
29. Koul, O. Biological activity of volatile di-*n*-propyl disulfide from seeds of neem, *Azadirachta indica* (Meliaceae), to two species of stored grain pests, *Sitophilus oryzae* (L.) and *Tribolium castaneum* (Herbst). *J. Econ. Entomol.* **2004**, *97*, 1142–1147. [CrossRef]
30. Kumar, A.; Kudachikar, V. Efficacy of aroma compounds for postharvest management of mango Anthracnose. *J. Plant Dis. Prot.* **2019**, *127*, 245–256. [CrossRef]
31. Valero, D.; Valverde, J.M.; Martínez-Romero, D.; Guillén, F.; Castillo, S.; Serrano, M. The combination of modified atmosphere packaging with eugenol or thymol to maintain quality, safety and functional properties of table grapes. *Postharvest Biol. Technol.* **2006**, *41*, 317–327. [CrossRef]
32. Herrera-Calderon, O.; Chacaltana-Ramos, L.J.; Huayanca-Gutiérrez, I.C.; Algarni, M.A.; Alqarni, M.; Batiha, G.E.-S. Chemical Constituents, In Vitro Antioxidant Activity and In Silico Study on NADPH Oxidase of *Allium sativum* L. (Garlic) Essential Oil. *Antioxidants* **2021**, *10*, 1844. [CrossRef] [PubMed]

*Article*

# Effect of Fiber Type and Content on Mechanical Property and Lapping Machinability of Fiber-Reinforced Polyetheretherketone

Shang Gao, Jialu Qu, Honggang Li and Renke Kang *

Key Laboratory for Precision and Non-Traditional Machining Technology of Ministry of Education, Dalian University of Technology, Dalian 116024, China; gaoshang@dlut.edu.cn (S.G.); 21904049@mail.dlut.edu.cn (J.Q.); superhonggang@mail.dlut.edu.cn (H.L.)
* Correspondence: kangrk@dlut.edu.cn; Tel.: +86-0411-8470-6059

**Abstract:** Polyetheretherketone (PEEK) is a novel polymer material with excellent material properties. The hardness and strength of PEEK can be further improved by introducing fiber reinforcements to meet the high-performance index of the aerospace industry. The machinability will be influenced when the material properties change. Therefore, it is crucial to investigate the influence of material properties of the fiber-reinforced PEEK on machinability. In this paper, the main materials include pure PEEK, short carbon-fiber-reinforced PEEK (CF/PEEK), and short glass-fiber-reinforced PEEK (GF/PEEK). The effects of the fiber type and mass fraction on the tensile strength, hardness, and elastic modulus of materials were discussed using the tensile test and nanoindentation experiments. Furthermore, the fiber-reinforced PEEK lapping machinability was investigated using lapping experiments with abrasive papers of different mesh sizes. The results showed that the hardness and elastic modulus of PEEK could be improved with fiber mass fraction, and the tensile strength of CF/PEEK can be improved compared with that of GF/PEEK. In terms of lapping ability, the material removal rates of the fiber-reinforced materials were found to be lower than the pure PEEK due to the higher hardness of the fiber. During the lapping process, the material removal methods mainly included the ductile deformation or desquamation of reinforcing fiber and ductile removal of the PEEK matrix. The lapped surface roughness of PEEK material can be improved by fiber reinforcement.

**Keywords:** polyetheretherketone; short fiber-reinforced; material property; lapping machinability

**Citation:** Gao, S.; Qu, J.; Li, H.; Kang, R. Effect of Fiber Type and Content on Mechanical Property and Lapping Machinability of Fiber-Reinforced Polyetheretherketone. *Polymers* **2022**, *14*, 1079. https://doi.org/10.3390/polym14061079

Academic Editors: Wei Wu, Hao-Yang Mi, Chongxing Huang, Hui Zhao and Tao Liu

Received: 22 February 2022
Accepted: 7 March 2022
Published: 8 March 2022

**Publisher's Note:** MDPI stays neutral with regard to jurisdictional claims in published maps and institutional affiliations.

## 1. Introduction

Polyetheretherketone (PEEK) is a novel crystalline thermoplastic polymer material that is widely used in aerospace, electronics, and medical industries for its excellent properties, such as low density and superior machinability [1–3]. However, PEEK has a few shortcomings, such as lower strength and lower hardness than most metal materials. These make it difficult to meet the higher performance requirements of certain industries. Therefore, the study about the performance of modified PEEK is of great significance. The mechanical properties of PEEK can be further improved through modification, and fiber-reinforced modification is a commonly used method to modify PEEK. The material properties of the fiber-reinforced PEEK vary with different fiber types. The fibers commonly used for PEEK reinforcement include carbon fiber, glass fiber, graphite fiber, and polytetrafluoroethylene (PTFE) fiber [4–7]. Carbon fiber has certain advantages, such as high strength, high modulus, small thermal expansion coefficient, and superior machinability [8,9]. On the other hand, glass fiber is often preferred due to its high stiffness, elastic modulus, and good load-bearing capacity [10]. Both carbon and glass fibers are widely used for reinforcement purposes [11–13]. However, despite materials possessing excellent properties, they also need to obtain better surface integrity through mechanical processing to fulfill the high requirements of the manufacturing industry [14]. Therefore, to expand the application fields of PEEK materials, it is crucial to study the material properties of the fiber-reinforced PEEK and analyze the influences of the change in the material properties on machinability.

The material properties and machinability of PEEK material change with fiber reinforcement, and the types and the mass fraction of fibers have diverse impacts on the PEEK matrix. The mechanical properties of PEEK can be enhanced with fiber reinforcing. Li et al. [9] compared the mechanical properties of PEEK and carbon-fiber-reinforced PEEK (CF/PEEK) and pointed out that carbon fiber can significantly improve the hardness, tensile, and compressive strength of PEEK. Zhang et al. [15] studied the material properties enhancement of the PEEK matrix by analyzing various fibers. The results showed that carbon fiber, glass fiber, and $TiO_2$ could effectively increase the tensile strength of the PEEK matrix. Although the material properties of PEEK have been improved by fiber reinforcement, PEEK material still needs to be mechanically processed to meet the higher surface integrity requirements of some industries. Therefore, it is of great significance to study the machinability of the fiber-reinforced PEEK materials. The commonly applied machining methods of modified PEEK materials include turning and grinding. Davim et al. [14] studied the influence of turning parameters on cutting force and surface roughness of PEEK and glass-fiber–reinforced PEEK (GF/PEEK) materials with PCD tools turning. The results showed that the cutting force decreased with the increase in cutting velocity and feed rate. The surface roughness of PEEK and GF/PEEK decreased with increasing cutting velocity and increased with the feed rate. The surface quality of pure PEEK was better than GF/PEEK with the same turning parameters. Ji et al. [16,17] worked on the nanomechanical properties and machinability with the single-point diamond turning of PEEK, CF/PEEK, and GF/PEEK materials. The results demonstrated that PEEK was a single-phase material with constant values of nano-hardness and modulus, whereas CF/PEEK and GF/PEEK were fiber composite materials with superior hardness and elastic modulus. The processing surface had poor uniformity of force in turning, which led to poorer turning processability as compared to the pure PEEK. Khoran et al. [18] investigated the grinding machinability of PEEK materials. They concluded that the ground temperature has a great influence on grinding surface morphology and force. The ground surface quality was profoundly affected by the cryogenic cooling that led to superior surface quality.

Although micron surface roughness of PEEK materials can be achieved by turning and grinding, the surface roughness after processing was bad due to the poor uniformity of the cutting force. Therefore, ultra-precision lapping is suitable for PEEK material processing due to its good uniformity and controllability of the lapping force, and improved machinability is achieved on the surface of the workpiece. Furthermore, ultra-precision lapping is an effective method to obtain high surface quality and accuracy [19]. However, there are only a few studies on the ultra-precision lapping process of the fiber-reinforced PEEK. Therefore, the lapping process based on analyzing the material properties of fiber-reinforced PEEK should be further researched to fully understand the process.

In this paper, the material properties and lapping machinability of fiber-reinforced PEEK has been studied. The effects of the type and mass fraction of modified fibers on the mechanical properties of PEEK have been investigated by analyzing the mechanical properties of PEEK that have been reinforced by carbon fiber and glass fiber. Finally, the lapping experiment was conducted to analyze the influences of property changes of materials on the material removal rate and surface roughness with abrasive paper lapping.

## 2. Materials and Methods

### 2.1. Materials

PEEK and its fiber-reinforced materials were provided by Nanjing Shousu Special Engineering Plastic Products Co., Ltd. (Nanjing, China). In this paper, five kinds of materials were investigated, which include the pure PEEK, carbon-fiber-reinforced PEEK with 10% mass fraction (CF10/PEEK) and 30% mass fraction (CF30/PEEK), glass-fiber-reinforced PEEK with 10% mass fraction (GF10/PEEK) and 30% mass fraction (GF30/PEEK). The average diameter and length of carbon fiber are 8 and 40 μm. The average diameter and length of glass fiber are 10 and 60 μm. The materials were prepared with molding.

*2.2. Methods*

2.2.1. Material Property Tests

Carbon fiber and glass fiber have excellent mechanical properties. The machinability and application of the fiber-reinforced PEEK are influenced by changing the material properties. Therefore, the tensile strength, hardness, and modulus of fiber-reinforced PEEK were tested in this paper to further analyze the influence of material properties on lapping machinability.

The tensile test was conducted at 298K with a tensile rate of 1mm/min using a material testing machine (5500A, Instron, Norwood, MA, USA) to measure the tensile stress curves of the fiber-reinforced PEEK. Flat dog-bone-shaped tensile specimens with a gauge length of 90 mm were made by wire cutting. The tensile specimens were ground on each side with SiC paper, resulting in a final specimen thickness of 4 mm and a gauge section width of 5 mm.

The nanoindenter (TI950, Hysitron, Eden Prairie, MN, USA) with a standard Berkovich indenter was used in this work to measure the load–displacement curves of the fiber-reinforced PEEK materials. The sample was a 15 mm × 15 mm × 5 mm square block made by wire cutting, and the sample surface was polished before the nanoindentation tests. To ensure that the data of the experiment are reliable, three points in a random site of the PEEK matrix on the samples were checked during the test for each indentation depth. The maximum load was 8 mN, with the loading for 10 s, holding for 5 s, and unloading for 10 s. Furthermore, nanoindentation was conducted on the pure PEEK material, carbon fiber, and glass fiber, respectively, with the same conditions.

2.2.2. Lapping Experiment Details

The surface roughness and the material removal rate (MRR) are the most relevant parameters to reflect the lapping machinability of materials. Thus, the MRR and surface roughness with abrasive paper lapping were studied in this work to analyze the influence of fiber types and mass fraction on the lapping machinability of fiber-reinforced PEEK.

All lapping experiments were conducted on an ultra-precision lapping machine (ZYP230, Shenyang Kejing Instrument Inc., Shenyang, China). The sample was a square block made by wire cutting with a size of 15 mm × 15 mm × 10 mm. The different mesh size of silicon carbide abrasive paper was applied in the lapping process to obtain the best surface roughness. During lapping, four samples for the same material were placed on a sample carrier in the once lapping process. A schematic diagram of the lapping experiment is shown in Figure 1.

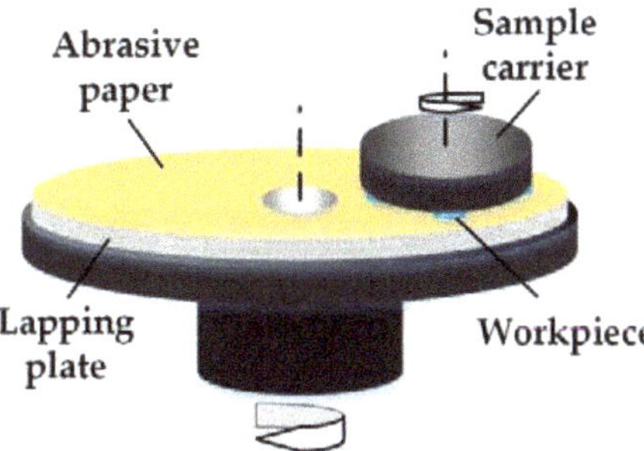


**Figure 1.** The lapping experimental setup.

The lapping parameters for the PEEK machinability analysis are presented in Table 1. The material removal rate (MRR) was calculated by the change of sample thickness per unit time and the sample thickness change during lapping was measured by a laser measuring device (Technological Gocator 2520, LMI, Mississauga, ON, Canada). The surface roughness with abrasive papers of different particle sizes was measured after the lapping process by Talysurf Profiler (CLI2000, Taylor Hobson Ltd., Leicester, UK), and the surface morphology was observed by SEM (SU820, Hitachi, Tokyo, Japan).

**Table 1.** Lapping parameters for fiber-reinforced PEEK lapping machinability analysis.

| Lapping Condition | Value |
| --- | --- |
| Abrasive | Silicon carbide abrasive paper |
| Mesh size | 180, 240, 320, 600, 1000, 2000 |
| Lapping plate speed (r/min) | 40 |
| Workpiece speed (r/min) | 63 |
| Lapping time (min) | 40 |
| Lapping pressure (KPa) | 11 |

## 3. Results

### 3.1. Material Properties of Fiber-Reinforced PEEK

#### 3.1.1. Nanomechanical Properties of Fiber-Reinforced PEEK

The components of the fiber-reinforced materials may have a great influence on the mechanical properties. The nanomechanical properties of five kinds of PEEK materials were studied in this work, and their results are exhibited in Figure 2. The nanoindentations in pure PEEK, CF/PEEK, and GF/PEEK show different load–displacement curves.

In Figure 2, the depth of carbon fiber and glass fiber were both smaller than the PEEK with the same load, which proved that the hardness of the fiber was higher than the PEEK. Furthermore, the depth of the fiber-reinforced PEEK matrix was smaller than the pure PEEK, and the influence of the indentation depth was more significant with an increased fiber mass fraction. The hardness and modulus could be calculated based on the Olive-Pharr method [20], and the results are shown in Figure 3. The hardness and modulus of the PEEK matrix and fiber are exhibited in Table 2. As shown in Figure 3 and Table 2, the hardness and modulus of carbon/glass fiber far exceeded that of the PEEK matrix. The hardness and modulus of the PEEK matrix were enhanced by the fiber due to its excellent mechanical properties, and the reinforcement was improved with the increased mass fraction of fiber.

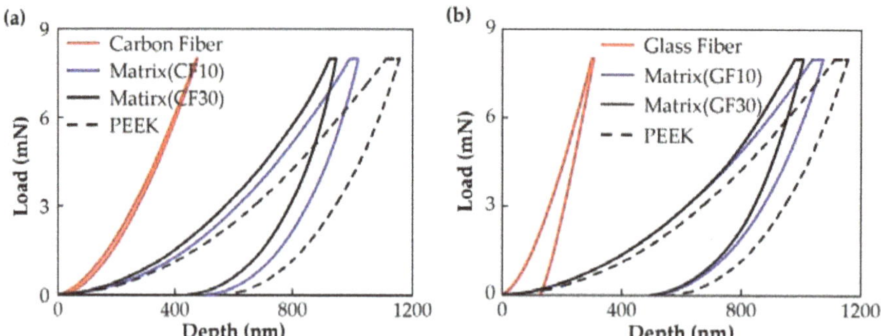


**Figure 2.** Load–depth (*P–h*) curves for (**a**) carbon-fiber-reinforced PEEK (CF/PEEK) materials and (**b**) glass-fiber-reinforced PEEK (GF/PEEK) materials from the nanoindentation tests with a maximum load of 8 mN. The *P–h* curves of the pure PEEK, the carbon fiber, and the glass fiber are displayed for comparative analysis.

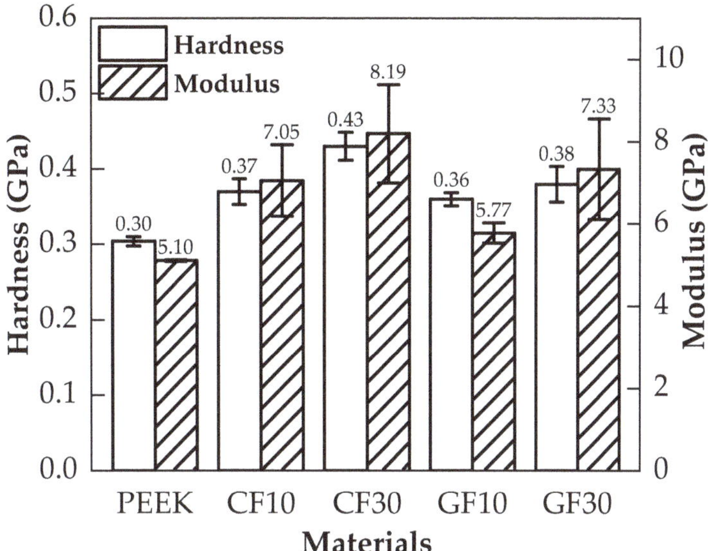

**Figure 3.** The average hardness and modulus of the pure PEEK and fiber-reinforced PEEK materials.

**Table 2.** The average hardness and modulus of the pure PEEK and carbon/glass fiber.

| Sample | PEEK | Carbon Fiber | Glass Fiber |
| --- | --- | --- | --- |
| Hardness (GPa) | 0.3 | 3.3 | 2.9 |
| Modulus (GPa) | 5.06 | 16.52 | 26.71 |

3.1.2. Tensile Mechanical Properties of Fiber-Reinforced PEEK

The tensile load–displacement curves of the pure PEEK and fiber-reinforced PEEK materials are shown in Figure 4. The tensile load–displacement curves demonstrated the different deformation phases of fiber-reinforced PEEK materials. In Figure 4, the load of pure PEEK increased with the increase in displacement in the elastic phase. When the load attained the maximum tensile load, the pure PEEK had a necked phenomenon and the load decreased to a constant value until the material failed at the fracture point. The ultimate tensile strength was the value of the maximum tensile load, and the tensile curves proved that the pure PEEK is a ductile material.

The tensile curves of CF/PEEK are represented in Figure 4b. In the elastic phase, the load increased with the displacement. However, necking of the CF/PEEK material occurred, and it fractured immediately when the load attained the ultimate tensile stress. The tensile length of the CF/PEEK fracture was smaller than that of pure PEEK. The tensile length of the material fracture was smaller and the ultimate tensile stress was higher with the increased carbon fiber mass fraction. The tensile curves proved that the CF/PEEK is a brittle material.

Figure 4c demonstrates the tensile curves of GF/PEEK. Similarly, the tensile load increased in the elastic phase until the ultimate tensile stress was attained. The tensile length of the material fracture and ultimate tensile stress became smaller as the glass fiber mass fraction increased. The GF/PEEK is a brittle material. The ultimate tensile stress of five kinds of PEEK materials is exhibited in Table 3. The CF/PEEK had the greatest tensile strength.

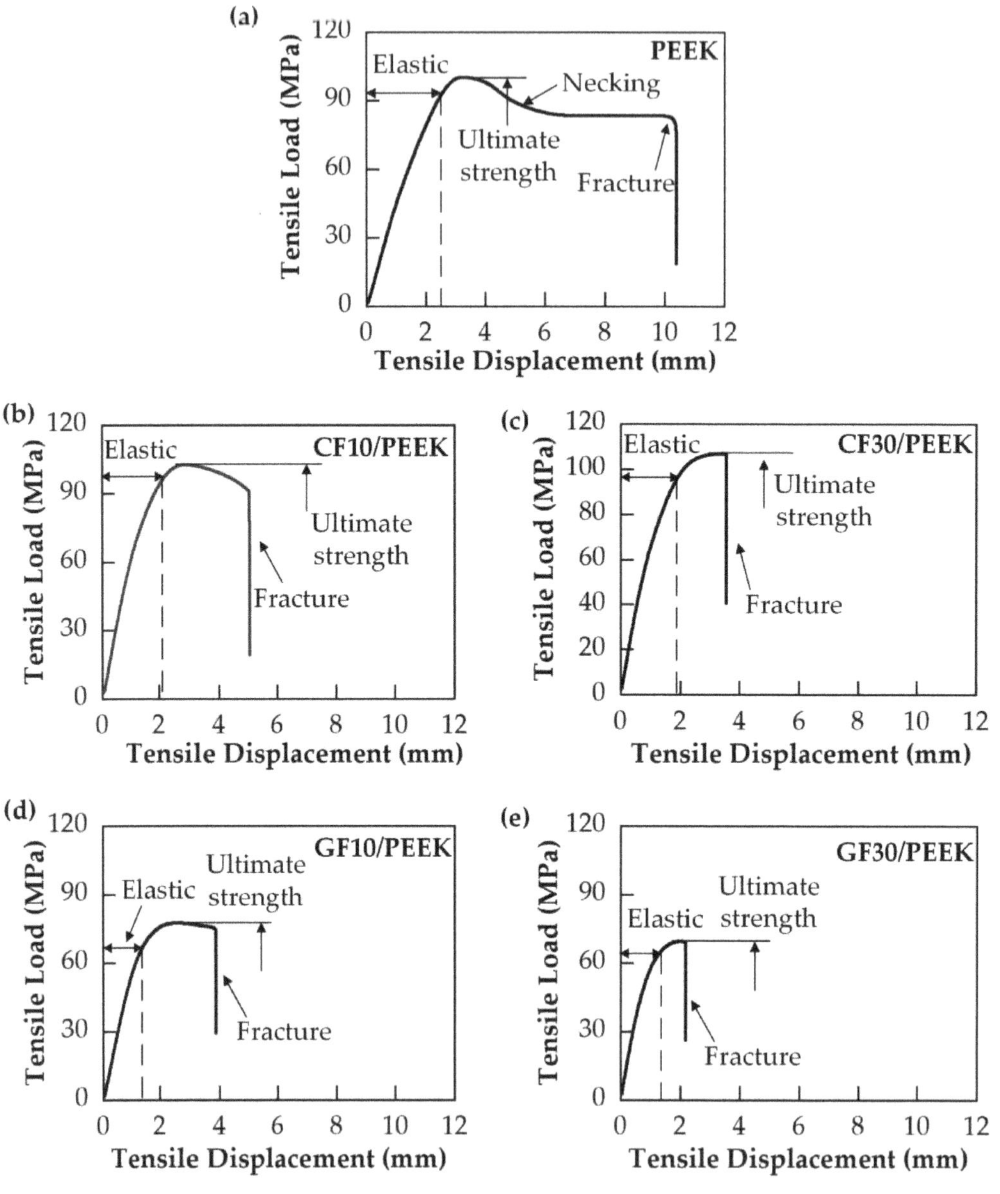


**Figure 4.** Tensile load–displacement curves of dog-bone-shaped specimens loaded in tension with tensile speeds of 1 mm/min for various materials: (**a**) PEEK; (**b**) carbon-fiber-reinforced PEEK with 10% fiber mass fraction (CF10/PEEK); (**c**) carbon-fiber-reinforced PEEK with 30% fiber mass fraction (CF30/PEEK); (**d**) glass-fiber-reinforced PEEK with 10% fiber mass fraction (GF10/PEEK); (**e**) glass-fiber-reinforced PEEK with 30% fiber mass fraction (GF30/PEEK).

**Table 3.** The ultimate tensile strength of the pure PEEK and fiber-reinforced PEEK materials.

| Materials | PEEK | CF10/PEEK | CF30/PEEK | GF10/PEEK | GF30/PEEK |
|---|---|---|---|---|---|
| Ultimate tensile strength (MPa) | 98.1 | 99.5 | 102.6 | 78.6 | 61.0 |

### 3.2. Lapping Processing Properties of Fiber-Reinforced PEEK

The material removal rate (MRR) reflects the degree of material removal difficulty during the lapping process. Taking the lapping experiment with abrasive paper of #240 mesh size as an example, Figure 5 shows the MRR of pure PEEK and fiber-reinforced PEEK materials. It can be seen in Figure 5 that the maximum MRR was of pure PEEK with the value of 17.4 µm/min. The MRR of CF10/PEEK and CF30/PEEK were 10.3 and 8.5 µm/min, respectively, whereas the MRR of GF10/PEEK and GF30/PEEK were 10.6 and 9.8 µm/min, respectively. As per the data, the MRR during the lapping process showed a decreasing trend when the fiber mass fraction increased.

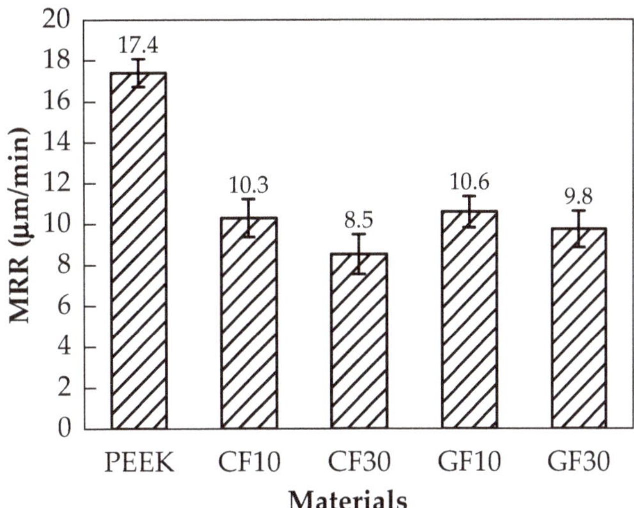

**Figure 5.** The material removal rates of the pure PEEK and the fiber-reinforced PEEK.

The surface roughness reflects the quality of the lapping process. The results of surface roughness after the lapping process are shown in Figures 6 and 7. The surface roughness of fiber-reinforced PEEK showed a downward trend with the particle size of the abrasive paper decreasing. When the mesh sizes of abrasive paper were small (#180, #240, #320), the lapped surface roughness was poor, and in addition, the downward trend of the surface roughness was obvious. When the abrasive paper of the #1000 mesh size was applied, the surface roughness improved and the downward trend flattened.

As shown in Figure 6a, with the same mesh size of the abrasive paper, the surface roughness of CF10/PEEK was lower than the GF10/PEEK, which was the same trend between the CF30/PEEK and GF30/PEEK. The CF/PEEK showed better lapping machinability than the GF/PEEK.

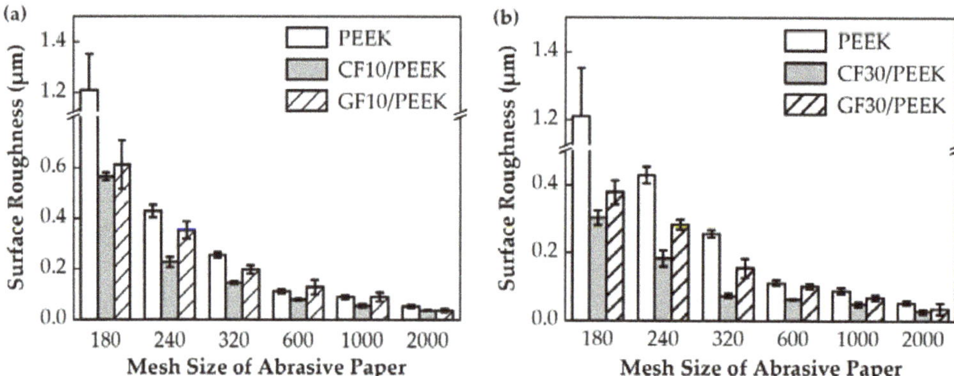

**Figure 6.** Comparative analysis on surface roughness of fiber types: (**a**) The surface roughness of the pure PEEK, CF10/PEEK, and GF10/PEEK, (**b**) the surface roughness of the pure PEEK, CF30/PEEK, and GF30/PEEK.

Figure 7 compares the surface roughness of different fiber mass fraction reinforced PEEK materials with the same type of fiber. As shown in Figure 7a, the surface roughness of CF30/PEEK was lower than that of CF10/PEEK. Similarly, the surface quality of GF30/PEEK was better than that of GF10/PEEK. The surface quality of fiber-reinforced PEEK improved as the fiber mass fraction increased. As per the data of MRR and surface roughness during the lapping process, the carbon fiber and glass fiber could improve the surface quality but decrease the MRR.

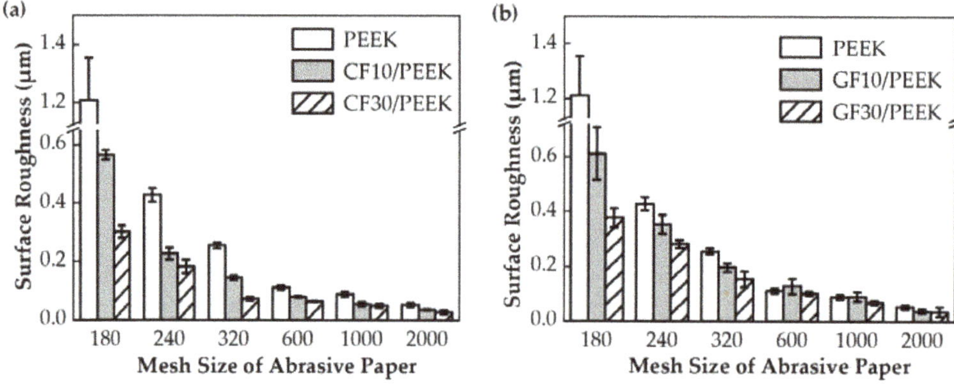

**Figure 7.** Comparative analysis on surface roughness of the fiber mass fraction: (**a**) The surface roughness of the pure PEEK, CF10/PEEK, and CF30/PEEK, (**b**) the surface roughness of the pure PEEK, GF10/PEEK, and GF30/PEEK.

## 4. Discussion

### 4.1. The Influence of Fiber Types on Lapping Machinability of Fiber-Reinforced PEEK Materials

The tensile strength, hardness, and elastic modulus of CF/PEEK materials were larger than the pure PEEK due to the extremely larger hardness and modulus of carbon fiber. The hardness and modulus of glass fiber also far exceeded that of pure PEEK, resulting in improvements in the hardness and modulus of GF/PEEK materials. In contrast, the average length and diameter ratio of glass fibers was too high, which resulted in an easier concentration and formation of defects inside the PEEK matrix and may reduce the tensile strength [21].

As shown in Figure 5, the MRRs of fiber-reinforced PEEK materials were smaller than that of the pure PEEK. During the lapping process, the interaction between abrasive particles and the workpiece could be divided into two categories. The first category was the interaction between abrasive particles and the PEEK matrix, and the second category was the interaction between abrasive particles and reinforcing fibers. The contact between the abrasive particles and the workpiece surface was irregular. Some abrasive particles were in contact with the PEEK matrix, and the others were in contact with the fibers. It can be observed in Table 2 that the hardness of carbon fiber or glass fiber far exceeded that of the pure PEEK. The high hardness of carbon fiber or glass fiber made it more difficult to remove than the pure PEEK during the lapping process, resulting in smaller MRRs of the fiber-reinforced PEEK materials than that of the pure PEEK. In the lapping process, the abrasive particle was coarse and exhibited poor height uniformity. Therefore, the surface roughness significantly decreased, resulting in the quick removal of the material. A rapid reduction in surface roughness would occur when the abrasives were courser. When the abrasive paper with the mesh size of #1000 was applied, the abrasive grains were finer. This improved the high equivalence more, and the pressure of a single abrasive particle on the fiber-reinforced PEEK materials became smaller. The MRR was then reduced, and the surface roughness tended to be better and constant. As shown in Figures 6 and 7, the PEEK had poorer surface quality than the fiber-reinforced PEEK materials with the same mesh size abrasive paper lapping. The lapped surface morphology of CF30/PEEK and GF30/PEEK are shown in Figure 8. The pure PEEK is a ductile material, and the fiber-reinforced PEEK is a brittle material (Figure 4). In Figure 8, there were many built-up edges and delamination on the PEEK matrix surface due to the ductile performance and low strength of the PEEK materials [22]. In the early period of lapping, there was only friction between the workpiece and abrasive particles with no material removal in the workpiece surface, resulting in ductile and elastic deformation. As the abrasive cutting process continued, the PEEK matrix was pressed by the abrasive particle and formed the scratches on the surface [23,24]. It can be observed in Figure 8 that there were many scratches and built-up edges on the lapped surface.

In Figure 8b, the scratch also occurred on the carbon fiber surface, but the scratch depth was shallower than the pure PEEK due to its high hardness. During the lapping process, the material removal methods of carbon fiber included cracks, scratches, and desquamations from the matrix [25]. The concaves formed by fiber desquamation were covered by the PEEK matrix as the lapping process continued; therefore, the fiber-reinforced PEEK had better surface quality than the pure PEEK with the same mesh size abrasive paper lapping.

In Figure 6, the CF/PEEK attained better surface quality than the GF/PEEK with the same fiber mass fraction. On the one hand, in Table 2, the hardness of carbon fiber was higher than the glass fiber. With the same mesh size abrasive paper lapping, the carbon fiber was more difficult to remove than the glass fiber, and the scratches on the carbon fiber were shallower than the glass fiber. Moreover, the average diameter and length of glass fiber were larger than that of carbon fiber, and in Figure 8c, the carbon fiber concaves formed by fiber desquamation were easier to cover by the PEEK matrix, resulting in better surface quality and a smaller MRR.

On the other hand, the reinforcement of fiber on the PEEK matrix varied with the fiber types. In Figure 3, the hardness and modulus of CF30/PEEK were higher than that of GF30/PEEK. Therefore, deformation of CF30/PEEK was more difficult, and the scratches of CF/PEEK would be shallower than the GF/PEEK. The CF/PEEK can attain better surface quality, whereas the CF/PEEK matrix attained higher hardness, which made it more difficult to be removed than the glass fiber, the MRR of CF/PEEK was smaller than that of GF/PEEK.

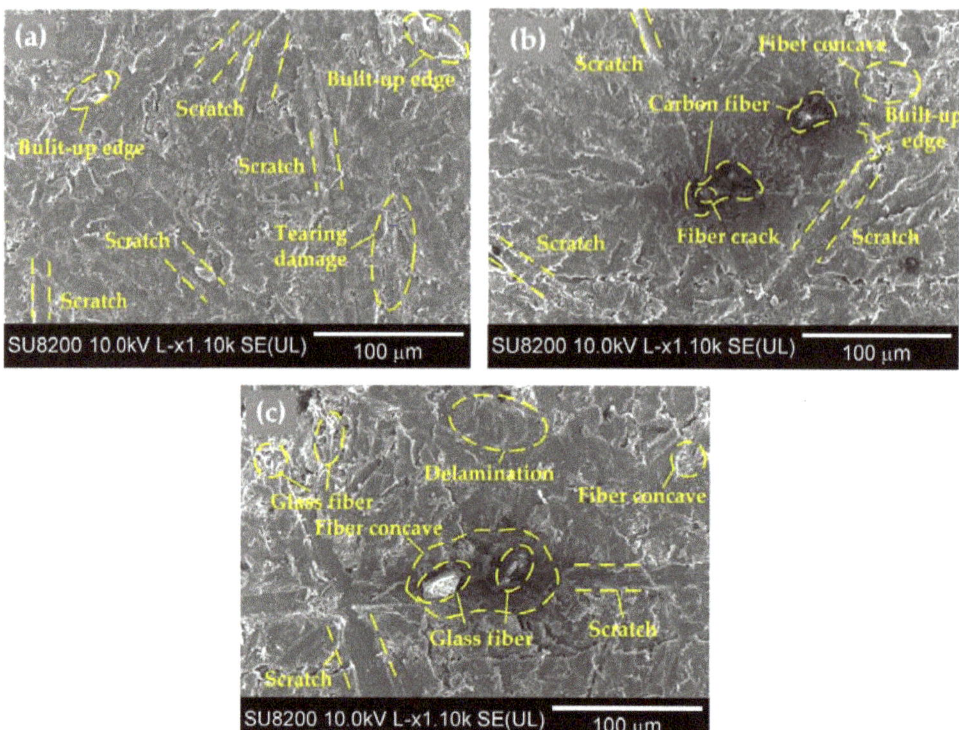



**Figure 8.** Surface morphology lapped with the #1000 mesh size silicon carbide abrasive paper of various fiber types of reinforced PEEK materials, (**a**) pure PEEK; (**b**) carbon-fiber-reinforced PEEK with the 30% fiber mass fraction; (**c**) glass-fiber-reinforced PEEK with the 30% fiber mass fraction.

*4.2. The Influence of Fiber Mass Fraction on Lapping Machinability of Fiber-Reinforced PEEK Materials*

In Figures 2 and 3, the hardness and modulus of the PEEK matrix improved significantly as the fiber mass fraction increased. The more carbon fiber or glass fiber mass fraction, the more the fiber was widely distributed on the PEEK matrix surface, which improved the hardness and modulus of fiber-reinforced PEEK materials. In Figure 4, the fiber-reinforced PEEK materials performed as brittle materials due to the brittleness of carbon and glass fiber. The tensile length of the fracture point decreased with the fiber mass fraction. The carbon fiber could improve the ultimate tensile strength of CF/PEEK. However, the length and diameter ratio of glass fiber was too large to form the defects inside the PEEK, resulting in the decrease in the ultimate tensile strength. Furthermore, the ultimate tensile strength decreased as the fiber mass fraction increased.

In Figures 5 and 7, the fiber-reinforced PEEK materials that had more fiber mass fraction could attain a smaller MRR and a better surface quality. In terms of MRRs of the same fiber type reinforced with PEEK of different fiber mass fractions, the more fiber mass fraction, the wider the interaction area between the abrasive particles and the fiber would be. Furthermore, the hardness of the PEEK matrix would be improved with an increased fiber mass fraction, resulting in the decreasing trend of the MRR due to the high hardness of the fiber.

Figure 9 shows the surface morphology of different carbon fiber mass fraction reinforced PEEK materials. As shown in Figure 2a, the nanoindentation curves of CF30/PEEK were above the CF10/PEEK, demonstrating that CF30/PEEK has a better ability to resist deformation. The scratches on the CF30/PEEK matrix were shallower than the CF10/PEEK. On the other hand, the carbon fiber could attain better surface quality than the PEEK matrix. Therefore, the improvement effect on the surface quality was more significant with an increased carbon fiber mass fraction.

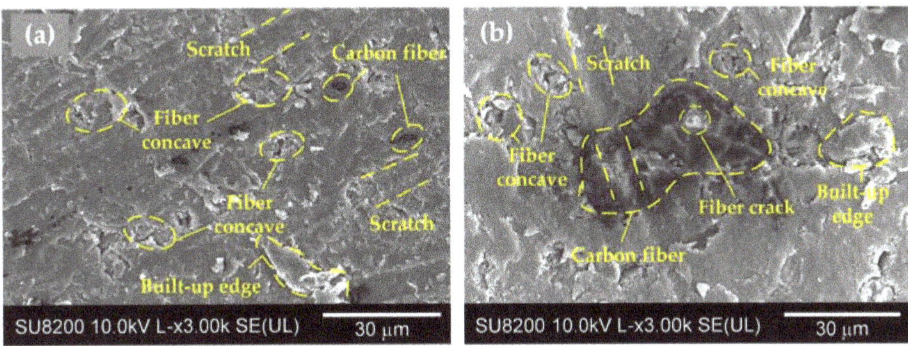


**Figure 9.** Surface morphology lapped with the #1000 mesh size silicon carbide abrasive paper of carbon-fiber-reinforced PEEK materials with different carbon fiber mass fractions, (**a**) carbon-fiber-reinforced PEEK with 10% carbon fiber mass fraction; (**b**) carbon-fiber-reinforced PEEK with 30% carbon fiber mass fraction.

The lapped surface morphology of GF10/PEEK and GF30/PEEK are shown in Figure 10. In Figures 2b and 3, the GF30/PEEK had superior mechanical properties to the GF10/PEEK. In Figure 10a, the scratches on the GF10/PEEK matrix were wider and more densely distributed, which led to poor surface quality, and there was tearing damage and material accumulation near the glass fiber concaves due to the ductile performance of the PEEK matrix. In Figure 10b, the glass fiber was distributed wider and the mechanical properties of PEEK matrix were improved significantly more than the CF10/PEEK. The scratches became shallow and narrow. The tearing damage phenomenon was greatly weakened and the surface quality improved.

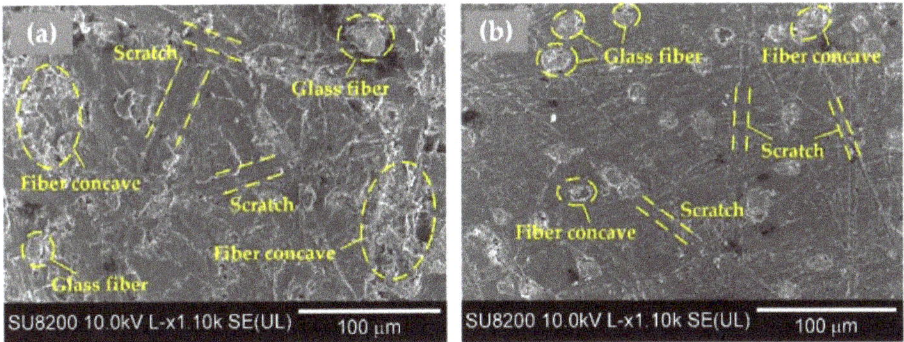


**Figure 10.** Surface morphology lapped with the #1000 mesh size silicon carbide abrasive paper of glass-fiber-reinforced PEEK materials with different glass fiber mass fractions, (**a**) glass-fiber-reinforced PEEK with 10% glass fiber mass fraction; (**b**) glass-fiber-reinforced PEEK with 30% glass fiber mass fraction.

## 5. Conclusions

In this paper, tensile tests, nanoindentation tests, and lapping experiments were carried out on five varieties of PEEK and its fiber-reinforced materials. The influences of the types and mass fraction of the modified fibers on the material properties were investigated in this work, including tensile strength, hardness, modulus, and lapping machinability. The following conclusions have been drawn from the results.

(1) In terms of material properties for CF/PEEK, carbon fiber has high hardness and modulus and is evenly distributed inside the material. Thus, the tensile strength, hardness, and modulus are improved. For GF/PEEK, the glass fiber is easy to concentrate and form defects inside the material due to its high hardness, modulus, and brittleness. This, in turn, decreases the tensile strength and improves the hardness and modulus. The influences on tensile strength and nanomechanical properties are more significant as the fiber mass fraction increases.

(2) In the lapping process, fiber-reinforced PEEK has better surface quality, and the MRR is lower than the pure PEEK due to the superior mechanical properties of the fiber. The lapped surface quality of CF/PEEK is better than GF/PEEK. Because the carbon fiber has higher hardness and modulus than the glass fiber, this results in a weaker deformation on the PEEK matrix surface.

(3) The higher the mass fraction is, the more the mechanical properties of fiber-reinforced PEEK are improved. As the fiber reinforcing can improve the lapped surface quality and hardness of the PEEK matrix, the fiber-reinforced PEEK with a higher fiber mass fraction has a better lapped surface quality and lower MRR.

**Author Contributions:** Conceptualization, S.G. and R.K.; methodology, S.G. and J.Q.; software, J.Q. and H.L.; investigation, S.G. and J.Q.; data curation, J.Q.; writing—original draft preparation, S.G.; writing—review and editing, H.L. and J.Q.; visualization, J.Q. and H.L.; supervision, R.K.; funding acquisition, S.G. and R.K. All authors have read and agreed to the published version of the manuscript.

**Funding:** This research was financially supported by the National Natural Science Foundation of China (51975091, 51991372, 51735004), the National Key Research and Development Program of China (2018YFB1201804-1), and the Lab of Space Optoelectronic Measurement and Perception (LabSOMP-2019-05).

**Institutional Review Board Statement:** Not applicable.

**Informed Consent Statement:** Not applicable.

**Data Availability Statement:** The data presented in this study are available on request from the corresponding author.

**Acknowledgments:** The authors would like to acknowledge the financial support from the National Natural Science Foundation of China, the National Key Research and Development Program of China and the Lab of Space Optoelectronic Measurement and Perception.

**Conflicts of Interest:** The authors declare no conflict of interest.

## References

1. Mate, F.; Gaitonde, V.N.; Karnik, S.R.; Davim, J.P. Influence of cutting conditions on machinability aspects of PEEK, PEEK CF 30 and PEEK GF 30 composites using PCD tools. *J. Mater. Process. Technol.* **2009**, *209*, 1980–1987. [CrossRef]
2. Najeeb, S.; Zafar, M.S.; Khurshid, Z.; Siddiqui, F. Applications of polyetheretherketone (PEEK) in oral implantology and prosthodontics. *J. Prosthodont. Res.* **2016**, *60*, 12–19. [CrossRef] [PubMed]
3. Arevalo, S.E.; Pruitt, L.A. Nanomechanical analysis of medical grade PEEK and carbon fiber-reinforced PEEK composites. *J. Mech. Behav. Biomed. Mater.* **2020**, *111*, 104008. [CrossRef] [PubMed]
4. Kang, H.; Qi, L.; Dang, H.; Jin, K.; Thomson, D.; Cui, H.; Li, Y. Biaxial tensile failure of short carbon-fibre-reinforced PEEK composites. *Compos. Sci. Technol.* **2021**, *208*, 108764. [CrossRef]
5. Davim, J.P.; Mata, F.; Gaitonde, V.N.; Karnik, S.R. Machinability evaluation in unreinforced and reinforced PEEK composites using response surface models. *J. Thermoplast. Compos. Mater.* **2010**, *23*, 5–18. [CrossRef]
6. Talbott, M.F.; Springer, G.S.; Berglund, L.A. The effects of crystallinity on the mechanical properties of PEEK polymer and graphite fiber reinforced PEEK. *J. Compos. Mater.* **1987**, *21*, 1056–1081. [CrossRef]

7.  Frick, A.; Sich, D.; Heinrich, G.; Lehmann, D.; Gohs, U.; Stern, C. Properties of melt processable PTFE/PEEK blends: The effect of reactive compatibilization using electron beam irradiated melt processable. *J. Appl. Polym. Sci.* **2013**, *128*, 1815–1827. [CrossRef]
8.  Qin, W.; Ma, J.; Liang, Q.; Li, J.; Tang, B. Tribological, cytotoxicity and antibacterial properties of graphene oxide/carbon fibers/polyetheretherketone composite coatings on Ti–6Al–4V alloy as orthopedic/dental implants. *J. Mech. Behav. Biomed. Mater.* **2021**, *122*, 104659. [CrossRef]
9.  Li, F.; Hu, Y.; Hou, X.; Hu, X.; Jiang, D. Thermal, mechanical, and tribological properties of short carbon fibers/PEEK composites. *High Perform. Polym.* **2018**, *30*, 657–666. [CrossRef]
10. Najim, A.S.; Adwaa, M. Studying mechanical properties specially fatigue behavior of (polyether ether ketone)/glass fiber composites in aerospace applications. *Appl. Mech. Mater.* **2014**, *666*, 8–16. [CrossRef]
11. Han, X.; Yang, D.; Yang, C.; Spintzyk, S.; Scheideler, L.; Li, P.; Li, D.; Geis-Gerstorfer, J.; Rupp, F. Carbon fiber reinforced PEEK composites based on 3D-printing technology for orthopedic and dental applications. *J. Clin. Med.* **2019**, *8*, 240. [CrossRef] [PubMed]
12. Zheng, B.; Deng, T.; Li, M.; Huang, Z.; Zhou, H.; Li, D. Flexural behavior and fracture mechanisms of short carbon fiber reinforced polyether-ether-ketone composites at various ambient temperatures. *Polymers* **2019**, *11*, 18. [CrossRef] [PubMed]
13. Díez-Pascual, A.M.; Ashrafi, B.; Naffakh, M.; González-Domínguez, J.M.; Johnston, A.; Simard, B.; Martínez, M.T.; Gómez-Fatou, M.A. Influence of carbon nanotubes on the thermal, electrical and mechanical properties of poly(ether ether ketone)/glass fiber laminates. *Carbon* **2011**, *49*, 2817–2833. [CrossRef]
14. Davim, J.P.; Reis, P.; Lapa, V.; Antonio, C.C. Machinability study on polyetheretherketone (PEEK) unreinforced and reinforced (GF30) for applications in structural components. *Compos. Struct.* **2003**, *62*, 67–73. [CrossRef]
15. Zhang, L.Z.; Li, M.; Hui, H. Study on mechanical properties of PEEK composites. In Proceedings of the 3rd International Conference on Manufacturing Science and Engineering, Xiamen, China, 27–29 March 2012. [CrossRef]
16. Ji, S.; Sun, C.; Zhao, J.; Liang, F. Comparison and analysis on mechanical property and machinability about polyetheretherketone and carbon-Fibers reinforced polyetheretherketone. *Materials* **2015**, *8*, 4118–4130. [CrossRef]
17. Ji, S.; Yu, H.; Zhao, J.; Liang, F. Comparison of mechanical property and machinability for polyetheretherketone and glass fiber-reinforced polyetheretherketone. *Adv. Mech. Eng.* **2015**, *7*, 1–7. [CrossRef]
18. Khoran, M.; Amirabadi, H.; Azarhoushang, B. The effects of cryogenic cooling on the grinding process of polyether ether ketone (PEEK). *J. Manuf. Process.* **2020**, *56*, 1075–1087. [CrossRef]
19. Ahn, Y.; Park, S.S. Surface roughness and material removal rate of lapping process on ceramics. *KSME Int. J.* **1997**, *11*, 494–504. [CrossRef]
20. Oliver, W.C.; Pharr, G.M. An improved technique for determining hardness and elastic modulus using load and displacement sensing indentation experiments. *J. Mater. Res.* **1992**, *7*, 1564–1583. [CrossRef]
21. Hou, L.; Qu, M.; He, J.; Sun, Y.; Wu, L.; Na, F. Study on properties of glass fiber reinforced PEEK composites with different length and diameter ratio. *Plast. Sci. Technol.* **2018**, *46*, 41–46. [CrossRef]
22. Quan, Y.; Yan, Y. Machined surface texture and roughness of composites. *Acta Mater. Compos. Sin.* **2001**, *5*, 128–132. [CrossRef]
23. Kobayashi, A.; Saito, K. On the cutting mechanism of high polymers. *J. Polym. Sci.* **1962**, *58*, 1377–1396. [CrossRef]
24. Rentsch, R.; Inasaki, I. Molecular dynamics simulation for abrasive process. *CIRP Ann.* **1994**, *43*, 327–330. [CrossRef]
25. Mionru, A.; Kazuo, N.; Wang, X.D. Study on the surface intergrity in aluminum reinforced by short fibre. *J. Jpn. Soc. Precis. Eng.* **1991**, *57*, 172–177. [CrossRef]

 *polymers*

*Article*

# Facile Preparation of Cellulose Fiber Reinforced Polypropylene Using Hybrid Filler Method

Safarul Mustapha [1], Jacqueline Lease [1], Kubra Eksiler [1], Siew Teng Sim [1], Hidayah Ariffin [2] and Yoshito Andou [1,3,*]

[1] Department of Biological Functions Engineering, Graduate School of Life Science and Systems Engineering, Kyushu Institute of Technology, 2–4 Hibikino, Wakamatsu, Fukuoka, Kitakyushu 808-0196, Japan; paroys_9@yahoo.com (S.M.); lease.jacqueline708@mail.kyutech.jp (J.L.); kubra.eksiler@gmail.com (K.E.); simsiewteng92@gmail.com (S.T.S.)

[2] Faculty of Biotechnology and Biomolecular Sciences, Universiti Putra Malaysia (UPM), Serdang 43400, Selangor, Malaysia; hidayah@upm.edu.my

[3] Collaborative Research Centre for Green Materials on Environmental Technology, Kyushu Institute of Technology, 2–4 Hibikino, Wakamatsu, Fukuoka, Kitakyushu 808-0196, Japan

* Correspondence: yando@life.kyutech.ac.jp

**Abstract:** Dried hybrid fillers comprised of silica/CNF were successfully synthesized in ethanol/water mixed solvents at room temperature without the usage of any precursor. The as-prepared fillers were incorporated with polypropylene (PP) as a polymer matrix through a twin-screw extruder. From surface morphology analysis, the agglomeration of the silica/CNF hybrid fillers was prevented in the PP matrix and they exhibited moderate transparency, around 17.9% and 44.6% T at 660 nm. Further, the chemical structures of the polymer composites were identified by Fourier transform infrared (FT-IR) analysis. According to thermogravimetric analysis (TGA), the insertion of silica as a co-filler to the PP matrix resulted in an increase in its degradation onset temperature and also thermal stability. In addition, the mechanical properties of the PP composites also increased after the blending process with the hybrid fillers. Overall, sample PP-SS/CNF exhibited the highest tensile strength, which was 36.8 MPa, or around 73.55% compared to the pristine PP. The improvements in tensile strength were attributed to good dispersion and enhanced efficiency of the stress transfer mechanism between the silica and the cellulose within the PP matrix. However, elongation of the sample was reduced sharply due to the stiffening effect of the filler.

**Keywords:** cellulose nanofiber; silica; polypropylene; composite; hybrid filler

**Citation:** Mustapha, S.; Lease, J.; Eksiler, K.; Sim, S.T.; Ariffin, H.; Andou, Y. Facile Preparation of Cellulose Fiber Reinforced Polypropylene Using Hybrid Filler Method. *Polymers* **2022**, *14*, 1630. https://doi.org/10.3390/polym14081630

Academic Editors: Wei Wu, Hao-Yang Mi, Chongxing Huang, Hui Zhao, Tao Liu and Carlo Santulli

Received: 1 March 2022
Accepted: 15 April 2022
Published: 18 April 2022

**Publisher's Note:** MDPI stays neutral with regard to jurisdictional claims in published maps and institutional affiliations.



## 1. Introduction

A polymer matrix composite is a material consisting of a polymer matrix with a re-inforcing dispersed phase. Incorporating inorganic fillers into a polymer matrix can give the composite unique properties such as rigidity, high thermal stability, good mechanical property, flexibility, and ductility [1]. Fillers with particle sizes in the 1–100 nm range are defined as nanofillers. Generally, nanofillers are categorized into three types based on their geometries: one-dimensional (1D, rod-like), two-dimensional (2D, platelet-like), and three-dimensional (3D, spherical) materials [2]. Important factors used to determine the reinforcing effect of these fillers are the polymer matrix's properties, the nature and type of filler, concentration of polymer and filler, and particle size as well as particle distribution [1]. Moreover, the most common nanofillers include metallic nanoparticles, polyhedral oligomeric silsesquioxane, carbon nanomaterials, graphite nanoplates, silica nanoparticles, and nanocellulose [3–5]. These nanofillers are inserted into polymer matrixes improve the mechanical properties, gas and solvent barrier properties, thermal degradability, and chemical resistance of the polymer [6,7]. In contrast, traditional micro-fillers can lead to polymer embrittlement, loss of transparency, and loss of lightness.

Cellulose fiber has drawn considerable interest as natural filler in polymer composites due to its superior mechanical properties, eco-friendliness, processability, biodegradability, biocompatibility, low toxicity, cost savings, and improved fuel efficiency [8,9]. However, it is difficult to disperse it properly in hydrophobic polymer because of their different surface properties. Fillers with various properties are used as hybrid fillers along with reinforcing fibers to further enhance physical and mechanical properties.

Researchers have utilized more than one filler material to investigate the synergistic effect of fillers on the final properties of polymer composites [10]. Anwer et al. [11] prepared nanocomposites using epoxy resin with carbon nanofibers (CaNFs), graphene nanoplatelets (GNPs), and a hybrid combination of CaNFs/GNPs as fillers. These composites were processed with and without the use of surfactants. It was proposed that GNPs prevented agglomeration of CaNFs during the process, leading to larger particle aspect ratios in the nanocomposite. In addition, Kwak et al. [12] successfully prevented the agglomeration of CaNFs using fish-derived gelatin during the dehydration process. According to Thomas et al. [13], a hybrid filler of carbon nanotubes (CNTs)/clay enhanced dispersion through the synergism and prevented agglomeration in the composite blend. A hybrid filler can improve crosslink density, tensile strength, and tear resistance due to the large contact area between clay and CNTs.

Typically, silica/cellulose-reinforced polymer composites have been prepared using the sol-gel method. Li et al. [14] synthesized cellulose nanocrystal/silica hybrids using TEOS as the silica precursor. This hybrid material was melt-blended with ultra-high molecular weight polyethylene (UHMWPE) polymer in a twin-screw extruder. The nanocomposite showed improvement in its flexural modulus and tensile and flexural strength. Although this sol-gel method is widely used to produce hybrid fillers, the precursors, such as TEOS, are relatively expensive and contain high amounts of embedded energy [3,15].

Herein, an ethanol/water mixed-solvent method was proposed as an accessible, fast, and low-cost protocol to prepare silica particles (SiPs) with cellulose nanofibers (CNFs) without chemical modification. Ethanol is miscible with water at any ratio, and the addition of ethanol to water can easily change its physicochemical properties. The in situ nucleation and growth of silica onto cellulose occurs rapidly in an aqueous solution. However, the addition of ethanol to water can restrain the diffusion of ions and the nucleation and growth of the silica. It is well known that rapid nucleation and slow growth favor the formation of particles with narrow size distribution. Thus, we chose ethanol/water mixed solvents as the solvent system.

Commercialized SiPs and CNFs were dispersed in the mixed solvent, and SiPs were deposited onto the CNF surface. During the solvent evaporation process, it was expected that the SiPs could prevent the agglomeration of CNFs. The interaction between the silica surface group and cellulose chain prevented immediate aggregation. To the best of our knowledge, there are no studies on incorporating cellulose fibers into hydrophobic polymers without surfactants or chemical modifications [3].

In this study, polypropylene (PP) was used as a polymer matrix. It is a widely utilized polymer with various advantages including low cost, superior transparency, good moisture barrier properties, and high recyclability compared to other polymers [16]. The composites, PP with fillers (SiPs, CNFs or SiP/CNF) were prepared by melt-blending in a twin-screw extruder. The effects of fillers and hybrid fillers in PP matrixes were investigated by the analysis of the matrixes' morphological, chemical, thermal, and mechanical properties. The performance of pulverized SiPs and pulverized CNFs as filler in PP composites was also identified.

## 2. Materials and Methods

### 2.1. Materials

Polypropylene (PP), as a polymer matrix, was obtained from Japan Polypropylene Corporation, Tokyo, Japan (Polypropylene FY-6, MFR 2.5 g/10 min, CrI 59.80%). It has melting point of 162 °C, crystallization temperature of 104 °C and degradation temperature of 461 °C. Cellulose nanofibers (5 wt% aqueous dispersion, DP 650) were provided from Sugino Machine Limited, Toyama, Japan. The CNFs were produced by a super high-pressure water jet system. SiP powders, Sylosphere 200 (SS, diameter 3.0 μm) and Sylophobic 200 (SP, diameter 3.9 μm) were supplied from Fuji Silysia Chemical, Aichi, Japan. The ethanol (>99.5%) was purchased from Wako Pure Chemical Industries Ltd., Osaka, Japan, and reverse osmosis (RO) water was used throughout the experiments.

### 2.2. Filler Preparation Procedure

The compositions of the reinforced PP created by using as-prepared filler and hybrid filler samples prepared in this study are listed in Table 1. The required amount of SiPs and CNFs was added into the ethanol/water mixed solvents and stirred for one hour. The mixture was then evaporated by using a rotary evaporator (Eyela N-1110, Tokyo Rikakiki Co., Ltd., Tokyo, Japan). Subsequently, it was dried under vacuum overnight before use.

**Table 1.** Sample composition of PP and filler.

| Sample | PP (wt%) | Sylosphere (wt%) | Sylophobic (wt%) | CNF (wt%) |
| --- | --- | --- | --- | --- |
| PP | 100 | - | - | - |
| PP-8.75 SS | 91.25 | 8.75 | - | - |
| PP-10 SS | 90 | 10 | - | - |
| PP-1.25 CNF | 98.75 | - | - | 1.25 |
| PP-SS/CNF | 90 | 8.75 | - | 1.25 |
| PP-8.75 SP | 91.25 | - | 8.75 | - |
| PP-8.75 pul.SP | 91.25 | - | 8.75 | - |
| PP-10 SP | 90 | - | 10 | - |
| PP-SP/CNF | 90 | - | 8.75 | 1.25 |

### 2.3. Composite Preparation

The desired amount of filler, hybrid filler, and PP polymer were fed into a twin-screw extruder with an ellipse-blade type screw L/D 0.5 (IMC-1979, Imoto Machinery Co., Toyama, Japan). The rotation speed, temperature, and residence time for melt-mixing were controlled at 40 rpm, 180 °C, and 5 min, respectively. The films were prepared using a hydraulic hot-press (IMC-180C, Imoto Machinery Co., Toyama, Japan) at 180 °C for 3 min under a pressure of 30 MPa, and then cooled at room temperature.

### 2.4. Characterization

The surface morphologies of SiPs were observed under a 3D laser scanning confocal microscope (LSCM) model VK-X 100 (Keyence Corporation, Osaka, Japan) under prescribed conditions of laser: red semiconductor laser, $\lambda$ = 658 nm, 0.95 mW, and pulse width of 1 ns, using a depth composition procedure.

The morphology of the fractured composites and dispersion state of the hybrid filler in the PP matrixes were investigated using a scanning electron microscope equipped with energy dispersive spectroscopy (SEM-EDS) (JCM 6000, JEOL Ltd., Tokyo, Japan). SEM images were obtained at 15 kV accelerating voltage and the PP composite samples were fractured under liquid nitrogen. Each sample was deposited on carbon tape and carbon-coated for 90 s before observation.

The optical properties of the PP composite films were determined by UV–Vis spectra. A rectangular piece of each film sample (4 cm × 4 cm) was directly mounted between two spectrophotometer magnetic cell holders. The transmittance spectra of the films were measured at selected wavelength ranges from 190 to 1000 nm using a UV–Vis spectrophotometer (GENESYS 50, Thermo Fisher Scientific, Waltham, MA, USA). The optical properties of the PP and PP composite films were characterized by the transmittance of visible (660 nm) regions.

Chemical analysis was carried out using a Fourier Transform Infrared (FT-IR) spectroscope Nicolet iD7 ATR (Thermo Fisher Scientific, Waltham, MA, USA). Each sample recording consisted of 16 scans recorded from 4000 to 400 cm$^{-1}$.

The PP composite films were cut into a rectangular shape of 40 mm × 5 mm × 0.5 mm. The mechanical properties of each PP composite were determined by an IMC-18E0 model machine (Imoto Machinery Co., Ltd., Kyoto, Japan) at a rate of 10 mm/min crosshead speed at 23 °C. Each measurement was carried out with five replicates.

The thermal properties of PP composites were characterized using an EXSTAR TG/DTA 7200 (SII Nanotechnology Inc., Chiba, Japan). The samples were scanned at a range of 30 to 550 °C at a constant heating rate of 10 °C/min under a continuous nitrogen flow rate of 100 mL/min.

## 3. Results and Discussion

### 3.1. Silica/CNF Filler Mechanism in PP Polymer

Fine particles aggregated due to van der Waals forces or chemical bonds [17], and CNFs agglomerated due to strong fiber–fiber hydrogen bonding coupled with their polar nature, especially in a non-polar polymeric environment [18]. We hypothesized that the silica particles would adhere to the CNFs' surface, which weakens the hydrogen bonds and prevents agglomeration. Figure 1 shows the interaction of the fine particles and CNFs if mixing in solvents. When the ratio of the fine particles is low, CNFs will agglomerate, and if too high, the fine particles will agglomerate. To further elucidate the respective mechanisms, SiPs were utilized and mixed with CNFs in ethanol/water mixed solvent. In this research, we only focus on the specific ratio of CNFs and SiPs first in order to investigate the basic interactions of the hybrid filler.

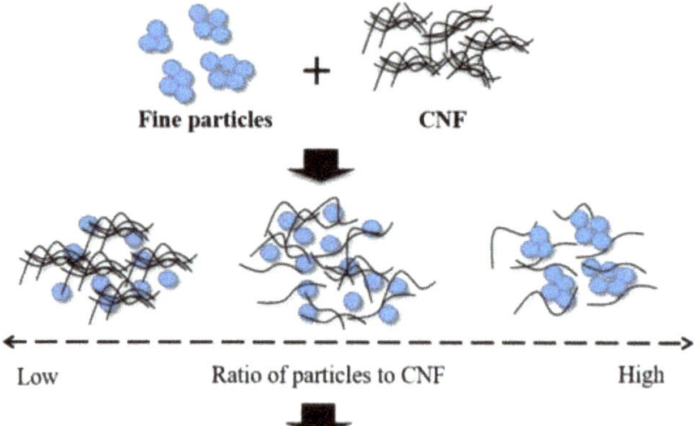


**Figure 1.** Interaction of fine particles and CNFs in ethanol/water mixed solvent.

Figure 2a,c shows the hybrid filler preparation in the ethanol/water mixed solvent and the possible hybrid filler mechanism in the hydrophobic polymer. Hydrophilic CNFs caused irreversible agglomeration during drying (Figure 2b), and due to the formation of additional hydrogen bonds between fibers, hydrophilic CNFs induced aggregation in the non-polar matrix [19]. Therefore, SiP powder was used to prevent the CNFs' aggregation from occurring when incorporated into the hydrophobic polymer. From the observation of the hot-press film, the hybrid filler prevented the aggregation of CNFs in PP composites, and a good dispersion rate and transparency were achieved (Figure 2c).

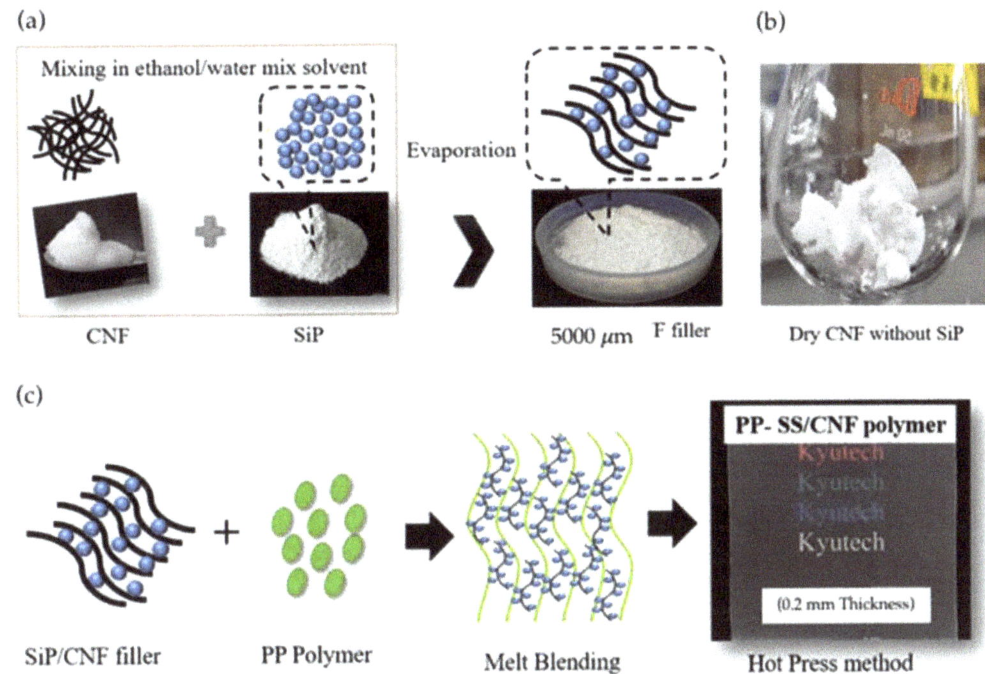



**Figure 2.** Images of SiP/CNF filler; (**a**) hybrid filler preparation, (**b**) agglomerated CNFs without SiPs and (**c**) hybrid filler mechanism in PP polymer.

Based on Figure 2a, the SiPs were mostly localized at each of the CNF fibers. Interaction between SiPs and CNFs was posited to be due to the presence of hydrogen bonds, due to the large surface areas and high amount of hydroxyl groups present in SiPs [20,21]. To further study the effect of hybrid fillers in hydrophobic polymer, PP polymer as a polymer matrix was melt-blended with fillers using a twin-screw extruder. Then, a hot-press sheet of the composite was prepared for further analysis. The chemical structures, thermal stabilities, and degree of substitution of MCC with different reaction temperatures and times were analyzed and investigated.

### 3.2. Surface Morphologies of the Hybrid Filler and PP Composite

Figure 3a,c presents the surface morphologies of dried SiP SS and SP under a laser microscope with 50× magnification. A SiP SS is a single-distributed spherical silica particle and has a smooth surface with average size of 3–4 μm, having features such as high mobility and outstanding dispersity. Meanwhile, A SiP SP has a rough surface and irregular shape with an average size of 2–7 μm. According to the manufacturer, SiP SP demonstrates hydrophobic properties by chemically replacing the hydroxy groups on the silica surface with organic silicone compounds.

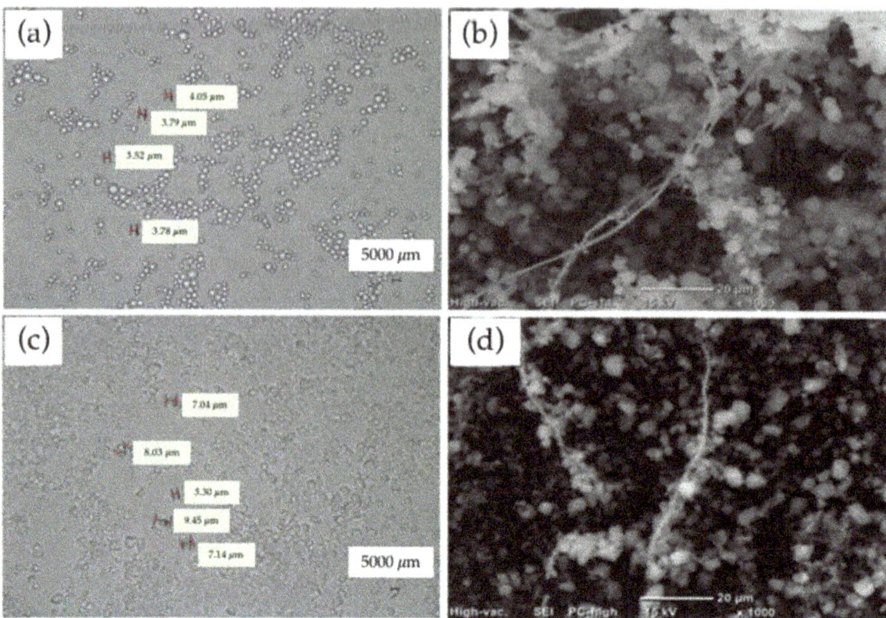

**Figure 3.** Morphological image of SiP particles: (**a**) SiP SS 50X, (**b**) SS/CNF 1000X, (**c**) SiP SP 50X, and SEM image of hybrid filler, (**d**) SP/CNF 1000X.

The structures of both hybrid fillers can be directly observed by using SEM. The SEM images in Figure 3b,d (SS/CNF and SP/CNF) illustrate that the SiPs were relatively homogeneously dispersed in the CNF matrix. It was verified that both types of SiPs were deposited predominantly on the surface of CNFs and prevented CNF agglomeration during the drying process. Both SiP SS and SP helped the dispersion of CNFs via a synergism effect. Silica not only offered a great surface area for coherence, but also provided aid in the dispersal of CNFs, which in turn reduced and minimized agglomeration. According to Sharip et al. [18], conventionally water-dispersed CNFs were employed to prepare a bio-nanocomposite, which required higher cost and energy. Hence, an ethanol/water mixed solvent method was used to address the setbacks of the conventional approach.

### 3.3. Optical Properties of the PP Composite with Fillers

The effect of different types of silica and CNFs on optical properties was elucidated. The PP composite was melt-blended through a twin-screw extruder. Figure 4 shows the PP composite after melt-blending and hot-pressing for analyses and the study of mechanical properties. Pristine PP (Figure 4a) acted as a reference. Figure 4b,c demonstrates the PP composite with SiP SS and SiP SP, while Figure 4e,f shows the PP-SS/CNF and PP-SP/CNF composites. Figure 4d presents the PP composite with only CNFs as the filler.

Based on observations, excellent dispersibility was noticed for the PP composite with SiPs and also the PP composite with SiPs/CNF. However, significant agglomeration or poor dispersion of CNFs can be observed in PP/CNF composites. This scenario was posited to be due to the different surface properties and the hydrophilic nature of CNFs. Inherently strong intra- and inter-molecular hydrogen bonds of CNFs led to a strong affinity to itself. Achieving good dispersion of fillers within polymer matrixes is the most critical factor in determining their resultant mechanical performance [22].

**Figure 4.** Digital images showing the transparency of (**a**) PP, (**b**) PP-8.75 SS, (**c**) PP-8.75 SP, (**d**) PP-1.25 CNF, (**e**) PP-SS/CNF, and (**f**) PP-SP/CNF.

UV-vis spectroscopy analysis was carried out to further support the optical transparency results of the PP nanocomposites. Figure 5 shows the UV-vis transmittance spectra of the PP composite with the hybrid fillers, and the percentages of visible light transmittance values of the PP composite are summarized in Table 2.

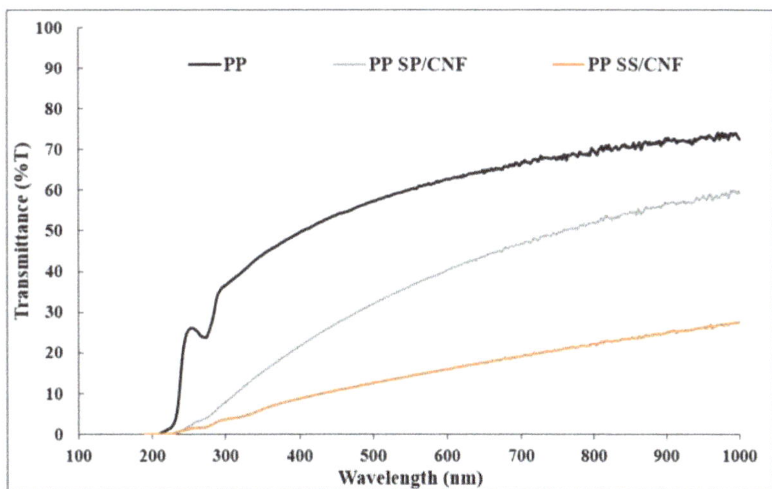


**Figure 5.** UV–vis transmittance spectra for PP and PP composite.

**Table 2.** Optical properties of PP and PP composite.

| Sample | Thickness (mm) | %T (Transmittance) (660 nm) |
|---|---|---|
| PP | 0.255 | 65.3 |
| PP SS/CNF | 0.446 | 17.9 |
| PP SP/CNF | 0.356 | 44.6 |

The UV-visible spectra of PP nanocomposites were analyzed in the wavelengths between 200 and 800 nm. Based on data given in Table 2, the transmittance of plain PP was about 65.3 %T at 660 nm. After adding the hybrid fillers, the transmittance value was reduced. The transmittance efficiency decreased due to blocking by the hybrid filler. The dispersity of the reinforcement will affect the transparency of the nanocomposite. The transmittance level of the PP SP/CNF (44.6 %T) was higher than the PP SS/CNF (19.9 %T), even provided the same hybrid filler loading level. This scenario was due to the higher hydrophobicity of SP making it easier to incorporate with the hydrophobic PP polymer matrix.

### 3.4. Chemical Structure of Composite Films

Functional group analysis of the PP and PP composites can be performed using FTIR spectroscopy. Figure 6 depicts the typical FT-IR spectra of pristine PP, SiPs, CNFs, and PP composite. The appearance of characteristic absorption bands of SiPs at 1054 and 1048 cm$^{-1}$ were assigned to the siloxane Si–O–Si bonds for SiP SS and SP, respectively. Furthermore, transmittance peaks at 794 and 797 cm$^{-1}$ indicated the presence of hydroxyl groups on the surface [23]. The C-H stretch vibration at 2966, 2898 and 1392 cm$^{-1}$ were the typical characteristics of polyalkyl siloxane on the SiP SP as hydrophobic surface treatments [24].

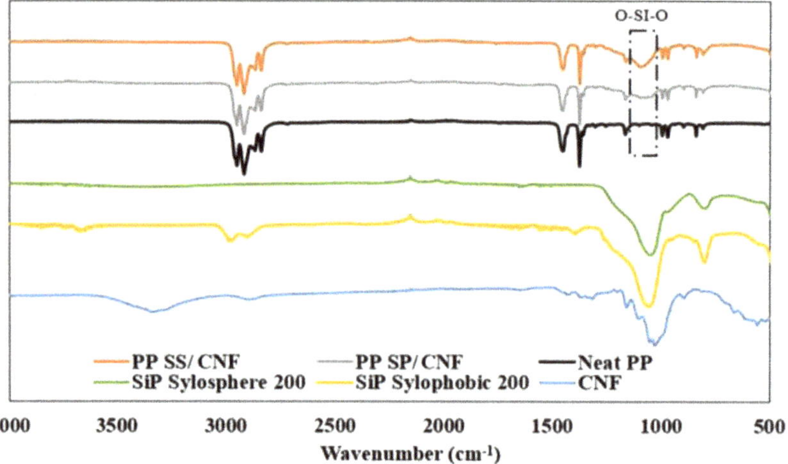


**Figure 6.** FTIR spectra of SiP Sylosphere 200, Sylophobic 200, CNF, and PP composites.

For the CNFs, the broad peak at 3400–3300 cm$^{-1}$ was attributed to the stretching vibration of O–H bonding from absorbed water molecules of the cellulose chains. This peak also included inter- and intra-molecular hydrogen bond vibrations in hydroxyl groups in cellulose I [25,26]. The peak at 2895 cm$^{-1}$ was assigned to the C–H stretching vibration of all hydrocarbon constituent in polysaccharides and the peak at 894 cm$^{-1}$ attributed to the β-glycosidic linkages of the cellulose chain [27].

Furthermore, the peaks at 2950–2835 cm$^{-1}$ were contributed by C–H stretching vibrations while peaks at 1450 and 1376 cm$^{-1}$ were assigned to CH$_2$ and CH$_3$ bending vibration in neat PP [28]. Further, the transmittance peak located at 840 cm$^{-1}$ was assigned to C–CH$_3$ stretching vibration [29]. This peak is a typical characteristic of PP polymers.

New transmittance peaks of the PP nanocomposites at 1087 and 1076 cm$^{-1}$ of the asymmetric vibration of O–Si–O bonds indicated that PP with SS/CNF and SP/CNF contained silica fillers. In addition, increasing peak intensity at 806 cm$^{-1}$ indicated stretching vibrations of Si–O. From the FT-IR spectra, it is confirmed that the hybrid fillers did not damage the composite materials via denaturing through the extrusion process. However, the characteristic bonds of the C–O group in CNFs were undetected due to the overlap with the O–Si–O band [30].

### 3.5. Mechanical Properties of PP Composite

Mechanical properties such as tensile strength and elongation are often used to measure the strength and elasticity of composite films. The representative tensile stress-strain curves of pure PP and its composites are shown in Figure 7 and their key mechanical properties are summarized in Table 3. The fillers incorporated into composite films significantly affected the mechanical properties of the composites due to the specific surface area and dispersibility of each filler. Improved mechanical properties can be achieved through an improved interface between the filler and polymer matrix [31].

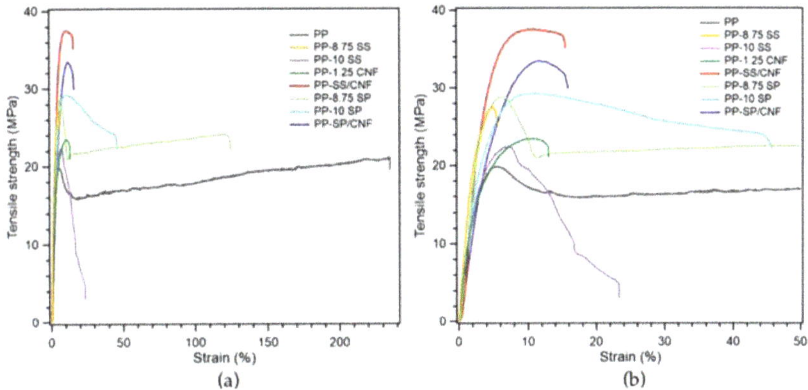


**Figure 7.** Mechanical properties of (**a**) stress-strain curves of PP composites and (**b**) an enlarged graph.

**Table 3.** Mechanical properties PP composites.

| Sample Code | Tensile Strength at Yield (MPa) | Tensile Elongation at Break (%) |
|---|---|---|
| PP | 20.78 ± 1.06 | 260.50 ± 27.03 |
| PP-8.75 SS | 27.80 ± 1.09 | 7.30 ± 1.10 |
| PP-10 SS | 23.53 ± 0.86 | 24.71 ± 12.20 |
| PP-1.25 CNF | 23.30 ± 2.22 | 12.17 ± 2.54 |
| PP-SS/CNF | 36.81 ± 1.56 | 20.87 ± 2.68 |
| PP-8.75 SP | 29.98 ± 2.94 | 149.20 ± 49.36 |
| PP-10 SP | 31.45 ± 1.34 | 47.40 ± 15.21 |
| PP-SP/CNF | 33.08 ± 0.47 | 14.57 ± 2.77 |

Figures 7 and 8 and Table 3 also show the effect of filler on the tensile strength at yield of PP composites compared to pristine PP polymer. The yield strength of PP increased after melt-blending with filler or hybrid filler. This scenario was due to the addition of filler into the matrix, which reduced the available free spaces and hence increased the stiffness of the composite. The filler linked the matrix together, leading to enhanced interaction between the reinforcement and the matrix. Once the load has been applied, the stress can easily be transferred from the polymer matrix to the reinforcing materials, thus improving the tensile strength of the polymer composite [32].

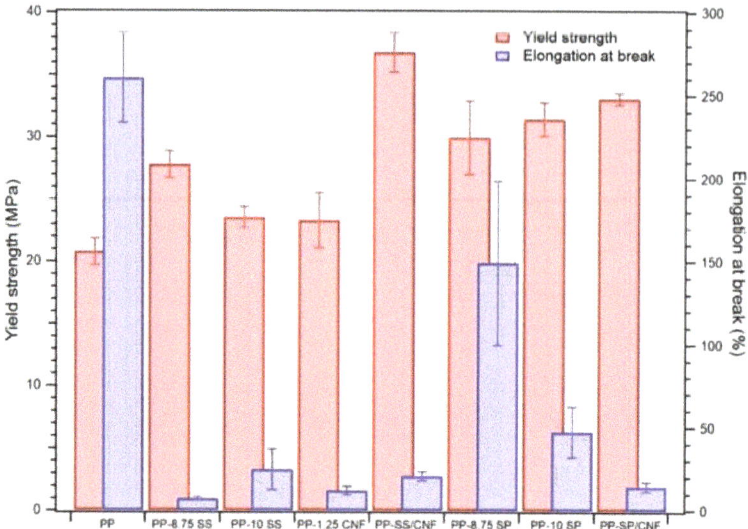

**Figure 8.** Mechanical properties of PP composites: bar charts of elongation at break and yield strength.

The tensile strength at yields of PP-8.75 SS and PP-10 SS was 27.80 and 23.53 MPa, respectively. Increasing the by-weight percentage of SS led to inferior mechanical properties. The agglomeration extent increased with the increase of nanofiller content and reduction of nanofiller size. When compared to SP, the particle size of SS is smaller overall, so aggregation has a higher chance to occur in PP/SS composites. Agglomeration decreased the effectiveness of nanoparticles in polymer matrix, which lastly resulted in the poor properties of the samples.

However, increasing the by-weight percentage of SP from 8.75 to 10 increased the tensile strength. The irregular shape and slightly larger particle size of the filler produced different properties. The surface roughness and irregular shape enhanced the interfacial properties between the matrix and reinforcement, such that the tensile strength still increased after the addition of 10 wt.% of SP. Both PP-SS/CNF and PP-SP/CNF samples depicted excellent yield strengths, which were 36.81 and 34.06 MPa, respectively. The synergy of the hybrid filler occurred due to the possibility of physical interaction between the functional groups of SiPs and CNFs. This functional group built the interfacial surface compatibility and increased the dispersibility between the CNFs and PP matrix. In addition, the homogeneity and matrix particle interactions were improved after adding the hybrid filler. Compared to the pristine PP (21.2 MPa), the tensile strength at yield of PP-SS/CNF and PP-SP/CNF samples showed 73.5% and 60.3% increments, respectively.

Furthermore, the elongation at break of PP composites was sharply reduced by the addition of filler due to the stiffening effect of SiPs and CNFs on the PP composite [31]. The presence of a rigid interface between SiPs and CNFs and the PP matrix decreased the deformability of the PP matrix, which led to more rigid and stiffer composites [28,33]. The PP-8.75 SP (201.70%) composites exhibited the highest elongation at break. SP is a hydrophobic filler, which can enhance the interfacial interaction between the filler and PP matrix and generate a stronger interfacial bonding. However, further study needs to be carried out to investigate the unexpected elongation behavior of sample PP-8.75 SP compared to other polymer composites.

### 3.6. Morphology of the Fracture Surface of PP Composites

PP composites were further analyzed with SEM-EDS to observe their cross-sectional fracture surface. First, each sample was submerged in liquid nitrogen for 5 min to ensure it was completely frozen. It was then removed from the liquid nitrogen and immediately snapped in half. The sample was dried to remove excess water. After freeze-fracture, the cross-section of the sample was carbon-coated and observed. Figure 9 depicts the SEM-EDS micrograph and quantitative analysis of elements present at the white spot. The elements are oxygen (O) and silica (Si). The SEM image also shows the typical fracture surfaces of neat PP and PP composite samples. This characterization technique can be used to identify the hybrid filler distribution in the polymer composite as well.

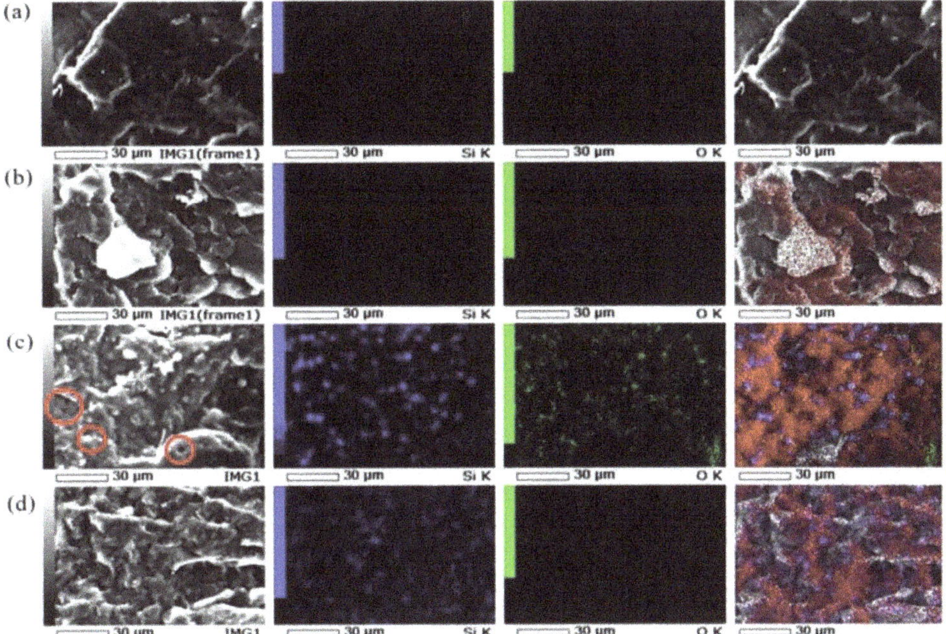

**Figure 9.** SEM images mapped with EDS analysis for distribution of oxygen (O) element represents CNF and silica (Si) element represent SiP: (**a**) PP polymer, (**b**) PP-1.25 CNF, (**c**) PP SS/CNF and (**d**) PP SP/CNF.

In Figure 9a, the rough surface observed on neat PP indicated the typical characteristic of the elastic behavior of PP [34] and the element mapping of Si and O was almost completely black in colour. In addition, it can be observed that CNFs were agglomerated in the polymer composite (white circle) in the EDS analysis and also SEM morphology, which is supported by the study of Yasim-Anuar et. al. [35]. According to the respective literature, the distribution of oxygen (O) represents CNFs (Figure 9b).

Moreover, the EDS images showed the samples PP SS/CNF and PP SP/CNF in Figure 9c,d to consist of uniform white spots of O and Si, which further confirm the good dispersion of the hybrid filler in the polymer composite. The oxygen element of PP nanocomposites increase compared with PP/CNF is due to Si-O-Si bonding formed after blending. These results are in line with the FT-IR assignment in the previous section.

Further, the SEM images present a uniform distribution of the hybrid fillers in a PP composite surface. Sample PP SS/CNF showed that the particulate fillers with a spherical and fibrous shape (red circle) were found in the composite, suggesting the applied stress effectively transferred to the filler from the polymer matrix by scattering the energy during crack propagation [36]. While on the sample PP SP/CNF, interconnected nanofibers formed with rough surfaces and irregular shapes were observed. SiP SP exhibited better interfacial bonding strength and filler dispersion in the PP composite. These results indicate robust interfacial adhesion between the PP matrix and filler due to their hydrophobicity compatibility.

### 3.7. Thermogravimetric Analysis (TGA) of PP Composite

Figure 10 shows the thermal stability of the neat PP, SiPs, CNFs, and PP composite. The TGA results show the neat PP and PP composite degradation in a single step of degradation. SiP SP had weight loss at 420 °C, attributed to the thermal decomposition of the functional group of polyalkyl siloxane. Above 450 °C, there was no weight loss observed for either SiP sample, indicating residual SiPs. Further, SiPs have high thermal stability at temperatures up to 800 °C [37]. For the CNF sample, weight loss started from 270 °C due to the thermal decomposition of cellulose [38]. The TGA result also showed the neat PP almost degraded without any char formation, with the residual of the original sample mass being only 1.3%.

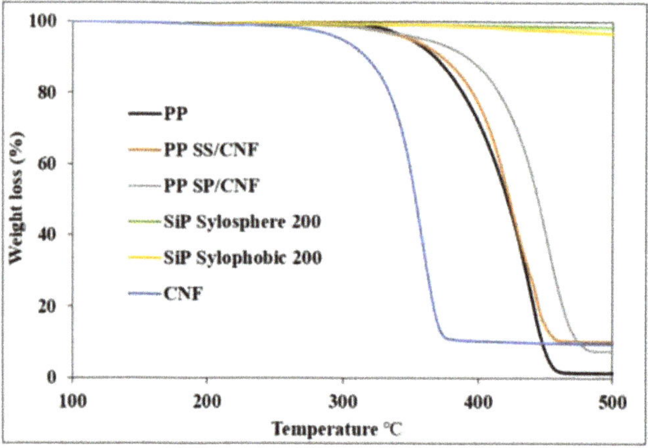


**Figure 10.** TGA curves of PP, SiP, CNF, PP SS/CNF, and PP SP/CNF.

Compared to the neat PP, the decomposition temperature of the PP composite increased significantly with the addition of hybrid filler, indicating that SiPs hindered thermal decomposition of PP composite. The degradation depends on the particle encapsulation and extent of interaction with matrix. Improvement in the thermal stability of PP composites was posited to be due to the excellent dispersion of the hybrid filler in the polymer matrix. Due to the stable thermal properties of the hybrid filler, chain scission process occurred at a higher temperature compared to pristine PP polymer [39]. Additionally, the weight residue at 500 °C of the PP composite showed the highest for sample PP SS/CNF at 10.28% residue, followed by CNF at 9.54% residue, and PP SP/CNF at 7.53% residue. The char yield increase after melt-blending with filler is attributed to the increased interactions between the filler and the matrix which resulted in the onset of thermal degradation.

Additionally, the DTA curves of the discussed composite PP SS/CNF and PP SP/CNF show that during their decomposition, endothermic reactions occurred (Figure 11). The DTA curve exhibited two endothermic events. The first is attributed to Tm at around 161 °C, corresponding to the melting of the crystalline phase in the studied polymer (PP). Other endothermic peaks were observed with decomposition maxima at 428 and 450 °C in higher temperature ranges. The thermal decomposition of PP SP/CNF is higher than PP SS/CNF.

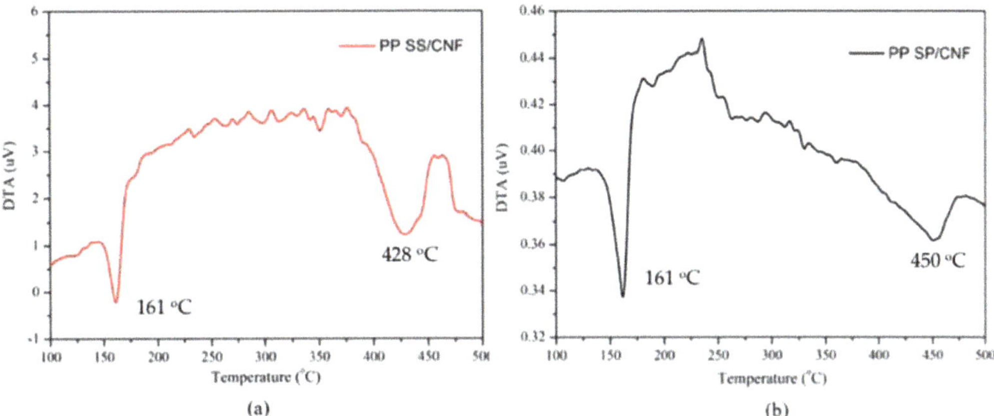

**Figure 11.** DTA curves of (**a**) PP SS/CNF, and (**b**) PP SP/CNF.

## 4. Conclusions

An environmentally benign ethanol/water mixed solvent method was used to prepare a PP-SiP/CNF composite. The method applied in this work is very convenient, requires less energy, and sidesteps chemical modification. It is noted that a synergistic effect of the hybrid filler occurred in this composition, showing a significant enhancement of the dispersion of CNFs in the hydrophobic polymer. Morphology results confirmed the deposition of SiPs onto the CNF surface, resulting in good dispersion of the hybrid filler in the PP matrix and preventing the agglomeration of CNFs. The incorporation of filler into the PP matrixes significantly increased the mechanical properties of the composites. Sample PP-SCNF exhibited higher tensile strength at 36.8 MPa (increments 73.55%) compared to the neat PP. However, its elongation sharply reduced due to the stiffening effect of the filler. The thermal stability of PP composites was improved by the incorporation of fillers as a physical barrier. The hybrid filler exhibited synergistic effects, especially for tensile strength, and proved to be more effective than single-filler systems. The simplicity of this strategy can be applied to other fillers such as metal oxide and graphene oxide to prepare hybrid fillers for designing polymer composites, especially to enhance the mechanical properties and thermal stability of a product.

**Author Contributions:** Conceptualization, Y.A.; Methodology, Y.A. Validation, S.M.; Formal analysis, S.M.; Investigation, S.M., K.E. and S.T.S.; Resources, Y.A.; Data curation, Y.A. and H.A.; Writing-original draft preparation, S.M.; Writing-review and editing, J.L. and Y.A.; Visualization, S.M.; Supervision, Y.A.; Project administration, Y.A.; Funding acquisition, Y.A. All authors have read and agreed to the published version of the manuscript.

**Funding:** This research was supported by aXis (Accelerating Social Implementation for SDGs Achievement) project funded by Japan Science and Technology Agency (JST), and Kyushu Institute of Technology (Kyutech) and Universiti Putra Malaysia (UPM) were supported by co-funding research program.

**Institutional Review Board Statement:** Not applicable.

**Informed Consent Statement:** Not applicable.

**Data Availability Statement:** The data presented are contained within the article.

**Conflicts of Interest:** The authors declare no conflict of interest.

## References


1. Dantas de Oliveira, A.; Augusto Gonçalves Beatrice, C. Chapter 6—Polymer nanocomposites with different types of nanofiller. In *Nanocomposites—Recent Evolution*; Sivasankaran, S., Ed.; IntechOpen: London, UK, 2019; pp. 103–128.
2. Rodríguez-Robledo, M.C.; González-Lozano, M.A.; Ponce-Peña, P.; Owen, P.Q.; Aguilar-González, M.A.; Nieto-Castañeda, G.; Bazán-Mora, E.; López-Martínez, R.; Ramírez-Galicia, G.; Poisot, M. Cellulose-silica nanocomposites of high reinforcing content with fungi decay resistance by one-pot synthesis. *Materials* **2018**, *11*, 1–13. [CrossRef] [PubMed]
3. Scaffaro, R.; Botta, L. Chapter 5—Nanofilled thermoplastic—Thermoplastic polymer blends. In *Nanostructured Polymer Blends*; Thomas, S., Shanks, R., Chandrasekharakurup, S., Eds.; Elsevier: London, UK, 2014; pp. 133–160.
4. Ajorloo, M.; Fasihi, M.; Khoramishad, H. The role of nanofiller size and polymer chain configuration on the properties of polypropylene/graphite nanoplates composites. *J. Taiwan Inst. Chem. Eng.* **2020**, *108*, 82–91. [CrossRef]
5. Vasquez-Zacarias, L.; Ponce-Peña, P.; Pérez-López, T.; Franco-Urquiza, E.A.; Ramirez-Galicia, G.; Poisot, M. Hybrid cellulose-silica materials from renewable secondary raw resources: An eco-friendly method. *Glob. Chall.* **2018**, *2*, 1700119. [CrossRef]
6. Rothon, R.; Dearmit, C. Particulate fillers, selection, and use in polymer composites. In *Polymers and Polymeric Composites: A Reference Series*; Rothon, R., Ed.; Springer: Berlin, Germany, 2017; pp. 3–27.
7. Dorigato, A.; Dzenis, Y.; Pegoretti, A. Nanofiller aggregation as reinforcing mechanism in nanocomposites. *Procedia Eng.* **2011**, *10*, 894–899. [CrossRef]
8. Pandey, J.K.; Takagi, H.; Nakagaito, A.N.; Kim, H.J. Handbook of polymer nanocomposites: Processing, performance and application. Volume C, Polymer nanocomposites of cellulose nanoparticles. In *Electrospun Cellulose Composite Nanofibers*; Abdul-Khalil, H.P.S., Davoudpour, A., Bhat, A.H., Rosamah, E., Tahir, P.M., Eds.; Springer: Berlin, Germany, 2015; Volume C, pp. 1–511.
9. Shubhra, Q.T.H.; Alam, A.K.M.M.; Quaiyyum, M.A. Mechanical properties of polypropylene composites: A review. *J. Thermoplast. Compos. Mater.* **2013**, *26*, 362–391. [CrossRef]
10. Nuruddin, M.; Hosur, M.; Gupta, R.; Hosur, G.; Tcherbi-Narteh, A.; Jeelani, S. Cellulose nanofibers-graphene nanoplatelets hybrids nanofillers as high-performance multifunctional reinforcements in epoxy composites. *Polym. Polym. Compos.* **2017**, *25*, 273–283. [CrossRef]
11. Anwer, M.A.S.; Wang, J.; Naguib, H.E. 1D/2D CNF/GNP Hybrid nanofillers: Evaluation of the effect of surfactant on the morphological, mechanical, fracture, and thermal characteristics of their nanocomposites with epoxy resin. *Ind. Eng. Chem. Res.* **2019**, *58*, 8131–8139. [CrossRef]
12. Kwak, H.W.; You, J.; Lee, M.E.; Jin, H.J. Prevention of cellulose nanofibril agglomeration during dehydration and enhancement of redispersibility by hydrophilic gelatin. *Cellulose* **2019**, *26*, 4357–4369. [CrossRef]
13. Thomas, S.; Abraham, J.; George, S.C.; Thomas, S. Role of CNT/clay hybrid on the mechanical, electrical and transport properties of NBR/NR blends. *Polym. Bull.* **2020**, *77*, 1–16. [CrossRef]
14. Li, Y.; He, H.; Huang, B.; Zhou, L.; Yu, P.; Lv, Z. In situ fabrication of cellulose nanocrystal-silica hybrids and its application in UHMWPE: Rheological, thermal, and wear resistance properties. *Polym. Compos.* **2018**, *39*, 1701–1713. [CrossRef]
15. Carter, C.B.; Norton, M.G. Sols, gels, and organic chemistry. In *Ceramics Materials*; Carter, C.B., Norton, M.G., Eds.; Springer: Berlin/Heidelberg, Germany, 2013; pp. 411–422.
16. Zu, L.; Li, R.; Jin, L.; Lian, H.; Liu, Y.; Cui, X. Preparation and characterization of polypropylene/silica composite particle with interpenetrating network via hot emulsion sol-gel approach. *Prog. Nat. Sci. Mater. Int.* **2014**, *24*, 42–49. [CrossRef]
17. Zare, Y. Study of nanoparticles aggregation/agglomeration in polymer particulate nanocomposites by mechanical properties. *Compos. Part A Appl. Sci. Manuf.* **2016**, *84*, 158–164. [CrossRef]
18. Sharip, N.S.; Ariffin, H.; Andou, Y.; Shirosaki, Y.; Bahrin, E.K.; Jawaid, M.; Tahir, P.M.; Ibrahim, N.A. process optimization of ultra-high molecular weight polyethylene/cellulose nanofiber bionanocomposites in triple screw kneading extruder by response surface methodology. *Molecules* **2020**, *25*, 4498. [CrossRef] [PubMed]
19. Dufresne, A. Nanocellulose: A new ageless bionanomaterial. *Mater. Today* **2013**, *16*, 220–227. [CrossRef]
20. Abbate Dos Santos, F.; Bruno Tavares, M.I. Development of biopolymer/cellulose/silica nanostructured hybrid materials and their characterization by NMR relaxometry. *Polym. Test.* **2015**, *47*, 92–100. [CrossRef]
21. Pouresmaeel-Selakjani, P.; Jahanshahi, M.; Peyravi, M. Synthesis of cellulose/silica nanocomposite through electrostatic interaction to reinforce polysulfone membranes. *Cellulose* **2017**, *24*, 1333–1353. [CrossRef]
22. Wen, X. One-pot route to graft long-chain polymer onto silica nanoparticles and its application for high-performance poly(l-lactide) nanocomposites. *RSC Adv.* **2019**, *9*, 13908–13915. [CrossRef]
23. Gieseking, B.; Jäck, B.; Preis, E.; Jung, S.; Forster, M.; Scherf, U.; Deibel, C.; Dyakonov, V. Effect of the modified silica nanofiller on the mechanical properties of unsaturated polyester resins based on recycled polyethylene terephthalate. *Adv. Energy Mater.* **2012**, *2*, 1477–1482. [CrossRef]
24. Deniz, S.; Arikan, B. Effect of silica type on superhydrophobic properties of pdms-silica nanocomposite coatings. *Int. J. Eng. Appl. Sci.* **2016**, *8*, 19–27.
25. Hishikawa, Y.; Togawa, E.; Kondo, T. Characterization of individual hydrogen bonds in crystalline regenerated cellulose using resolved polarized FTIR spectra. *ACS Omega* **2017**, *2*, 1469–1476. [CrossRef]
26. Sahlin, K.; Forsgren, L.; Moberg, T.; Bernin, D.; Rigdahl, M.; Westman, G. Surface treatment of cellulose nanocrystals (CNC): Effects on dispersion rheology. *Cellulose* **2018**, *25*, 331–345. [CrossRef]

27. Lamaming, J.; Hashim, R.; Sulaiman, O.; Leh, C.P.; Sugimoto, T.; Nordin, N.A. Cellulose nanocrystals isolated from oil palm trunk. *Carbohydr. Polym.* **2015**, *127*, 202–208. [CrossRef] [PubMed]
28. Chun, K.S.; Husseinsyah, S.; Osman, H. Utilization of cocoa pod husk as filler in polypropylene biocomposites: Effect of maleated polypropylene. *J. Thermoplast. Compos. Mater.* **2015**, *28*, 1507–1521. [CrossRef]
29. Gopanna, A.; Mandapati, R.N.; Thomas, S.P.; Rajan, K.; Chavali, M. Fourier transform infrared spectroscopy (FTIR), raman spectroscopy and wide-angle x-ray scattering (WAXS) of polypropylene (PP)/cyclic olefin copolymer (COC) blends for qualitative and quantitative analysis. *Polym. Bull.* **2019**, *76*, 4259–4274. [CrossRef]
30. Jia, N.; Li, S.M.; Ma, M.G.; Zhu, J.F.; Sun, R.C. Synthesis and characterization of cellulose-silica composite fiber in ethanol/water mixed solvents. *BioResources* **2011**, *6*, 1186–1195.
31. Islam, S.; Masoodi, R.; Rostami, H. The effect of nanoparticles percentage on mechanical behaviour of silica-epoxy nanocomposites. *J. Nanosci.* **2013**, *2013*, 10. [CrossRef]
32. Hu, Y.; Ding, J.; Chen, Y. Effects of nanofiller geometries and interfacial properties on the mechanical performance of polymer nanocomposites—A numerical study. *Polym. Polym. Compos.* **2021**, *29*, S19–S35. [CrossRef]
33. Singh, K.; Ohlan, A.; Saini, P.; Dhawan, S.K. Poly (3,4-ethylenedioxythiophene) $\gamma$-Fe$_2$O$_3$ polymer composite—Super paramagnetic behavior and variable range hopping 1D conduction mechanism—Synthesis and characterization. *Polym. Adv. Technol.* **2008**, *19*, 229–236. [CrossRef]
34. Luna, I.Z.; Dam, K.C.; Chowdhury, A.M.S.; Gafur, M.A.; Khan, N.; Khan, R.A. Physical and thermal characterization of alkali treated rice husk reinforced polypropylene composites. *Adv. Mater. Sci. Eng.* **2015**, *2015*, 7. [CrossRef]
35. Yasim-Anuar, T.A.T.; Ariffin, H.; Norrrahim, M.N.F.; Hassan, M.A.; Andou, Y.; Tsukegi, T.; Nishida, H. Well-dispersed cellulose nanofiber in low density polyethylene nanocomposite by liquid-assisted extrusion. *Polymers* **2020**, *12*, 927. [CrossRef]
36. Fu, S.Y.; Feng, X.Q.; Lauke, B.; Mai, Y.W. Effects of particle size, particle/matrix interface adhesion and particle loading on mechanical properties of particulate-polymer composites. *Compos. Part B Eng.* **2008**, *39*, 933–961. [CrossRef]
37. Lin, B.; Zhou, S. Poly(ethylene glycol)-grafted silica nanoparticles for highly hydrophilic acrylic-based polyurethane coatings. *Prog. Org. Coat.* **2017**, *106*, 145–154. [CrossRef]
38. Yasim-anuar, T.A.T.; Sharip, N.S.; Megashah, L.N.; Ariffin, H.; Nor, N.A.M. Cellulose nanofibers from waste waper and their utilization as reinforcement materials in poly((r)-3 hydroxybutyrate-co-(r)-3 hydroxyhexanoate bionanocomposite. *Pertanika J. Sci. Technol.* **2020**, *28*, 259–272. [CrossRef]
39. Alaei, M.H.; Mahajan, P.; Brieu, M.; Kondo, D.; Rizvi, S.J.A.; Kumar, S.; Bhatnagar, N. Effect of particle size on thermomechanical properties of particulate polymer composite. *Iran. Polym. J.* **2013**, *22*, 853–863. [CrossRef]

*Article*

# A Morphological Study of Dynamically Vulcanized Styrene-Ethylene-Butylene-Styrene/Styrene-Butylene-Styrene/MethylVinylSilicon Rubber Thermoplastic Elastomer

Chunxu Zhao, Xiaohan Chen [†] and Xian Chen *,[†]

Room 602, Yifu Science and Technology Building, Wangjiang Campus, Sichuan University, Chengdu 610065, China; 2020222030015@stu.scu.edu.cn (C.Z.); xc2549@columbia.edu (X.C.)
* Correspondence: chen.xian@scu.edu.cn; Tel.: +86-028-85468166
† These authors contributed equally to this work.

**Abstract:** In this work, we prepared thermoplastic silicone rubber (TPSiV) by dynamically vulcanizing different relative proportions of methyl vinyl silicone rubber (MVSR), styrene ethylene butene styrene block copolymer (SEBS), and styrene butadiene styrene block copolymer (SBS). The compatibility and distribution of the MVSR phase and SEBS/SBS phase were qualitatively characterized by Fourier transform infrared spectroscopy (FTIR) and scanning electron microscopy (SEM) tests on TPSiV. Subsequently, the backscattered electron signal image was analyzed using a colorimeter, and it was found that the size of the interface layer between the MVSR phase and the SEBS-SBS phase could be quantitatively characterized. This method overcomes the defect of the etching method, which cannot quantitatively analyze the size of the compatible layer between the two polymers. The final experiment proved that the two phases in TPSiV exhibited a "sea-island" structure, in which the MVSR phase acted as a dispersed phase in the SEBS-SBS phase. In addition, the addition of the silane coupling agent KH-907 (γ-isocyanatopropyltriethoxysilane) improved the mechanical properties of TPSiV, increasing the tensile strength by about 40% and the elongation at break by 30%. The permanent tensile deformation increase rate was about 15%. Through the quantitative measurement of the compatible layer, it was found that KH-907 could increase the thickness of the interface layer between the MVSR phase and the SEBS-SBS phase by more than 30%, which explained why the silane coupling agent KH-907 improved the mechanical properties of TPSiV at the micro level.

**Keywords:** thermoplastic silicone rubber; backscattered electrons; compatibility layer; scanning electron microscope; dynamic vulcanization

**Citation:** Zhao, C.; Chen, X.; Chen, X. A Morphological Study of Dynamically Vulcanized Styrene-Ethylene-Butylene-Styrene/Styrene-Butylene-Styrene/MethylVinylSilicon Rubber Thermoplastic Elastomer. *Polymers* **2022**, *14*, 1654. https://doi.org/10.3390/polym14091654

Academic Editors: Wei Wu, Hao-Yang Mi, Chongxing Huang, Hui Zhao and Tao Liu

Received: 16 March 2022
Accepted: 15 April 2022
Published: 20 April 2022

**Publisher's Note:** MDPI stays neutral with regard to jurisdictional claims in published maps and institutional affiliations.

## 1. Introduction

Rubber is a class of polymers that is commonly used in daily life, and vulcanization is a necessary process to impart various properties to rubber [1,2]. Dynamic vulcanization refers to the melt blending of rubber and thermoplastic polymer at a high temperature and then vulcanizing the rubber under the action of a cross-linking agent, thereby obtaining a vulcanized rubber phase with a size in the micron level and uniformly dispersed in the thermoplastic polymer. Compared with other rubber vulcanization methods, dynamic vulcanization has the advantages of greatly improving the performance of blended thermoplastic elastomers, reducing equipment investment, and improving efficiency [3]. Since Gessler first proposed the concept of dynamic vulcanization in 1962, researchers have successively discovered polypropylene (PP)/ethylene propylene diene rubber (EPDM) [4], nitrile rubber (NBR)/polylactic acid (PLA) [5], and other dynamic vulcanization systems.

Silicone rubber is a polymer material with a silicon–oxygen bond as the main chain and phenyl, vinyl, and other groups as side chains and has excellent comprehensive properties [6]. At present, researchers often improve some properties of silicone rubber by blending it with different inorganic substances or polymers. In recent years, the development of special functions, such as high-voltage electrical properties, thermal conductivity, flame retardant properties, and antistatic properties of silicone rubber, has been a trend [7,8]. In addition to silicone rubber, thermoplastic elastomer is widely used because it has the physical and mechanical properties of vulcanized rubber and the processing properties of thermoplastic plastics [9,10]. As one of the most important types of thermoplastic elastomers, SBS is often used for asphalt modification [11]. Since the aging resistance of SBS is not ideal, SEBS obtained by hydrogenating SBS came into being, and it is often used in situations where high aging resistance is required [12]. In this paper, TPSiV prepared by melt blending MVSR and SEBS/SBS is also a material that can combine the biocompatibility of MVSR with the excellent mechanical properties of SEBS/SBS. This TPSiV can be used for medical elastomers and other applications. Before this, the study of this blend system was rarely reported.

Although a variety of materials with excellent properties can be obtained by melting and blending various polymers, because most polymers have poor compatibility with each other, solving the compatibility between polymers and enhancing the interaction between different phase interfaces has become the focus of researchers. Among many solutions, adding a silane coupling agent to the polymer blend system is one of the commonly used methods [13,14]. Therefore, in this paper, we chose the silane coupling agent KH-907 as the compatibilizer of TPSiV.

Scanning electron microscopy (SEM) is one of the most commonly used methods to characterize polymer compatibility and phase interface conditions [15]. The characterization of copolymer compatibility and phase interface by scanning electron microscopy can be roughly divided into two categories. One is to directly observe the surface with obvious phase separation by observing the difference in the phase separation before and after adding the compatibilizer as evidence of the enhanced interaction and improved compatibility of the two-phase interface [16–18]. The other is to dissolve the material section in a good solvent of a phase, dissolve and remove the phase, observe the section of the material, and measure the strength of the interaction between the phases by the roughness of the surface of the undissolved phase, to prove that a substance has an obvious effect on the improvement of the compatibility between the two phases [19–21]. However, this method is only applicable to the situation where one phase is completely dissolved by the solvent, and the other phase is not dissolved by the solvent at all, which is difficult in some dynamic vulcanization blend systems. In addition, SEM is often used in conjunction with Transmission Electron Microscopy (TEM) and Energy Dispersive X-Ray Spectroscopy (EDX) to characterize more information on the surface of the material [22,23]. However, TEM testing has high requirements for samples, and it is necessary to find suitable dyes.

The SEM test described above usually uses the secondary electron signal [24]. In addition to the secondary electron signal, the backscattered electron signal reflected from the sample surface can also be analyzed to obtain information about the sample surface [25,26]. This characterization method is usually used to observe the changes in alloy phases [27–29] and the phase distribution of cement cross-sections [30–32]. However, there are few reports on the characterization of dynamic vulcanization blend systems. Therefore, we applied the backscattered electron signal characterization method to the dynamic vulcanization system and quantitatively characterized the phase interface between the dispersed phase and the continuous phase by using the method combined with the colorimeter, which can avoid many shortcomings of the etching method. For example, it is impossible to effectively prove that the etched phase is completely etched, and the unetched phase collapses and easily adheres to the cross-section, which affects the observation and cannot quantitatively characterize the phase interface between the dispersed phase and the continuous phase. At the same time, this method can also observe and analyze the dynamic vulcanization blend system without obvious phase separation.

## 2. Experimental

### 2.1. Materials

To prepare thermoplastic silicone rubber, we used the following materials: methyl vinyl silicone rubber (HCR 9670 UE, Lanxingxinghuo Silicone Co., Ltd., Jiujiang, China), SEBS (YH-602T, Baling Petrochemical Branch of Sinopec Group, Yueyang, China), SBS (D1155JOP, Kraton, Belpre, OH, USA), white oil (250N, Wanghai Petrochemical Co., Ltd., Taizhou, China), silane coupling agent KH-907 ($\gamma$-isocyanatopropyl triethoxysilane, Jessica Chemical Co., Ltd., Hangzhou China), platinum catalyst (Daxi Chemical Raw Materials Co., Ltd., Guangzhou, China), and hydrogen-containing silicone oil (hydrogen content 1.6%, Xinglongda New Materials Co., Ltd., Jinan, China).

### 2.2. Sample Preparation

First, the raw materials were weighed according to the weights shown in Table 1. Then, to evenly disperse the white oil 250N in the sample, we first mixed the white oil 250N with SEBS and SBS and stirred for 5 min before letting it stand. After SEBS fully absorbed the white oil, we added the pre-mixed mixture to the torque rheometer (chamber temperature 170 °C, rotation speed 70 rpm). When the torque value was stable, we continued to add the methyl vinyl silicone rubber. We waited for the torque value to stabilize and then added the silane coupling agent, hydrogen-containing silicone oil, and platinum catalyst in sequence to allow for the dynamic vulcanization of the mixture in the cavity. When the torque value rose due to the cross-linking of the mixture in the cavity and then stabilized, the blend was taken out and placed on a flat vulcanizing machine to be pressed for tableting (pressing time was 5 min) to obtain a sample tablet with a thickness of 2 mm.

**Table 1.** TPSiV dynamic vulcanization experimental formula (unit: phr) [a].

| Entry | MVSR | SEBS | SBS | KH-907 | Platinum Catalyst |
|---|---|---|---|---|---|
| 1 | 40 | 30 | 30 | 1 | 0.5 |
| 2 | 50 | 25 | 25 | 1 | 0.5 |
| 3 | 60 | 20 | 20 | 1 | 0.5 |
| 4 | 70 | 15 | 15 | 1 | 0.5 |
| 5 | 40 | 30 | 30 | 0 | 0.5 |
| 6 | 50 | 25 | 25 | 0 | 0.5 |
| 7 | 60 | 20 | 20 | 0 | 0.5 |
| 8 | 70 | 15 | 15 | 0 | 0.5 |

[a] Unless otherwise specified, all samples contain 10 units of white oil and 1 unit of hydrogen-containing silicone oil.

### 2.3. Scanning Electron Microscope Test

The eight groups of prepared samples were cut into strips with a size of $10 \times 2 \times 2$ mm$^3$, and they were placed in liquid nitrogen to freeze for 5 min. Then, the test strips were taken out with tweezers and broken in the middle, and the cross-sections of the samples were treated with gold spraying. Finally, the cross-sections after gold spraying were observed by a scanning electron microscope (Thermo Fisher Scientific, Helios G4 UC, Waltham, MA, USA) to obtain backscattered electron (BSE) images.

### 2.4. Fourier Transform Infrared Spectroscopy Test

The prepared samples were tested with a Fourier transform infrared spectrometer (Bruker Scientific Technology Co., Ltd., Beijing, China). The detection mode was ATR, the resolution was 4 cm$^{-1}$, the test wavenumber range was 400–4000 cm$^{-1}$, and the number of scans was 16.

### 2.5. Etching Method to Test the Cross-Sectional Morphology of the Samples

The eight groups of prepared samples were cut into strips with a size of $10 \times 2 \times 2$ mm$^3$, and they were placed in liquid nitrogen to freeze for 5 min. Then, the test strips were taken out with tweezers and broken in the middle. The broken samples were completely immersed in chloroform for 72 h (the samples were taken out and replaced with a new chloroform solvent every 24 h to completely dissolve the etched MVSR phase). After 72 h, the samples were taken out from the chloroform solvent and placed in a fume hood for 12 h to allow the solvent on the surface of the samples to evaporate completely. Finally, the cross-sections were sprayed with gold, and we observed them with a scanning electron microscope (SEM) to obtain the secondary electron images.

### 2.6. Compatibility Layer Quantitative Characterization Test

The 1000-times backscattered electron images obtained by the scanning electron microscope test were cut by one-eighth according to the method shown in Figure 1 (image pixel: $3840 \times 2160$ dpi), and then the obtained images were printed on photo paper by a high-definition printer. The pixel dimensions of the printer were $1200 \times 1200$ dpi, and the size of the photo paper was $420 \times 297$ mm. Finally, according to the method shown in Figure 2, the light passage of the colorimeter was oriented and shifted quantitatively to determine the size of the compatibility layer between the MVSR phase and the SEBS/SBS phase.

**Figure 1.** Printing method of BSE diagrams.

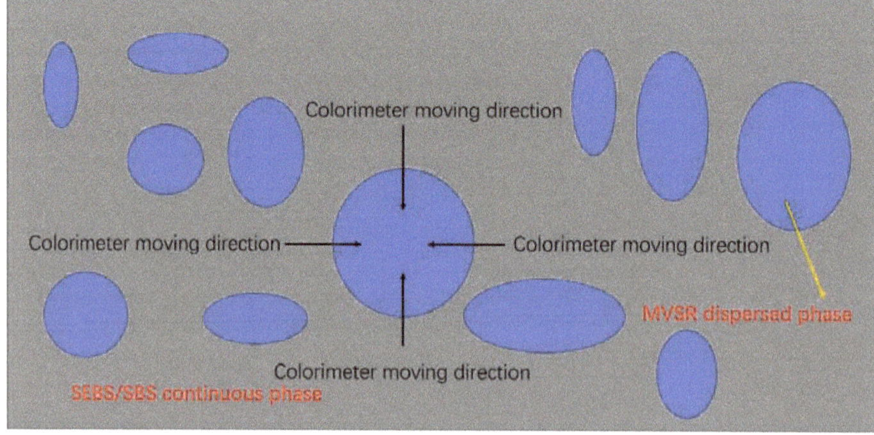


**Figure 2.** The test method for the colorimeter to test the size of the compatibility layer in TPSiV.

*2.7. Mechanical Property Test*

According to GB/T 528-2009 standards, five dumbbell-shaped samples with a gauge length of 20 mm, a width of 4 mm, and a thickness of 2 mm were cut from the eight groups of samples prepared in Section 2.2. The tensile strength and elongation at the break of the samples were tested by a tensile testing machine (AGS-J, Shimadzu Corporation, Kyoto, Japan), and the tensile rate was 200 mm min$^{-1}$. Then, the samples were pulled off and left to stand for five minutes, and the permanent tensile deformation was tested after the standing was completed. Finally, the arithmetic means of the test results of the five dumbbell-shaped specimens were taken for each mechanical property test result.

## 3. Results and Discussions

The scanning electron microscope (SEM) is a common characterization method to observe the microscopic information of polymer materials. By analyzing the electron signals excited by the interaction of incident electrons with the sample, information about the surface of the sample can be obtained. The excited signal when the incident electron interacts with the sample is shown in Figure 3 [24]. In this work, the condition of the sample surface is mainly characterized by analyzing the secondary electron signal and the backscattered electron signal.

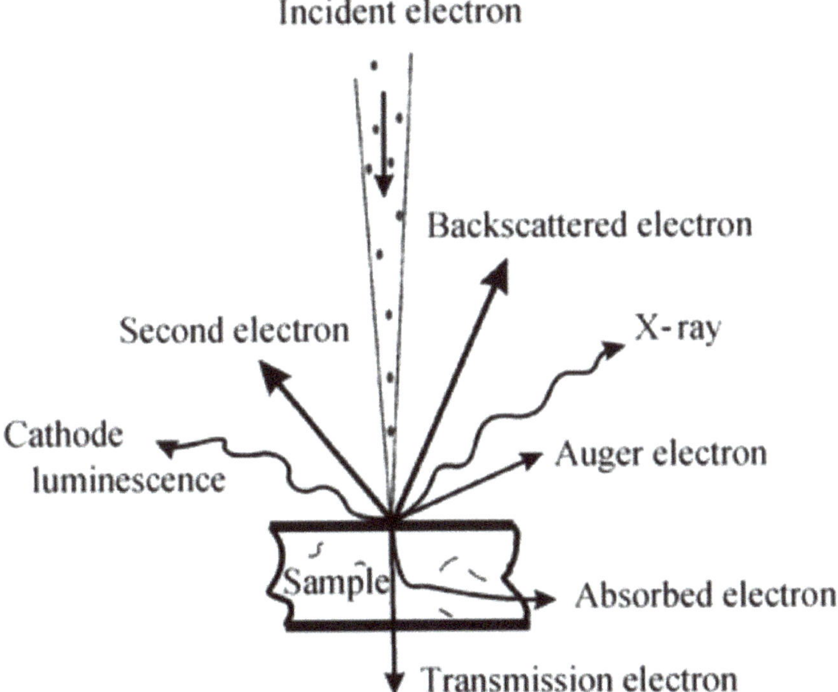


**Figure 3.** Signals excited when the incident electron interacted with the sample.

*3.1. Analysis of Secondary Electron Signal Results of TPSiV Made of MVSR and SEBS-SBS with Different Relative Contents*

Figure 4A–H are the SEM images of samples 1–8 after being etched by chloroform, during which the MVSR phase was dissolved by the solvent. As the content of silicone rubber increases, the diameter of the pores left by the etching and dissolution of the silicone rubber gradually increases. By observing the marked parts in Figure 4A–D, we can find that with the increase of the content of silicone rubber, the diameter of the pores left by the etching and the dissolution of the silicone rubber gradually increases. However, by comparing the secondary electron signal results of the silane coupling agent KH-907 (for example, Figure 4A,E), we found that the silane coupling agent KH-907 could not distinguish the MVSR and SEBS-SBS interface. Therefore, it cannot explain the improvement of mechanical properties after adding the silane coupling agent KH-907. Except for Figure 4A,B, it is difficult to distinguish the dispersed phase from the continuous phase. Moreover, we observed the marked parts in Figure 4D,H and found that the unetched phase was attached to the cross-section. Since the hardness of the SEBS-SBS phase is relatively small when the relative content of silicone rubber increases, the etching degree of the etched phase also increases, and the remaining SEBS-SBS phase will easily adhere to the cross-section and affect the observation (for example, Figure 4G,H).

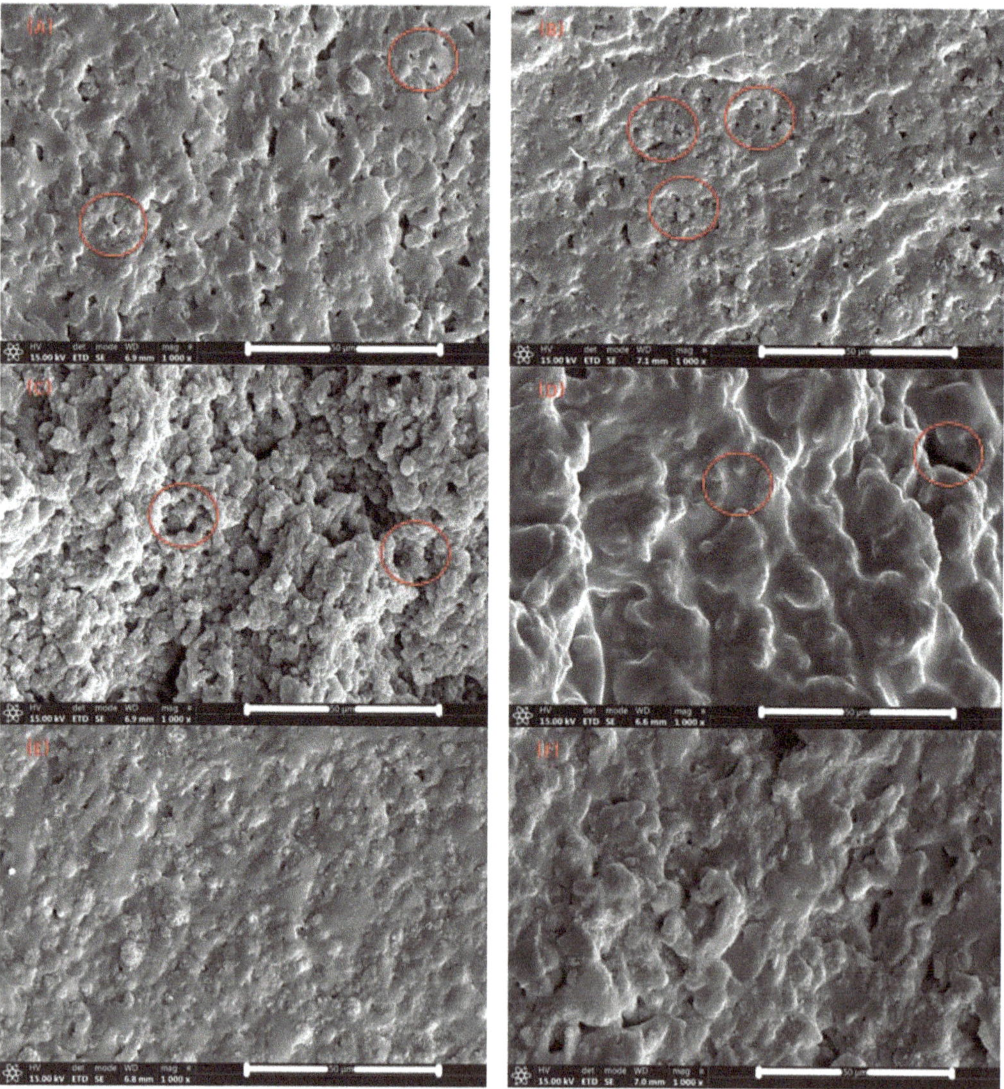

**Figure 4.** *Cont.*

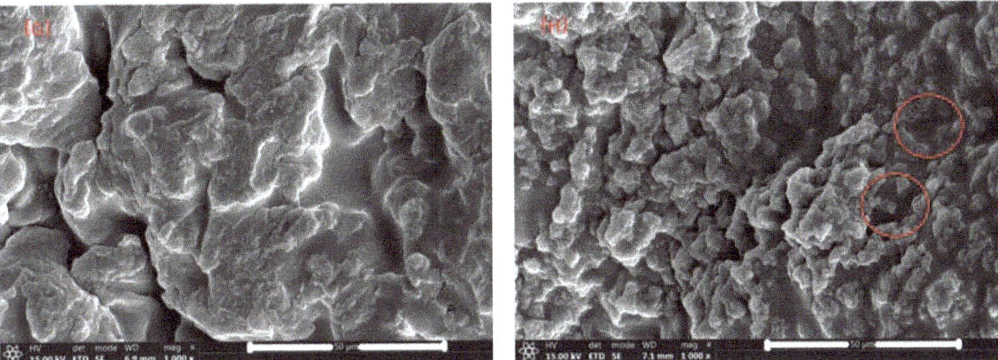

**Figure 4.** SEM diagrams obtained by etching method (**A–H**) correspond to samples 1–8, respectively.

*3.2. Analysis of the Backscattered Electron Signal of TPSiV Made of MVSR and SEBS-SBS with Different Relative Contents*

Among many electronic signals, backscattered electrons are the part of primary electrons reflected from the surface of the sample after the elastic and inelastic scattering of the incident electrons from colliding with the atoms on the surface of the sample. Elastically backscattered electrons are incident electrons reflected from the nucleus of the sample after a single or few large-angle elastic scatterings, and their energy does not change. We usually put reflected electrons with slightly varying energies in this category as well. Inelastic backscattered electrons are those incident electrons that are eventually reflected from the sample surface after tens or hundreds of inelastic collisions. Backscattered electrons are reflected in an irregular direction. However, their number is related to the angle of incidence and the average atomic number Z of the sample. The larger the value of Z, the stronger the signal strength of the backscattered electrons. Experiments show that when the energy of incident electrons is 10–40 keV, the backscattering coefficient $\eta$ of the sample increases with the increase of element atomic number Z. The relationship between the backscattering coefficient $\eta$ and the backscattered electron signal intensity $i_b$ and the incident electron intensity $i_p$ is as follows:

$$\eta = \frac{i_b}{i_p} \tag{1}$$

When the atomic number is greater than 10, the backscattering coefficient $\eta$ has the following quantitative relationship with the atomic number Z:

$$\eta = \frac{\ln Z\left(\overline{Z}\right)}{6} - \frac{1}{4} \tag{2}$$

Figure 5 shows the relationship between the atomic number Z and the backscattering coefficient $\eta$. From Figure 5, we can see that for elements with Z < 20, $\eta$ increases rapidly and linearly with the increase in Z. Therefore, the MVSR phase (the main constituent element is silicon) and the SEBS-SBS phase (the main constituent element is carbon) in TPSiV can be distinguished in the backscattering signal diagram, and the difference in the backscattering coefficient $\eta$ of the two phases is about 0.08 [25].

The larger the atomic number Z is, the stronger the backscattered signal is. Therefore, in Figure 4, the brighter area is the MVSR phase, and the darker area is the SEBS-SBS phase. From Figure 6, we can observe that the atomic number contrast of the two phases in TPSiV achieves the effect of distinguishing them. According to the marks in Figure 6A,B, we also found that when the silicone rubber content was low, a distinct "sea-island" structure was formed between the dispersed phase and the continuous phase. However, as the silicone rubber content increases, especially when the amount of silicone rubber reaches 70 phr, through the observation of the marks in Figure 6D,H, there is a tendency for a co-continuous phase in some regions, and it is difficult to distinguish between the continuous phase and the dispersed phase, which may affect some properties of the material. Through the image analysis system built into the software used when taking the SEM image, we obtained the size distribution of the dispersed phase, as shown in Figure 7. By comparing the dispersed phase size distribution of samples with the same content of silicone rubber and different content of silane coupling agent KH-907 (for example, samples 1 and 5), we found that adding silane coupling agent KH-907 would make the area of the dispersed phase smaller. We think that it may be that the promotion of interfacial interaction by silane coupling agent KH-907 improves the dispersion of the dispersed phase, but this needs to be further verified by other quantitative characterization methods.

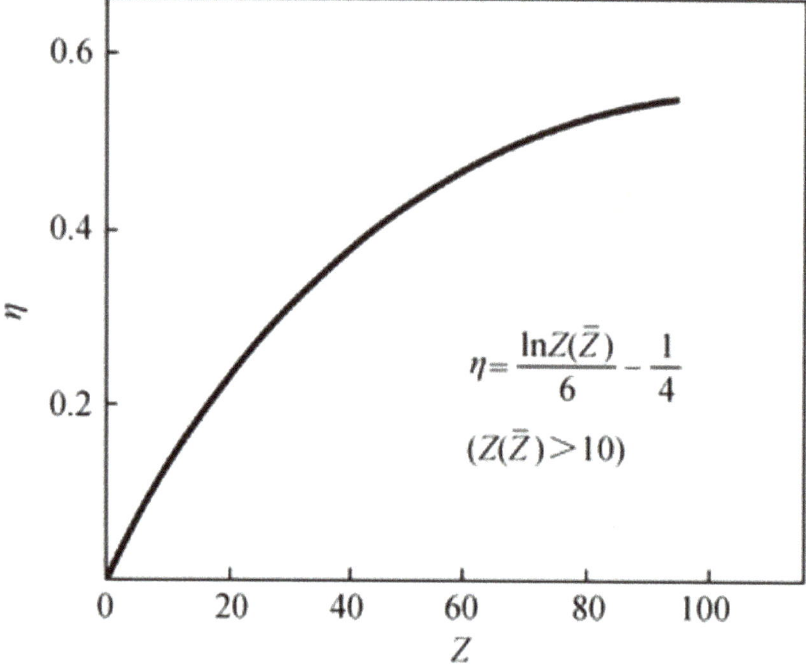



**Figure 5.** Relationship between atomic number $Z$ and backscattering coefficient $\eta$.

*3.3. Characterization of the Compatibility between MVSR and SEBS-SBS by Fourier Transform Infrared Spectroscopy*

Fourier transform infrared spectroscopy is one of the common methods used to qualitatively characterize the compatibility of two polymers after blending. According to the information provided in Table 2, for MVSR, we usually take 1260 and 1010 cm$^{-1}$ as characteristic peaks.

**Table 2.** Infrared characteristic peaks of MVSR and SEBS/SBS.

| Abbreviation | Peak Position (cm$^{-1}$) | Corresponding Groups and Vibration Types |
| --- | --- | --- |
| MVSR | 2965 | Stretching vibration of CH$_3$ on siloxane |
| | 1260 | Stretching vibration of Si(CH$_3$)$_2$ |
| | 1010 | Stretching vibration peak of Si-O-Si |
| SEBS/SBS | 2920, 2850 | Stretching vibration of CH$_3$ group in SEBS/SBS |
| | 760, 700 | Characteristic vibration of monosubstituted benzene |

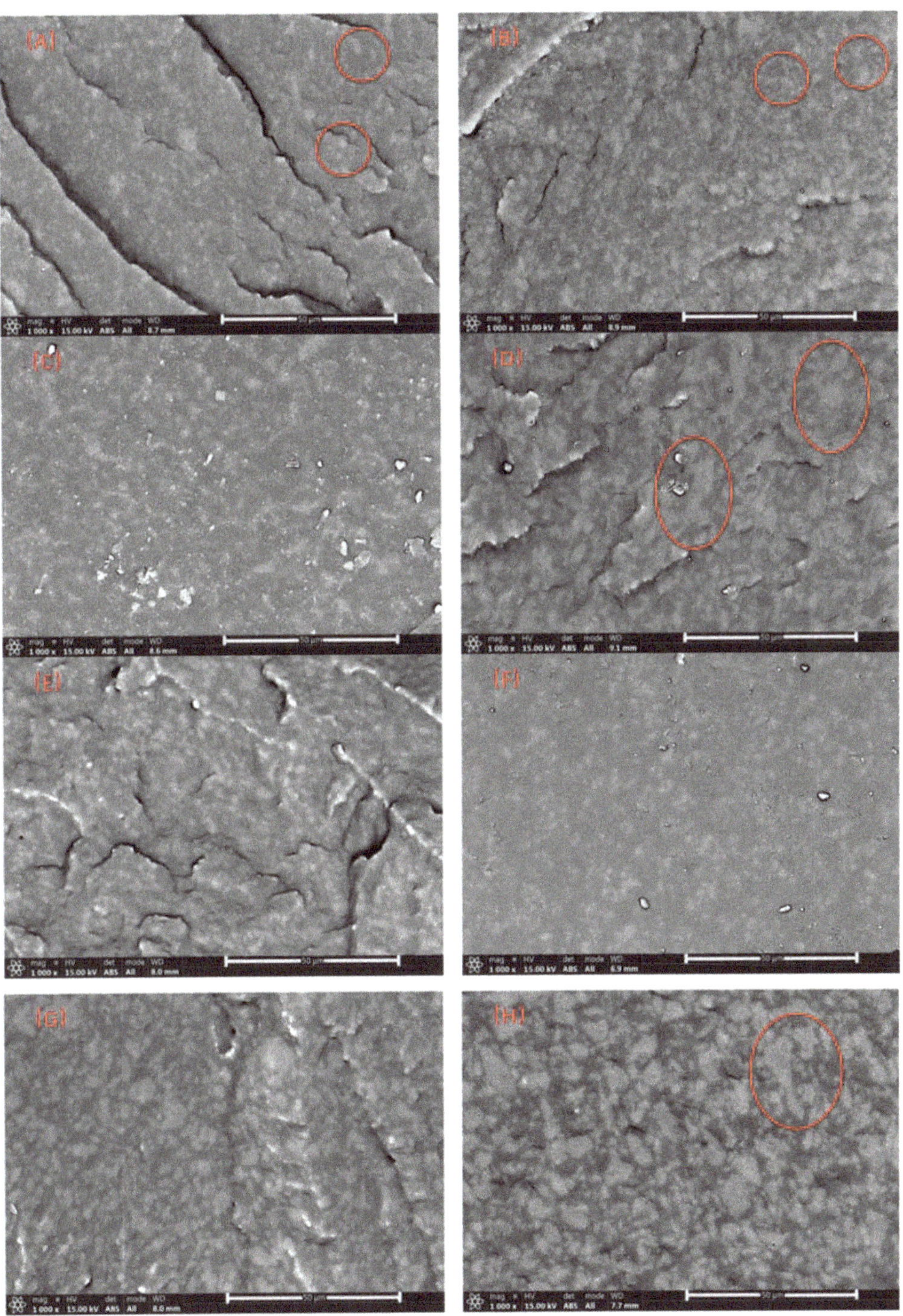

**Figure 6.** BSE diagrams of TPSiV, (**A–H**) correspond to samples 1–8, respectively.

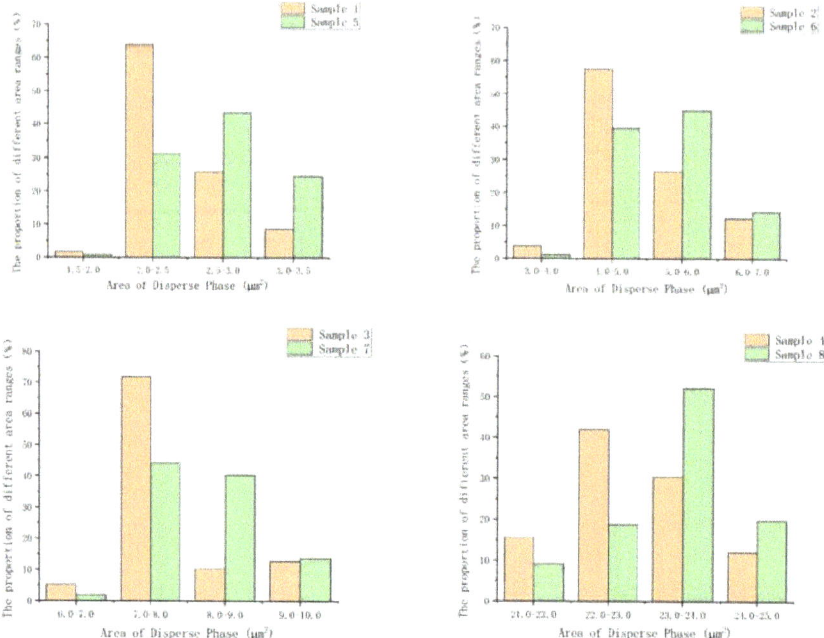

**Figure 7.** Bar diagrams of the size distribution of the dispersed phase in the backscattered signal diagrams.

According to Figures 8 and 9 and Table 3, we found that no matter whether the silane coupling agent KH-907 was added or not, it can bring about the shift in characteristic peaks, and with the decrease in the relative content of silicone rubber, the wavenumber shift is more obvious. However, after adding silane coupling agent KH-907, the value of the wavenumber shift is larger. For example, comparing pure MVSR with samples 1 and 5, the stretching vibration peak of $Si(CH_3)_2$ in sample 1 is shifted by eight wavenumbers compared to the same characteristic peak of pure MVSR, while the stretching vibration peak of $Si(CH_3)_2$ in sample 5 is shifted by six wavenumbers compared to the same characteristic peak of pure MVSR.

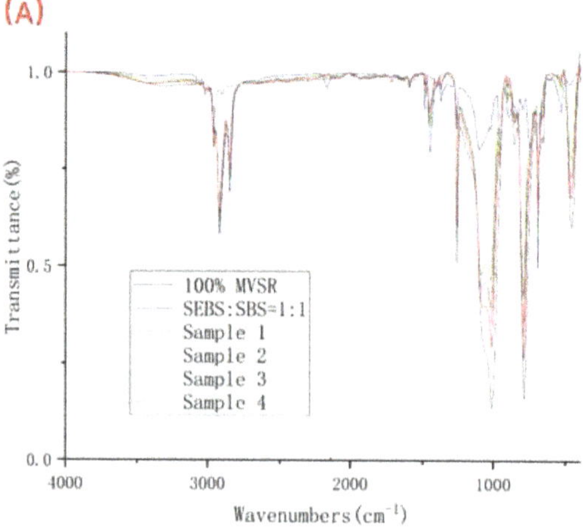


**Figure 8.** *Cont.*

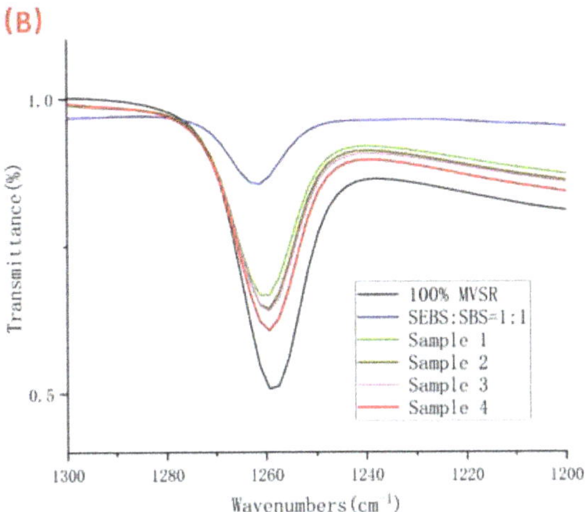

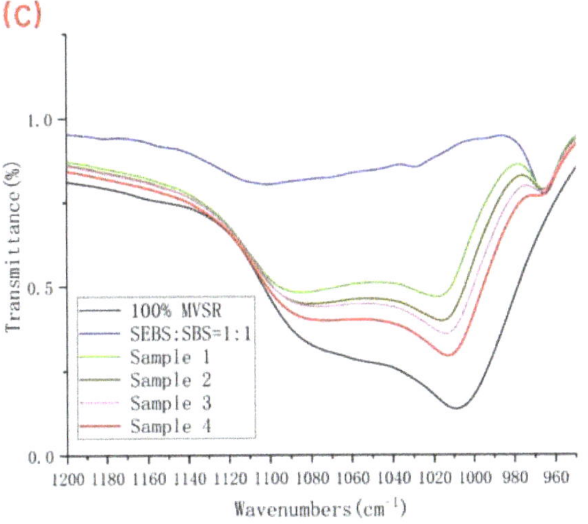

**Figure 8.** FI-IR diagrams of 100% MVSR, 50% SEBS + 50% SBS and samples 1–4. (**A**) Complete spectrum; (**B**) Enlarged detail near wavenumber 1260 cm$^{-1}$; (**C**) Enlarged detail near wavenumber 1010 cm$^{-1}$.

**Table 3.** Peak wave numbers of FI-IR diagrams of 100% MVSR, 50% SEBS + 50% SBS and samples 1–8.

| Abbreviation | Peak Position (cm$^{-1}$) | Abbreviation | Peak Position (cm$^{-1}$) |
|---|---|---|---|
| MVSR | 1010, 1259 | 50%SEBS + 50%SBS | 1261 |

| Sample No. | Peak position (cm$^{-1}$) | Sample No. | Peak position (cm$^{-1}$) |
|---|---|---|---|
| 1 | 1018, 1261 | 5 | 1016, 1261 |
| 2 | 1016, 1261 | 6 | 1014, 1260 |
| 3 | 1014, 1260 | 7 | 1014, 1259 |
| 4 | 1014, 1259 | 8 | 1012, 1259 |

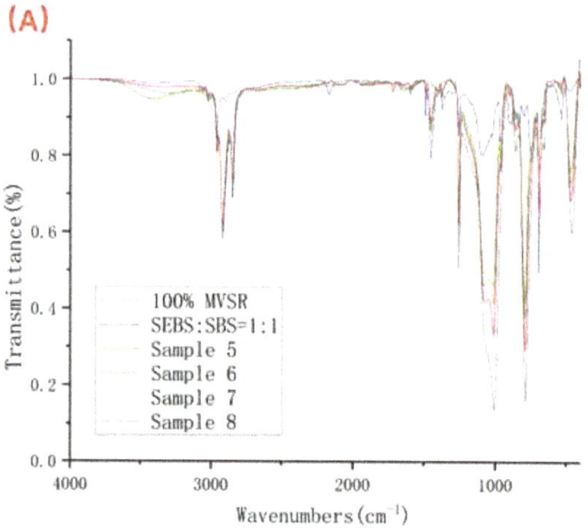

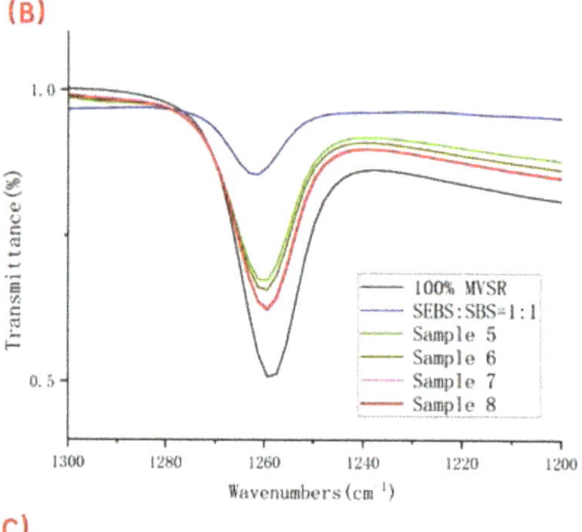

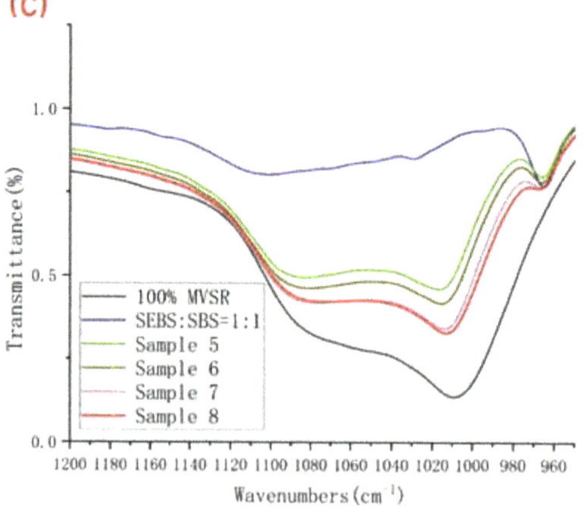

**Figure 9.** FI-IR diagrams of 100% MVSR, 50% SEBS + 50% SBS and samples 5–8. (**A**) Complete spectrum; (**B**) Enlarged detail near wavenumber 1260 cm$^{-1}$; (**C**) Enlarged detail near wavenumber 1010 cm$^{-1}$.

Usually, we use the compatibility results obtained by infrared spectroscopy as the basis for qualitative analysis and do not use the absolute magnitude of the wavenumber offset as the basis for the quantitative comparison of the compatibility of the two phases. The infrared spectrum results in this experiment can only show that the blend system of MVSR and SEBS/SBS has certain compatibilities, and we cannot further draw a quantitative conclusion regarding the compatibility of MVSR and SEBS/SBS.

### 3.4. Analysis of the Results of the Quantitative Characterization Test of the Compatibility Layer

We usually use three parameters, $L$, $A$ and $B$, to represent the chromaticity value of the color of the object. The size of these three values represents the color space coordinates of a certain color, and every color has a unique color space coordinate value. Among them, $L$ stands for lightness and darkness, black and white; $A$ stands for red and green; and $B$ stands for yellow and blue. Since the BSE image obtained by scanning electron microscopy is black and white—that is, the values of $A$ and $B$ are both 0—in the experiment, we only measure the chromaticity value of the color of a certain point in the BSE image by the $L$ value.

We printed the 1000 x BSE diagrams on A3 size (420 × 297 mm) photo paper with a high-definition printer, as shown in Figure 1, and moved the colorimeter, as shown in Figure 2. In each direction, we recorded the $L$ value every time the colorimeter moved by 1 mm and took the arithmetic average of the thickness of the compatible layer obtained from each direction as the final experimental result.

The field of view of the BSE diagram is not related to the size of the photo paper, so we can first find the field of view of the BSE diagram. According to the size of the scale bar in the BSE map and the measurement of the vernier caliper, we can calculate that the field of view of the BSE diagram was 27.19 × 18.54 μm. Next, as shown in Figure 1, we divided the BSE diagram into eight parts, randomly selected one of them, and printed it out on photo paper. According to the segmentation method in Figure 1, we can calculate that the field of view of the divided diagrams was 6.80 × 9.27 μm. We converted this field of view with the A3 size photo paper (420 × 297 mm), and we found that 1 mm length of photo paper represents 0.0162 μm of the diagram length and 1 mm width of photo paper represents 0.031 μm of the diagram width.

Before the start of the experiment, we tested the $L$ value of the MVSR and SEBS/SBS phases. We found that the $L$ value of the MVSR phase was between 70 and 72, and the $L$ value of the SEBS/SBS phase was between 47 and 50. Therefore, we inferred that the distance with the $L$ value of 50–70 can be regarded as the thickness of the compatibility layer between the MVSR phase and the SEBS/SBS phase.

According to the data in Table 4, with the increase in the relative content of MVSR, the thickness of the compatibility layer between the MVSR and the SEBS/SBS phases gradually decreased. Since the thickness of the compatibility layer can be used as a measure of the degree of interaction between the MVSR and the SEBS/SBS phases, combined with the statistical data in Figure 7, we believe that the dispersed phase size of TPSiV with a high proportion of MVSR was larger, and the "sea-island" structure tended to disappear because the force of the two phases was reduced. By comparing the thicknesses of the compatibility layers of several groups with the same MVSR content, with the variable only being the content of silane coupling agent KH-907 (for example, samples 1 and 5), we found that the thickness of the compatibility layer significantly increased after adding silane coupling agent KH-907, which shows that the silane coupling agent can promote the interaction between the MVSR and the SEBS/SBS phases. This can serve as a microscopic explanation for the improved macroscopic properties after adding the silane coupling agent KH-907. It can be seen that this characterization method overcomes the disadvantage that the etching method and infrared spectroscopy cannot quantitatively characterize the thickness of the compatibility layer and also avoids the defect that when the MVSR content is high, and the etching method is used, the non-etched phase collapses and then adheres to the section.

**Table 4.** Compatibility layer thickness of samples 1–8.

| Sample No. | Thickness of Compatible Layer (µm) | Sample No. | Thickness of Compatible Layer (µm) |
|---|---|---|---|
| 1 | 0.5616 | 5 | 0.4143 |
| 2 | 0.4439 | 6 | 0.3304 |
| 3 | 0.3518 | 7 | 0.2632 |
| 4 | 0.1276 | 8 | 0.0963 |

*3.5. Analysis of the Experimental Results of the Mechanical Property Test*

Table 5 shows the test results of the mechanical properties of samples 1–8. From this, we can easily find that with the increase in the silicone rubber content, the mechanical properties of TPSiV are significantly reduced. It is more obvious when it increases to 70%. This is because the mechanical properties of silicone rubber itself are poor and far inferior to SEBS and SBS. With the obvious increase in the content of silicone rubber, the material itself gradually showed properties closer to that of silicone rubber. In addition, according to Figure 6D,H, the partial disappearance of the "sea-island" structure in TPSiV is also one of the reasons for the obvious deterioration of the mechanical properties of the sample with a silicone rubber content of 70 phr.

**Table 5.** Mechanical properties of samples 1–8.

| Sample No. | Tensile Strength (MPa) | Break Elongation (%) | Tensile Set Rate (%) |
|---|---|---|---|
| 1 | 9.94 | 664.74 | 17.74 |
| 2 | 8.83 | 583.71 | 17.89 |
| 3 | 7.17 | 456.06 | 24.58 |
| 4 | 4.09 | 380.01 | 29.09 |
| 5 | 7.16 | 516.34 | 20.82 |
| 6 | 6.52 | 442.34 | 21.53 |
| 7 | 5.03 | 352.80 | 28.92 |
| 8 | 2.94 | 296.37 | 34.21 |

For a material, tensile strength, elongation at break and tensile set are the three most critical mechanical properties. Using the data shown in Table 5, we can analyze the performance change after adding the silane coupling agent KH-907. We can calculate the ratio of improvement in tensile strength and elongation at break by:

$$T = \frac{W_2 - W_1}{W_1} \tag{3}$$

where $T$ is the improvement ratio of tensile strength and elongation at break, $W_1$ is the tensile strength and elongation at break before adding silane coupling agent KH-907, and $W_2$ is the tensile strength and elongation at break after adding silane coupling agent KH-907.

Since the smaller the value of the tensile set, the better the performance of the material, we use the following formula to calculate the improvement ratio of the tensile set:

$$R = \frac{M_1 - M_2}{M_1} \tag{4}$$

where $R$ is the improvement ratio of the tensile set, $M_1$ is the tensile set before adding silane coupling agent KH-907, and $M_2$ is the tensile strength and elongation at break after adding silane coupling agent KH-907.

By substituting the data of Table 5 into Equations (3) and (4), we obtain Table 6. According to Table 6, we found that the addition of silane coupling agent KH-907 significantly improved the mechanical strength of TPSiV; the tensile strength is increased by about 40%, the elongation at break is increased by about 30%, and the tensile set is increased by about

15%. Since we have quantitatively characterized the thickness of the compatibility layer above, we believe that the reason for the improvement of mechanical properties comes from the increase in the thickness of the compatibility layer. When the material is stretched, the interaction between the MVSR phase and SEBS/SBS phase will be closer. This closer interaction gives the material better mechanical properties and resilience.

**Table 6.** The improvement rate of mechanical properties of TPSiV after the addition of silane coupling agent KH-907.

| Silicone Rubber Content (phr) | Tensile Strength Improvement Rate (%) | The Rate of Increase in Elongation at Break (%) | Tensile Set Improvement Rate (%) |
|---|---|---|---|
| 40 | 38.83 | 28.74 | 14.79 |
| 50 | 35.43 | 31.96 | 16.91 |
| 60 | 42.55 | 29.27 | 15.01 |
| 70 | 39.12 | 28.22 | 14.97 |

## 4. Conclusions

We prepared a thermoplastic silicone rubber from methyl vinyl silicone rubber, SEBS and SBS using a torque rheometer. By taking the backscattered electron diagram of the material, we found that the phase distribution of the material could be characterized and analyzed using this method. The MVSR phase is the dispersed phase, the SEBS/SBS phase is the continuous phase, and the two have a "sea-island" structure. When the MVSR content reaches 70 phr, the "sea-island" structure tends to disappear. After adding silane coupling agent KH-907, the size of the MVSR phase will decrease. According to the results obtained by infrared spectroscopy, we found that there is a certain degree of compatibility between the MVSR phase and the SEBS/SBS phase, but this can only be used as a basis for qualitative analysis. By using a colorimeter to analyze the printed backscattered electron diagrams, we found that the size of the compatibility layer between the two phases can be quantitatively characterized. Moreover, the silane coupling agent KH-907 can increase the thickness of the compatibility layer between the two phases. In the subsequent mechanical property test, we also found that silane coupling agent KH-907 improved the mechanical properties of TPSiV, in which the increase rate of tensile strength is about 40%, the increase rate of elongation at break is about 30%, and the increase rate of the tensile set is about 15%. The increase in the thickness of the compatibility layer can provide a microscopic explanation for the improvement of mechanical properties.

**Author Contributions:** Methodology, C.Z.; writing—original draft, C.Z.; validation, C.Z.; writing—review & editing, C.Z.; visualization, C.Z.; data curation, X.C. (Xiaohan Chen) and X.C. (Xian Chen). All authors have read and agreed to the published version of the manuscript.

**Funding:** This research received no external funding.

**Institutional Review Board Statement:** Not applicable.

**Informed Consent Statement:** Not applicable.

**Data Availability Statement:** The data presented in this study are available on request from the corresponding author.

**Conflicts of Interest:** The authors declare no conflict of interest.

## References

1. Kruzelak, J.; Sykora, R.; Hudec, I. Sulphur and peroxide vulcanisation of rubber compounds—Overview. *Chem. Pap.* **2016**, *70*, 1533–1555. [CrossRef]
2. Scheutz, G.M.; Lessard, J.J.; Sims, M.B. Adaptable Crosslinks in Polymeric Materials: Resolving the Intersection of Thermoplastics and Thermosets. *J. Am. Chem. Soc.* **2019**, *141*, 16181–16196. [CrossRef] [PubMed]
3. Bhattacharya, A.B.; Naskar, K. Automotive applications of thermoplastic vulcanizates. *J. Appl. Polym. Sci.* **2020**, *137*, 49181. [CrossRef]

4.	Naskar, K. Thermoplastic elastomers based on PP/EPDM blends by dynamic vulcanization. *Rubber Chem. Technol.* **2007**, *80*, 504–519. [CrossRef]

5.	Samthong, C.; Kunanusont, N.; Deetuam, C. Effect of acrylonitrile content of acrylonitrile butadiene rubber on mechanical and thermal properties of dynamically vulcanized poly(lactic acid) blends. *Polym. Int.* **2019**, *68*, 2004–2016. [CrossRef]

6.	Shit, S.C.; Shah, P. A Review on Silicone Rubber. *Natl. Acad. Sci. Lett.* **2013**, *36*, 355–365. [CrossRef]

7.	Lee, M.; Naila, D.; Sadanand, S. Characterization of Mechanical and Dielectric Properties of Silicone Rubber. *Polymers* **2021**, *13*, 1831.

8.	Renard, C.; Han, P. Remarkably improved electromechanical actuation of polyurethane enabled by blending with silicone rubber. *RSC Adv.* **2017**, *7*, 22900–22908. [CrossRef]

9.	Thongpin, C.; Buaksuntear, K.; Tanprasert, T. Bio-based thermoplastic vulcanizates from natural rubber (bioplastic/NR). *IOP Conf. Ser. Mater. Sci. Eng.* **2020**, *773*, 012045. [CrossRef]

10.	Hsissou, R.; Seghiri, R.; Benzekri, Z. Polymer composite materials: A comprehensive review. *Compos. Struct.* **2021**, *262*, 113640. [CrossRef]

11.	Ravindranath, S.; Islam, S.K.S. Performance Deterioration of SBS-Modified Asphalt Mix: Impact of Elevated Storage Temperature and SBS Concentration of Modified Binder. *J. Mater. Civ. Eng.* **2022**, *34*, 04021475.

12.	Ribeiro de Oliveira, C.I.; Guimaraes Rocha, M.C. Morphological, mechanical, and thermal properties of PP/SEBS/talc composites. *J. Thermoplast. Compos. Mater.* **2019**, *35*, 281–299. [CrossRef]

13.	Aziz, T.; Jamil, M.I.; Iqbal, M. Recent Progress in Silane Coupling Agent with Its Emerging Applications. *J. Polym. Environ.* **2021**, *29*, 3427–3443. [CrossRef]

14.	Anbupalani, M.S.; Venkatachalam, C.D. Influence of coupling agent on altering the reinforcing efficiency of natural fibre-incorporated polymers—A review. *J. Reinf. Plast. Compos.* **2020**, *39*, 520–544. [CrossRef]

15.	Mueller, E.; Meffert, M.; Gerthsen, D. Versatile application of a modern scanning electron microscope for materials characterization. *J. Mater. Sci.* **2020**, *55*, 13824–13835.

16.	Leila, B.; Alireza, K.; Mohammad, A.B. Correlation between viscoelastic behavior and morphology of nanocomposites based on SR/EPDM blends compatibilized by maleic anhydride. *Polymer* **2017**, *113*, 156–166.

17.	Subhan, S.; Gisèle, B.; Philippe, C. Dynamically Cured Poly(vinylidene fluoride)/Epoxidized Natural Rubber Blends Filled with Ferroelectric Ceramic Barium Titanate. *Compos. Part A Appl. Sci. Manuf.* **2017**, *93*, 107–116.

18.	Kerim, K.; Yasin, K.; Tayfun, Ü. Mechanical, thermo-mechanical and water uptake performance of wood flour filled polyurethane elastomer eco-composites: Influence of surface treatment of wood flour. *Holzforschung* **2019**, *73*, 401–407.

19.	Kana, M.; Hiromu, S. Tensile properties and interfacial adhesion of silicone rubber/polyethylene blends by reactive blending. *J. Appl. Polym. Sci.* **2018**, *135*, 46192.

20.	Andrea, K.; István, Z.H.; Tamás, B. Thermoplastic Dynamic Vulcanizates with In Situ Synthesized Segmented Polyurethane Matrix. *Polymers* **2019**, *11*, 1663.

21.	Chen, Y.K.; Zhou, G.; Cao, L.M. Novel fluorosilicone thermoplastic vulcanizates prepared via core-shell dynamic vulcanization: Effect of fluororubber/silicone rubber ratio on morphology, crystallization behavior, and mechanical properties. *Polym. Adv. Technol.* **2018**, *25*, 30–45. [CrossRef]

22.	Wang, Y.H.; Gong, Z.; Chen, Y.K. Poly (vinylidene fluoride)/fluororubber/silicone rubber thermoplastic vulcanizates prepared through core-shell dynamic vulcanization: Formation of different rubber/plastic interfaces via controlling the core from "soft" to "hard". *Mater. Chem. Phys.* **2017**, *195*, 123–131. [CrossRef]

23.	Husain, J.; Raghu, N.; Reddy, N. Synthesis, Characterization and Studies on Polyaniline/Nanocomposites Thin Films. *J. Phys. Conf. Ser.* **2020**, *1495*, 012008. [CrossRef]

24.	Seiler, H. Secondary-electron emission in the scanning electron-microscope. *J. Appl. Phys.* **1983**, *54*, R1–R18. [CrossRef]

25.	Goldstein, J.I.; Newbury, D.E.; Michael, J.R. *Scanning Electron Microscopy and X-ray Microanalysis*; Springer: Berlin/Heidelberg, Germany, 2017.

26.	Czernuszka, J.T.; Boyes, E.D.; Hirsch, P.B. Imaging of dislocations using backscattered electrons in a scanning electron-microscope. *Philos. Mag. Lett.* **1990**, *62*, 227–232. [CrossRef]

27.	Oliveira, J.P.; Curado, T.M. Gas tungsten arc welding of as-rolled CrMnFeCoNi high entropy alloy. *Mater. Des.* **2020**, *189*, 108505. [CrossRef]

28.	Azimi-Roeen, G.; Kashani-Bozorg, S. Effect of multi-pass friction stir processing on textural evolution and grain boundary structure of Al-Fe$_3$O$_4$ system. *J. Mater. Res. Technol.* **2020**, *9*, 1070–1086. [CrossRef]

29.	Ebied, S.; Hamada, A. Study on Hot Deformation Behavior of Beta Ti-17Mo Alloy for Biomedical Applications. *JOM* **2022**, *74*, 494–505. [CrossRef]

30.	Duraman, S.B.; Richardson, I.G. Microstructure & properties of steel-reinforced concrete incorporating Portland cement and ground granulated blast furnace slag hydrated at 20 degrees C. *Cem. Concr. Res.* **2020**, *137*, 106193.

31.	Gonzalez-Lopez, L.; Ventura, H.; Ardanuy, M. Surface modification of flax nonwovens for the development of sustainable, high performance, and durable calcium aluminate cement composites. *Compos. Part B Eng.* **2020**, *191*, 107955. [CrossRef]

32.	Sadighpour, L.; Mostafavi, A.S. Comparison of Bond Strength in the Different Class of Resin Cements to Cast and CAD/CAM Co-Cr Alloys. *Int. J. Dent.* **2021**, *2021*, 7843979. [CrossRef] [PubMed]

*polymers*

MDPI

*Article*

# A New Analytical Model for Deflection of Concrete Beams Reinforced by BFRP Bars and Steel Fibres under Cyclic Loading

Haitang Zhu [1,2], Zongze Li [2,*], Qun Chen [2], Shengzhao Cheng [3,4], Chuanchuan Li [2] and Xiangming Zhou [5]

[1] School of Civil Engineering, Henan University of Engineering, Zhengzhou 451191, China; htzhu@zzu.edu.cn
[2] School of Water Conservancy Engineering, Zhengzhou University, Zhengzhou 450001, China; chenqun9958@163.com (Q.C.); lichuanchuan1004@126.com (C.L.)
[3] China Construction Seventh Engineering Division, Co., Ltd., Zhengzhou 450004, China; chengshengzhao@aliyun.com
[4] Installation Engineering Co., Ltd. of CSCEC 7th Division, Zhengzhou 450004, China
[5] Department of Civil & Environmental Engineering, Brunel University London, Uxbridge UB8 3PH, UK; xiangming.zhou@brunel.ac.uk
* Correspondence: lizongze@gs.zzu.edu.cn

Citation: Zhu, H.; Li, Z.; Chen, Q.; Cheng, S.; Li, C.; Zhou, X. A New Analytical Model for Deflection of Concrete Beams Reinforced by BFRP Bars and Steel Fibres under Cyclic Loading. *Polymers* 2022, 14, 1797. https://doi.org/10.3390/polym14091797

Academic Editors: Wei Wu, Hao-Yang Mi, Chongxing Huang, Hui Zhao and Tao Liu

Received: 13 April 2022
Accepted: 26 April 2022
Published: 28 April 2022

Publisher's Note: MDPI stays neutral with regard to jurisdictional claims in published maps and institutional affiliations.

**Abstract:** Basalt-fiber-reinforced plastic-bars-reinforced concrete beams (i.e., BFRP-RC beams) usually possess significant deformations compared to reinforced concrete beams due to the FRP bars having a lower Young's modulus. This paper investigates the effects of adding steel fibers into BFRP-RC beams to reduce their deflection. Ten BFRP-RC beams were prepared and tested to failure via four-point bending under cyclic loading. The experimental variables investigated include steel-fiber volume fraction and shape, BFRP reinforcement ratio, and concrete strength. The influences of steel fibers on ultimate moment capacity, service load moment, and deformation of the BFRP-RC beams were investigated. The results reveal that steel fibers significantly improved the ultimate moment capacity and service load moment of the BFRP-RC beams. The deflection and residual deflection of the BFRP-RC beams reinforced with 1.5% by volume steel fibers were 48.18% and 30.36% lower than their counterpart of the BFRP-RC beams without fibers. Under the same load, the deflection of the beams increased by 11% after the first stage of three loading and unloading cycles, while the deflection increased by only 8% after three unloading and reloading cycles in the second and third stages. Finally, a new analytical model for the deflection of the BFRP-RC beams with steel fibers under cyclic loading was established and validated by the experiment results from this study. The new model yielded better results than current models in the literature.

**Keywords:** cyclic loading; deflection; BFRP-RC beams; steel fiber; analytical model

## 1. Introduction

The corrosion of steel bars in RC structures shortens the service life of RC structures and significantly increases maintenance costs. Over the past decades, fibre-reinforced polymer (FRP) bars have been used extensively in the construction industry as a new type of reinforcement that replaces steel bars to solve corrosion problems [1,2]. Compared with steel bars, FRP bars are corrosion-free, magnetically transparent, and lighter but with higher tensile strength [3]. These important features enable FRP-RC structures to withstand various complex and corrosive environments with desirable performances. Based on the raw materials used for manufacturing FRPs, FRPs are divided into four categories, including basalt-fiber-reinforced plastic (BFRP), aramid-fiber-reinforced plastic (AFRP), glass-fiber-reinforced plastic (GFRP) and carbon-fiber-reinforced plastic (CFRP). CFRP has a higher Young's modulus and tensile strength than any other FRP, but its high price hinders its wider applications in construction. Although GFRP and AFRP are less expensive, their alkali resistance is poor, leading to a large degree of strength degradation when reinforcing concrete with alkalinity of pH 12~13. In this regard, BFRP is now used in more applications in construction due to its relatively low cost, good thermal resistance, and excellent freeze/thaw resistance [4]. At the same time, the bond performance between BFRP bars and concrete is also better than that between GFRP bars and concrete. More importantly, BFRP is made of volcanic basalt, which is widely recognized as a type of green construction material with high sustainability credentials; therefore, BFRP has greater application prospects [5]. However, the Young's modulus of FRP bars is lower than that of steel bars, which leads to FRP-RC structures usually possessing larger crack widths and deflection than RC structures [6–8]. In addition, FRP-RC structures are prone to brittle failure because FRP bars are usually brittle while steel bars are ductile.

To solve the above problems, researchers have proposed various ways to improve the structural performance of FRP-RC beams, which include composite reinforcement composed of steel core and FRP to reinforce RC beams [9], increasing the transverse reinforcement ratio for FRP-RC beams [10], grouting FRP bars in corrugated sleeves to reinforce RC beams [11] and adding fibers as additional reinforcement into FRP-RC beams [12]. Previous studies have demonstrated that fibers can improve ductility and strain-hardening of cementitious composite mortars and grouts [13]. Adding discrete fibers into a concrete matrix is the most effective way to improve the serviceability performances of FRP-RC structures. More importantly, the bridging effect of fibers leads to the pseudo-ductile behavior of FRP-RC structures [14–29]. Chellapandian et al. [14] investigated the effects of adding macro-synthetic fibers into concrete on the cracking, stiffness, and deformability of GFRP-RC beams. They concluded that the fibers improved the post-cracking behavior of GFRP-RC beams with their deformation largely enhanced by adding only 1% by volume of steel fibers, which also transformed the GFRP-RC beams from brittle flexure–shear failure to ductile flexural failure with higher pseudo-ductility. Filipe et al. [15] found that the use of fibers enhanced the stiffness of FRP-RC members and helped to reduce crack spacing and width. The same findings were also obtained by other researchers [16,17]. Ibrahim and Eswari [18] investigated the strength and ductility of GFRP = laminated RC beams incorporated with various amounts of discrete steel fibers. Their study revealed that incorporating steel fibers can effectively improve the strength and ductility of FRP-RC beams. Issa et al. [19] concluded that polypropylene fibers, glass fibers, and steel fibers all improved the ductility of FRP-RC beams, in particular, in the case of steel fibers, which increased the ductility of the beams by 277.8%. Short discrete fibers not only improved the tensile properties of concrete but also improved the shear capacity of concrete beams [20–23]. Zhu et al. [24] studied the effects of partially steel fibers reinforced concrete (SFRC) on the flexural behavior of FRP-RC beams. The results suggest that compared with full section SFRC beams, partially reinforced SFRC beams cannot provide a better performance, and steel fibers helped to reduce the deflection of the FRP-RC structures. Similar findings were also reported by other researchers. In summary, previous studies have systematically studied the flexural behaviors of FRC beams strengthened with FRP bars under static loading, including crack behaviors, ductility, deflection, and ultimate moment capacity.

However, for practical purposes, RC beams always bear cyclic loading rather than static loading [30]. The flexural behaviors of RC beams under static loading are different from those under cyclic loading, so it is imperative to investigate the effects of the deterioration of concrete and FRP bars on the flexural performances of FRP-RC beams under cyclic loading. Zhu et al. [27] studied the influence of steel fiber on the bearing capacity of the BFRP-RC beams, and the results showed that steel fiber can help to increase the ultimate compressive strain of concrete so as to increase the bearing capacity of the BFRP reinforced concrete beams, and the calculation method of flexural capacity of the BFRP bars and steel-fiber-reinforced concrete beams was established. Li et al. [28,29] analyzed the influence of steel fibers on crack width and ductility of the BFRP-RC beams; the results showed that steel fibers were beneficial in reducing crack width and increasing the ductility of the BFRP-RC beams. The calculation method of maximum crack width and the evaluation method of ductility were put forward. However, the research on the dRC beams with steel fibers under cyclic load is still rare in literature.

The purpose of this research is to study the volume fraction and type of steel fibers on the deformation and flexural behavior of the BFRP-RC beams. In addition, the effects of concrete strength and BFRP reinforcement ratio on the deformation and flexural behavior of the BFRP-RC beams were also investigated. Ten beams were prepared and loaded via four-point bending under cyclic loading until failure, which included nine beams reinforced by both BFRP bars and steel fibers and one BFRP-RC beam without steel fibers as a reference case. The responses of the beams under cyclic loading were compared and analyzed from the aspects of failure mode, ultimate moment capacity, service load moment, load-deflection relation, envelop curves, residual deflection, and stiffness degradation. In addition, a new analytical model for the deformation of the BFRP-RC beams with steel fibers was proposed. Compared with other models, the newly proposed model in this paper accords better to experimental results.

## 2. Materials and Methods

### 2.1. Material Properties

In this study, three types of steel fibers were used as reinforcement for concrete, which were differentiated in the number of hook-ends, length, diameter, and tensile strength, but with the same aspect ratio (i.e., fiber length-to-diameter ratio). Based on the number of hook-ends, the three types of fibers were named 3D, 4D, and 5D, respectively. Three kinds of steel fibers were produced by Bekaert company in Shanghai, China. The steel fibers having only one hook-end were denoted as 3D fibers, as shown in Figure 1a; those with one and a half hook-ends were named 4D fibers, as shown in Figure 1b; and those with two steel fiber hook-ends were called 5D fibers, as shown in Figure 1c. The physical properties and dimensions of the three types of steel fibers are shown in Table 1. It can be seen that the 4D and 5D fibers both had a length of 60 mm, which was longer than that, i.e., 35 mm, of 3D fibers. In addition, 5D fibers had the highest tensile strength among the three types of fibers.

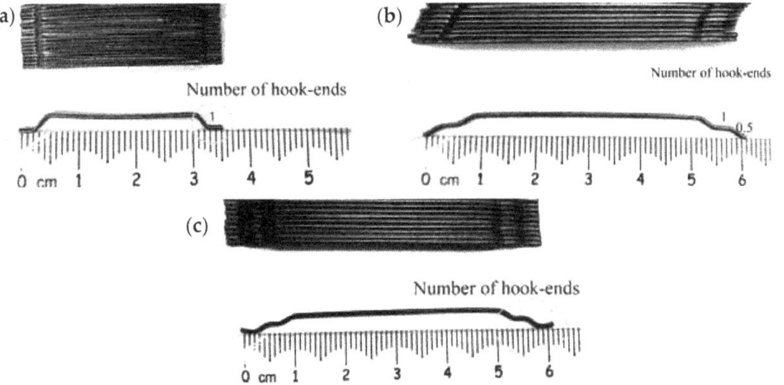


**Figure 1.** Three types of steel fibers: (**a**) 3D; (**b**) 4D; (**c**) 5D.

**Table 1.** Physical properties and dimensions of steel fibers used in this research.

| Types | $l_{sf}$ (mm) | $d_{sf}$ (mm) | $l_{sf}/d_{sf}$ | $f_{t,sf}$ (MPa) | $E_{sf}$ (GPa) | Number of Hook-Ends |
|---|---|---|---|---|---|---|
| 3D | 35 | 0.55 | 65 | 1345 | 200 | 1 |
| 4D | 60 | 0.90 | 65 | 1600 | 200 | 1.5 |
| 5D | 60 | 0.90 | 65 | 2300 | 200 | 2 |

Note: $l_{sf}$ is the length of steel fibers; $d_{sf}$ is the diameter of steel fibers; $f_{t,sf}$ is the tensile strength of steel fibers; $E_{sf}$ is the Young's modulus of steel fibers.

The BFRP bars were produced by Jiangsu lvcaigu new material technology development Co., Ltd. The BFRP bars used as the reinforcement had a diameter of 12 or 14 mm. Their tensile strength and Young's modulus were measured conforming to Chinese standard GB/T 30022-2013 [31], with the average of the results of five samples taken as the relevant representative value as shown in Table 2.

**Table 2.** Tensile properties of the BFRP bars.

| Types | Diameter (mm) | Tensile Strength ($f_{fu}$) (MPa) | Young Modulus ($E_f$) (GPa) | Yield Strength |
|---|---|---|---|---|
| BFRP | 12 | 1080 | 47.0 | — |
| BFRP | 14 | 1060 | 46.5 | — |

In this study, two grades of concrete were prepared with the characteristic strength of 30 and 60 MPa, respectively. Mix proportions of the two grades of concrete were designed according to JG/T 472–2015 [32], as shown in Table 3. All beam specimens are labelled as "BN-CN-VN-SN", where "BN" represents the BFRP reinforcement ratio in percentage, "CN" refers to concrete grade, "VN" refers to steel-fiber volume fraction in percentage and "SN" denotes the type of steel fibers used. For example, "B0.56C60V1.0S3" refers to the beam specimen with a BFRP reinforcement ratio of 0.56% (i.e., 2Φ12), concrete strength of 60 MPa, the steel-fiber volume fraction of 1.0%, and 3D steel fibers.

**Table 3.** Concrete mixtures (in kg/m$^3$) of the specimens.

| Beams | Water | Cement | Sand | Steel Fiber | Coarse Aggregate | Polycarboxylate Superplasticizer |
|---|---|---|---|---|---|---|
| B0.56C60V1.0S3 | 172 | 521.2 | 669.3 | 78.5 (3D) | 1013.5 | 5.212 |
| B0.77C60V1.0S3 | 172 | 521.2 | 669.3 | 78.5 (3D) | 1013.5 | 5.212 |
| B1.15C60V1.0S3 | 172 | 521.2 | 669.3 | 78.5 (3D) | 1013.5 | 5.212 |
| B1.65C60V1.0S3 | 172 | 521.2 | 669.3 | 78.5 (3D) | 1013.5 | 5.212 |
| B1.15C60 | 172 | 521.2 | 648.6 | — | 1058.2 | 2.606 |
| B1.15C60V0.5S3 | 172 | 521.2 | 658.9 | 39.3 (3D) | 1035.9 | 4.170 |
| B1.15C60V1.5S3 | 172 | 521.2 | 679.6 | 117.8 (3D) | 991.1 | 7.297 |
| B1.15C60V1.0S4 | 172 | 521.2 | 669.3 | 78.5 (4D) | 1013.5 | 5.212 |
| B1.15C60V1.0S5 | 172 | 521.2 | 669.3 | 78.5 (5D) | 1013.5 | 5.212 |
| B1.15C30V1.0S3 | 215 | 330.8 | 706.2 | 78.5 (3D) | 1124.0 | 0 |

To ensure the slump of both grades of concrete falling between 50 and 70 mm, a polycarboxylate superplasticizer was added to the concrete mixture. Natural gravels with particle size ranging between 5 and 20 mm was used as coarse aggregates when making concrete. The fine aggregate used was river sand, and its particle size was less than 5 mm. The physical properties of aggregate are shown in Table 4. Grade 42.5R ordinary Portland Cement conforming to Chinese standard GB175-2007 [33] was used as the binder in preparing concrete; its physical properties are shown in Table 5. Six $150 \times 150 \times 150$ mm$^3$ concrete cubes and six $150 \times 150 \times 300$ mm$^3$ concrete prisms were also prepared alongside each beam specimen for measuring compression strength and splitting tensile strength of concrete.

**Table 4.** Physical properties of aggregate.

| Aggregate | Specific Gravity | Water Absorption | Fineness Modulus | Free Moisture Content | Graded Zone |
|---|---|---|---|---|---|
| Fine aggregate | 2.60 | 1.01% | 2.78 | 2% | II |
| Coarse aggregate | 2.74 | 0.30% | 7.5 | NIL | NIL |

**Table 5.** Physical properties of the used cement.

| Compressive Strength/MPa | | Flexural Strength /MPa | | Setting Time/min | | Specific Surface Area m$^2$/kg |
|---|---|---|---|---|---|---|
| 3 d | 28 d | 3 d | 28 d | Initial Setting Time | Final Setting Time | |
| 27.8 | 46.8 | 5.6 | 8.5 | 122 | 232 | 345 |

### 2.2. Test Beams

The flexural behaviors of nine BFRP-RC with steel fibers were investigated, together with that of one BFRP-RC beam without steel fibers, which was also investigated as a reference point. All ten beam specimens had the same sizes, i.e., 150 mm wide (b) × 300 mm deep (h) × 2100 mm long (l), and were tested via four-point bending under cyclic loading until failure, as shown in Figure 2. ACI 440 1R-15 [34] stipulates that FRP-RC beams should be designed with concrete crushing failure, and its reinforcement ratio should be greater than 1.4 times of the balanced reinforcement ratio ($\rho_{fb}$), which can be obtained via Equation (1).

$$\rho_{fb} = 0.85\beta_1 \frac{f'_c}{f_{fu}} \frac{E_f \varepsilon_{cu}}{E_f \varepsilon_{cu} + f_{fu}} \tag{1}$$

where $\beta_1$ is a factor, which can be calculated by Equation (2).

$$\beta_1 = 0.85 - 0.05 \times (\frac{f'_c - 28}{7}) \geq 0.65 \tag{2}$$

The balanced reinforcement ratio of all ten beam specimens was calculated by Equation (1). The design BFRP reinforcement ratios were 0.56% (in the case of 2Φ12 FRP bars), 0.77% (in the case of 2Φ14 FRP bars), and 1.15% (in the case of 3Φ14 FRP bars), respectively. The reinforcement ratios of all beam specimens were higher than 1.4 times the balanced reinforcement ratio. Four levels of steel-fiber volume fractions were investigated, which are 0%, 0.5%, 1.0% and 1.5% respectively. Among them, three types of steel fibers were used as reinforcement when the steel-fiber volume fraction was 1%. Steel stirrups with a diameter of 10 mm and spacing of 75 mm were placed along the whole length of all beam specimens, which ensures that shear failure will not occur. The concrete cover was 25 mm. Figure 2 depicts the details of the beam specimens. Table 6 lists technical details of all beam specimens as well as the actual mechanical properties of concrete making the specimens.

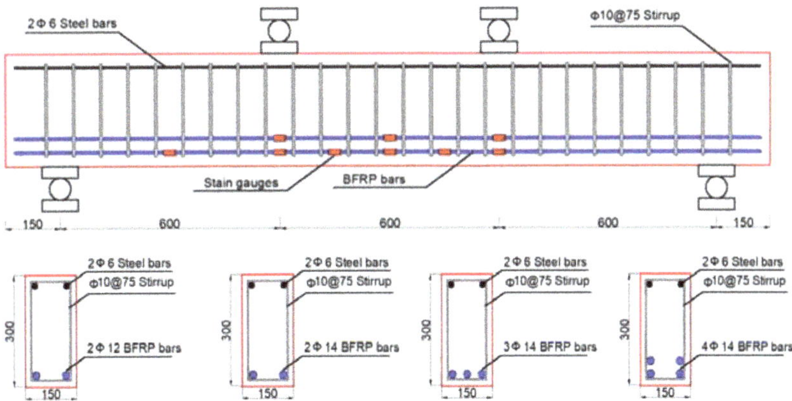


**Figure 2.** Specimen details (all dimensions in millimeters).

**Table 6.** Technical details and actual mechanical properties of concrete for all specimens.

| Beams | $\rho_f$ (%) | $\rho_{sf}$ (%) | Steel Fiber Shapes | $f_{cu,k}$ (MPa) | Actual Physical Properties of Concrete | | | |
|---|---|---|---|---|---|---|---|---|
| | | | | | $f_{cu}$ (MPa) | $f_t$ (MPa) | $f_c'$ (MPa) | $E_c$ (GPa) |
| B0.56C60V1.0S3 | 0.56 | 1.0 | 3D | 60 | 60.16 | 5.59 | 48.13 | 41.30 |
| B0.77C60V1.0S3 | 0.77 | 1.0 | 3D | 60 | 74.99 | 5.70 | 52.45 | 42.70 |
| B1.15C60V1.0S3 | 1.15 | 1.0 | 3D | 60 | 81.47 | 6.60 | 65.18 | 42.40 |
| B1.65C60V1.0S3 | 1.65 | 1.0 | 3D | 60 | 76.47 | 5.84 | 61.18 | 42.40 |
| B1.15C60 | 1.15 | 0 | — | 60 | 74.54 | 3.56 | 59.63 | 41.62 |
| B1.15C60V0.5S3 | 1.15 | 0.5 | 3D | 60 | 69.00 | 4.88 | 51.75 | 41.00 |
| B1.15C60V1.5S3 | 1.15 | 1.5 | 3D | 60 | 81.47 | 5.17 | 65.18 | 42.23 |
| B1.15C60V1.0S4 | 1.15 | 1.0 | 4D | 60 | 83.89 | 5.83 | 62.92 | 42.38 |
| B1.15C60V1.0S5 | 1.15 | 1.0 | 5D | 60 | 79.14 | 5.51 | 63.31 | 43.02 |
| B1.15C30V1.0S3 | 1.15 | 1.0 | 3D | 30 | 44.00 | 3.36 | 34.00 | 35.00 |

*2.3. Experiment Setup and Procedure*

From Figure 3a, four-point bending tests under cyclic loading mode were carried out on beams using a 2000 kN Hydraulic Press Machine (HPM) together with a load-distribution steel beam. Seven linear voltage differential transformers (LVDTs) were mounted at both supports, midspan, both loading points, and the other two quartile spans of the pure bending zone of a BFRP-RC beam. To capture the strain of the BFRP bars during loading, nine electrical strain gauges with the size of $3 \times 2$ mm$^2$ were attached to the bottom of the BFRP bars, and their distribution on the beam was depicted in Figure 2. Nineteen $\pi$-type strain gauges were attached to the top, bottom, and front surfaces of each BFRP-R beam specimen to capture its strain evolution during loading, as shown in Figure 3b. The crack width of concrete at BFRP bar levels was observed by the ZBL-F120 crack width gauge.

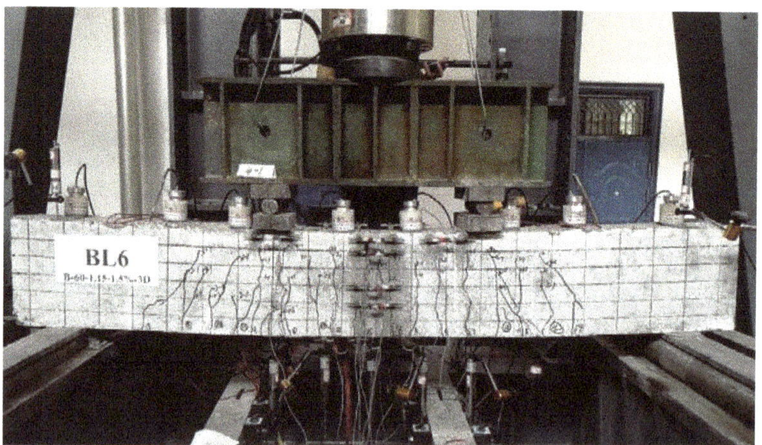


(**a**) Beam loading setup

**Figure 3.** *Cont.*

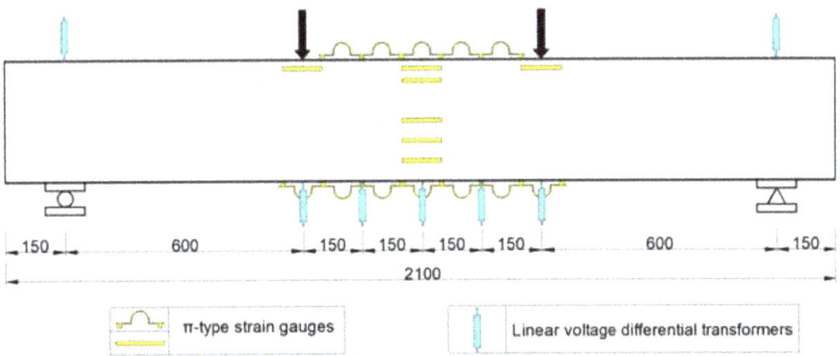



(**b**) Details of beam instrumentation

**Figure 3.** Test setup (dimensions in millimeters).

The unloading–reloading protocol is depicted in Figure 4. First, the beams were loaded at a rate of 0.5 kN/min until cracking; subsequently, the displacement of the hydraulic pressure head was increased every 6 mm (i.e., actuator) as the loading grade; for example, the displacement of the first loading grade actuator was 6 mm, the displacement of the second loading grade actuator was 12 mm, and so on. At each loading grade, the loading–unloading cycles were done three times until the test beam failed.

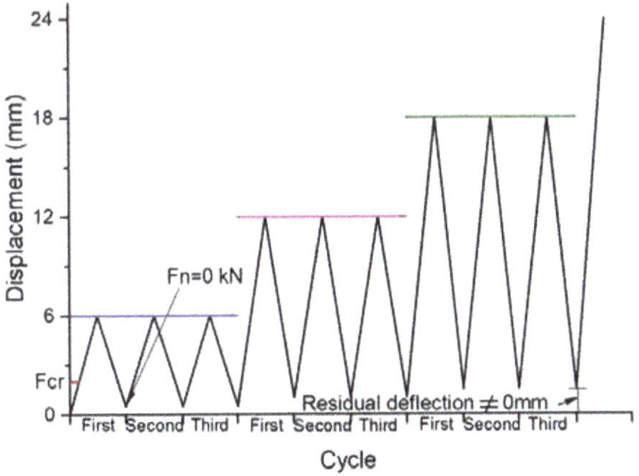



**Figure 4.** Unloading–reloading process. Note: $F_{cr}$ is the cracking load; $F_n$ is the load applied to a beam.

## 3. Results and Discussion

This section presents the experimental results of the nine BFRP-RC beams with steel fibers and the one BFRP-RC beam without steel fibers in terms of failure mode, cracking load, service load moment, ultimate moment capacity, cracking moment, load-deflection evolution, envelope curve, residual deflection, and stiffness degradation. The cracking moment ($M_{cr}$) of a BFRP-RC beam was defined as the moment when the stress of the BFRP bars quickly increased or the initial concrete cracking occurred. The stabilized moment ($M_s$) referred to the moment when no new cracks appeared. Table 7 lists the experimental results for all beams tested.

**Table 7.** Experimental results for all beams.

| Beams | Failure Modes | $M_{cr}$ (kN·m) | $M_s$ (kN·m) | $M_u$ (kN·m) | $\Delta_{max}$ (mm) | $\omega_{100}$ kN (mm) | Number of Cracks |
|---|---|---|---|---|---|---|---|
| B0.56C60V1.0S3 | BFRP bars rupture | 13.50 | 21.07 | 51.85 | 32.23 | 0.72 | 7 |
| B0.77C60V1.0S3 | BFRP bars rupture | 14.10 | 23.40 | 73.28 | 35.23 | 0.70 | 8 |
| B1.15C60V1.0S3 | Concrete crushing | 14.25 | 25.74 | 101.34 | 44.32 | 0.37 | 10 |
| B1.65C60V1.0S3 | Concrete crushing | 15.00 | 28.23 | 101.43 | 46.83 | 0.35 | 10 |
| B1.15C60 | Concrete crushing | 9.30 | 18.14 | 93.48 | 44.31 | 0.75 | 7 |
| B1.15C60V0.5S3 | Concrete crushing | 13.50 | 21.27 | 94.92 | 46.03 | 0.52 | 9 |
| B1.15C60V1.5S3 | Concrete crushing | 16.50 | 27.93 | 106.77 | 44.42 | 0.33 | 11 |
| B1.15C60V1.0S4 | Concrete crushing | 15.00 | 25.50 | 103.53 | 46.04 | 0.35 | 10 |
| B1.15C60V1.0S5 | Concrete crushing | 15.00 | 26.82 | 104.37 | 45.50 | 0.34 | 11 |
| B1.15C30V1.0S3 | Concrete crushing | 9.75 | 22.90 | 80.50 | 46.45 | 0.50 | 10 |

Note: $M_{cr}$ is the cracking moment; $M_s$ is the stabilized moment; $M_u$ is the ultimate moment capacity of a beam; $\Delta_{max}$ is the deflection when the ultimate moment capacity is reached; $\omega_{100}$ kN is the crack width of a beam at 100 kN.

### 3.1. Failure Modes, Service Load Moment, and Ultimate Moment Capacity

Although the BFRP reinforcement ratios of the beam specimens were all greater than 1.4 times the balanced reinforcement ratio as recommended by ACI 440.1R-15, the beam specimens tested in this research exhibited two distinguished failure modes, which are concrete crushing and BFRP bar rupturing. Figure 5 depicts the two failure modes. BFRP bar rupturing occurred in specimens B0.56C60V1.0S-3 and B0.77C60V1.0S-3 (see Figure 5a), while all other beam specimens failed by concrete crushing. As can be seen from Table 7, the number of cracks of specimens B0.56C60V1.0S-3 and B0.77C60V1.0S-3 was less than that of other beam specimens, but their crack width was larger. For the beams that failed by BFRP bar rupture, the stiffness of the beams decreased rapidly after cracking, and the deflection increased sharply. Then the bearing capacity decreased suddenly before the ultimate failure, and BFRP bars were broken, which was companied by a loud sound. Beams exhibited no ductility under this failure mode, which shall be avoided in design. A beam that failed by concrete crushing is shown in Figure 5b. It can be seen that under such a failure mode, multiple cracks but with smaller widths occurred, and horizontal cracks were observed at the top of the beam section. As observed, the beam experienced the following cracking process before ultimate failure: first, small horizontal cracks appeared one by one at the top of the beam section; then, the small horizontal cracks gradually connected and formed a crack, which led to the bulge of concrete in the compression zone; finally, the ultimate moment capacity of the beam was reached. Therefore, the FRP-RC beams with concrete crushing exhibited good ductility [19,27].

The maximum crack width of FRP-RC beams under service load moment was larger than that of RC beams due to FRP bars possessing an anticorrosion property. The CSA code [35] recommends that the maximum crack width of FRP-RC beams in outdoor and indoor service environments shall be less than 0.5 mm and 0.7 mm, respectively. Coastal engineering structures, bridges, and other infrastructure, which frequently experience cyclic loading, are usually constructed in an outdoor service environment. Therefore, the service load moment ($M_{ser}$) of an FRP-RC beam was defined as the moment when the maximum crack width reached 0.5 mm. Figure 6 illustrates the service load moment ($M_{ser}$) and ultimate moment capacity ($M_u$) of all beams. It can be seen from Figure 6a that service load moment and ultimate moment capacity increases with the BFRP reinforcement ratios, but the influence of the BFRP reinforcement ratio on the ultimate moment capacity of beams with BFRP bars rupture was significantly higher than that of beams with concrete crushing. The reason is that the ultimate moment capacity of the beams with BFRP bars rupture was determined by the BFRP reinforcement ratio, while the ultimate moment capacity of beams with concrete crushing was dictated by concrete performances. Steel fibers made a significant contribution to improving the performance of concrete, including enhancing concrete's tensile strength, ultimate compressive strain, and bond strength, which is beneficial for improving the service load moment and ultimate moment capacity of the beams. The service load moment and ultimate moment capacity of the beam with 1.5% by volume steel fibers were 103.3% and 14.2%, respectively, higher than their counterparts of those beams without steel fibers, as shown in Figure 6b. Steel fibers significantly improved the serviceability of the beams under cyclic loading. From Figure 6c, the service performance and ultimate moment capacity of the beams increased with the increase in fiber length and the number of fiber hook-ends. Compared with 3D and 4D steel fibers reinforced beams, the load moment and ultimate moment capacity of beams with 5D steel fibers were higher. The concrete strength also significantly affects the flexural performances of the beams. As can be seen from Figure 6d, the service load moment and ultimate moment capacity of the beams with high strength concrete (i.e., Grade 60) were 17.1% and 25.9%, respectively, higher than those with low strength concrete (i.e., Grade 30).

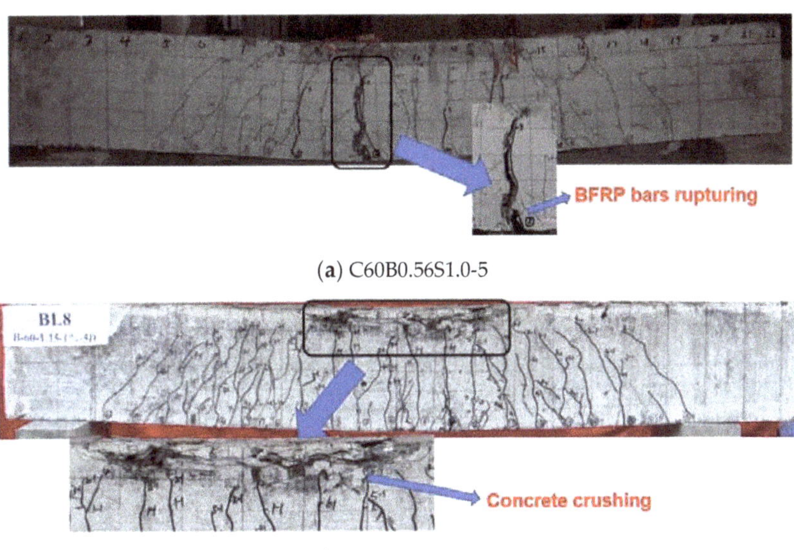


(**a**) C60B0.56S1.0-5

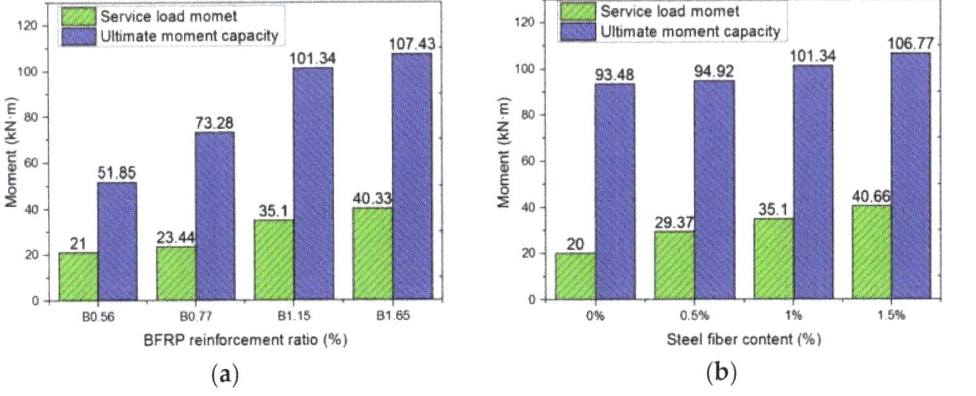


(**b**) C60B1.15S1.0-5

**Figure 5.** Failure modes of beams: (**a**) BFRP bars rupture; (**b**) concrete crushing.

**Figure 6.** *Cont.*

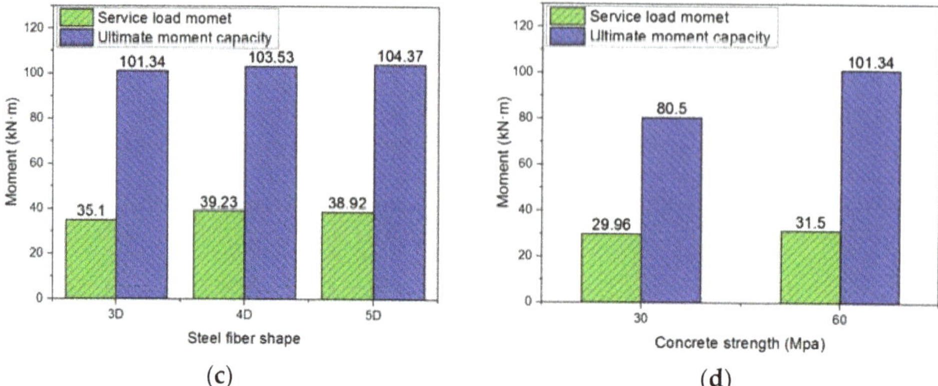



(**c**)  (**d**)

**Figure 6.** Service load moment and ultimate moment capacity of beams with respect to (**a**) BFRP reinforcement ratio; (**b**) steel fiber content; (**c**) steel fiber shape; (**d**) concrete strength.

### 3.2. Load-Deflection Curve and Envelope Curve

Load-deflection curves of beams under cyclic loading were commonly used for examining their flexural behaviours, from which the envelope curve, energy dissipation, residual deflection, stiffness, etc., was able to be derived. The envelope curve was the curve connecting the peak load of all cycles in the load-deflection curve of a beam under cyclic loading. The enclosed area in the load-deflection curve after the unloading–reloading cycle represented the energy dissipation of the beam under this unloading–reloading cycle. The residual deflection was defined as the irrecoverable deflection of a beam after the load was unloaded to 0. The load degradation coefficient meant the reduction coefficient of the peak load at the same displacement in different unloading–reloading cycles. Figure 7 presents the load-deflection curves of all beams tested in this research. The red curves, blue curves, and pink curves indicate the first cycle, the second cycle, and the third cycle envelope curves, respectively, of the load-deflection curves. The envelope curve was also an important index for studying the flexural performance of a beam under cyclic loading. From Figure 7, it is obvious that the load-deflection curves of all beams demonstrate the identical characteristics, i.e., all load-deflection curves increased linearly after cracking; the residual deflection increased with unloading–reloading cycles, especially at larger deflection; the peak load and energy consumption of the beam under the same deflection decreased gradually with the increase of loading–unloading cycles. Figure 8 reproduces the first cycle envelope curves for all beams.

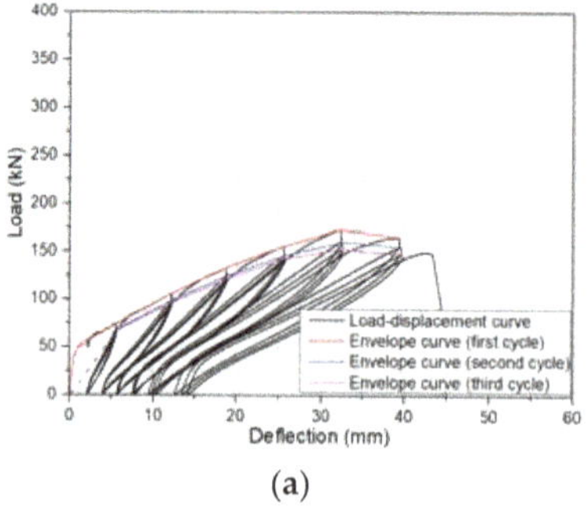



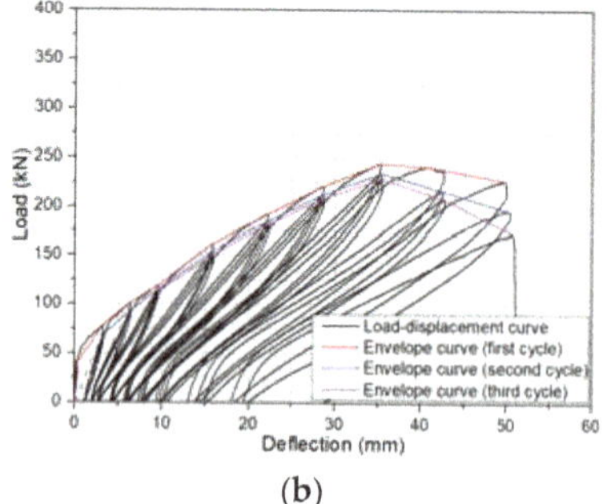



(**a**)  (**b**)

**Figure 7.** *Cont.*

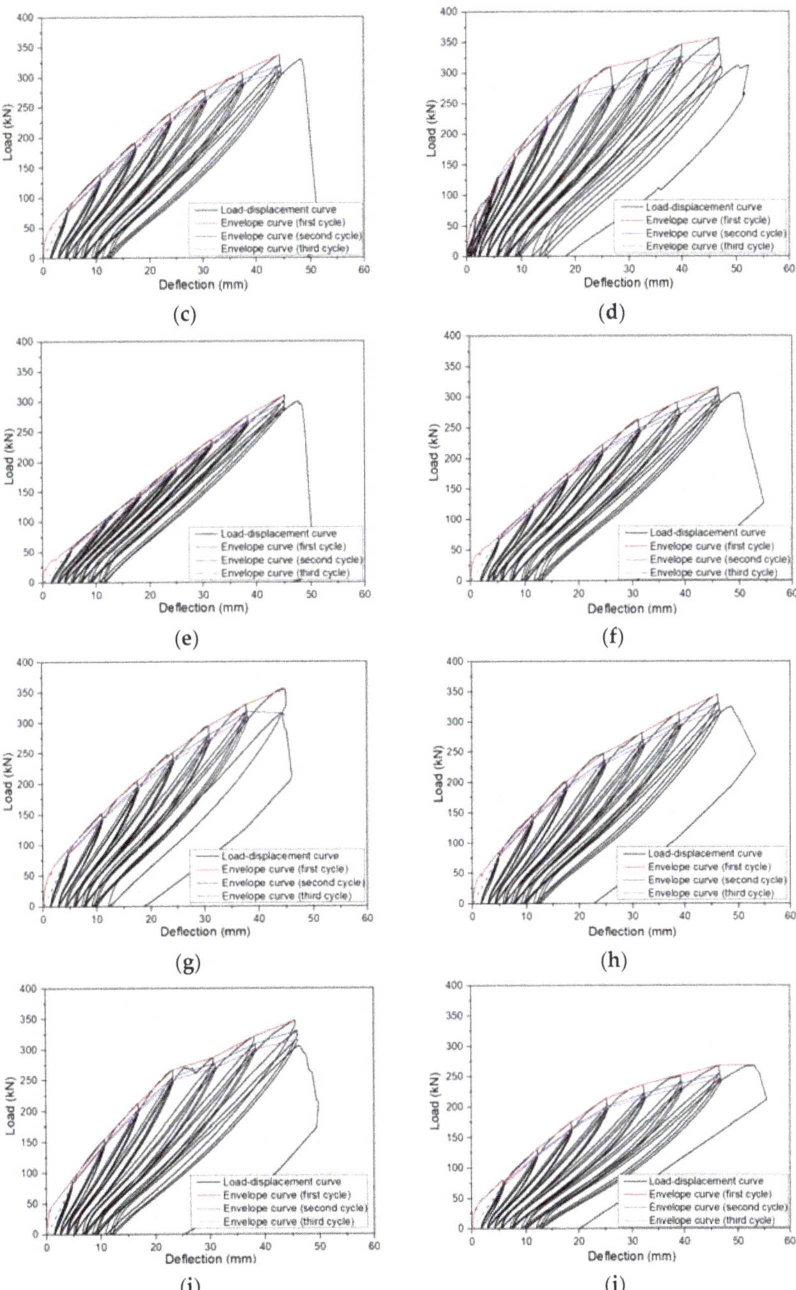

**Figure 7.** Load-deflection curves for all beams tested: (**a**) B0.56C60V1.0S3; (**b**) B0.77C60V1.0S3; (**c**) B1.15C60V1.0S3; (**d**) B1.65C60V1.0S3; (**e**) B1.15C60; (**f**) B1.15C60V0.5S3; (**g**) B1.15C60V1.5S3; (**h**) B1.15C60V1.0S4; (**i**) B1.15C60V1.0S5; (**j**) B1.15C30V1.0S3.

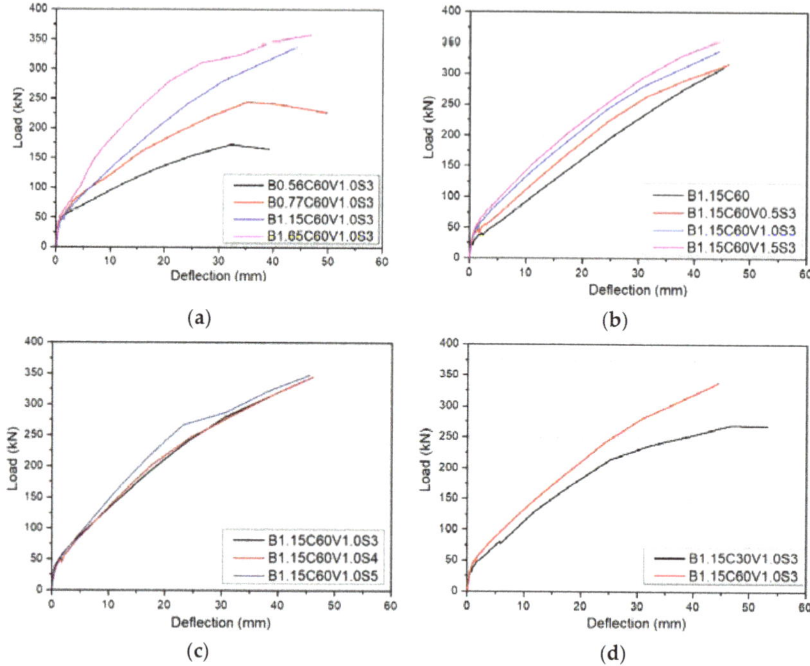

**Figure 8.** Envelope curve (first cycle) for beams with different: (**a**) BFRP reinforcement ratio; (**b**) steel-fiber volume fraction; (**c**) steel fiber shape; and (**d**) concrete strength.

### 3.2.1. Number of Unloading–Reloading Cycles

From Figure 7, it can be found that unloading–reloading cycles at the same stroke displacement significantly reduced the peak load of a beam. However, the peak load reduction rate decreased with the increase of unloading–reloading cycles. According to the experimental results, the average peak load of the second cycle was 3%~12% lower than that of the first cycle, while the average peak load of the third cycle was only 1%~5.38% lower than that of the second cycle. This was due to greater damage to the beams caused by the increase in loading in the first cycle, leading to the increase in crack width and height, the decrease of the effective area of concrete, and hence the reduction of stiffness. The peak load of the second and third unloading–reloading cycles was lower than that of the first unloading–reloading cycle. But the width and height of cracks after the second and the third unloading–reloading cycles were comparable to those after the first unloading–reloading cycle. The decrease in stiffness was only related to the internal damage and bond between concrete and BFRP bars. Therefore, the reduction rate of peak load was decreased with the increase of unloading–reloading cycles. For example, for beam B1.15C60V1.0S3 with a stroke displacement of 6 mm, the peak load degradation coefficient after the second and the third unloading–reloading cycles were 7.12% and 1.08%, respectively.

More importantly, the deflection of all beams increased with the increase of the number of unloading–reloading cycles under the same applied load. Table 8 shows the deflections of the beams at the first cycle and the deflections after three unloading–reloading cycles under the same applied load. It can be seen from Table 8 that after three loading and unloading cycles of the first stage under the same applied load, the deflections of the beams increased by 11% on average, but after three loading and unloading cycles of the second and third stages under the same applied load, the deflections of the beams increased by only 8% on average. The reason was that the skeleton curves of the beams were bilinear; due to the lower elastic modulus of the BFRP bars, the stress of the BFRP bars increased rapidly after concrete cracking, resulting in a large increase of the deflection after three loading and unloading cycles of the first stage. Therefore, the deflections of the beams under cyclic loading can be calculated by the following formula.

$$\Delta' = \Delta \times (1 + 11\%) \times (1 + 8\%)^{n-1} \tag{3}$$

where $\Delta_n'$ is the deflection of a beam after three cycles under cyclic loading, $\Delta$ is the deflection of the beam under static loading, and $n$ is the loading grade under cyclic loading ($n \geq 1$).

**Table 8.** Beam's deflections at the first cycle and after three loading–unloading cycles under the same applied load.

| Beams | Load (kN) | $\Delta_1$ (mm) | $\Delta_{1'}$ (mm) | $\Delta_1/\Delta_{1'}$ | Load (kN) | $\Delta_2$ (mm) | $\Delta_{2'}$ (mm) | $\Delta_2/\Delta_{2'}$ | Load (kN) | $\Delta_3$ (mm) | $\Delta_{3'}$ (mm) | $\Delta_3/\Delta_{3'}$ |
|---|---|---|---|---|---|---|---|---|---|---|---|---|
| B0.56C60V1.0S3 | 74.43 | 5.57 | 6.50 | 1.17 | 106.19 | 12.20 | 13.56 | 1.11 | 133.17 | 18.85 | 20.72 | 1.10 |
| B0.77C60V1.0S3 | 78.53 | 3.59 | 4.20 | 1.17 | 119.00 | 9.54 | 10.51 | 1.10 | 160.93 | 15.74 | 17.39 | 1.10 |
| B1.15C60V1.0S3 | 85.80 | 4.63 | 5.15 | 1.11 | 140.40 | 10.82 | 11.75 | 1.09 | 191.50 | 17.27 | 18.50 | 1.07 |
| B1.65C60V1.0S3 | 130.90 | 5.77 | 6.60 | 1.14 | 169.40 | 8.76 | 9.50 | 1.08 | 229.00 | 14.87 | 16.19 | 1.09 |
| B1.15C60 | 60.00 | 5.33 | 5.86 | 1.10 | 105.32 | 11.79 | 12.62 | 1.07 | 150.39 | 18.27 | 19.11 | 1.05 |
| B1.15C60V0.5S3 | 71.00 | 5.02 | 5.60 | 1.12 | 124.90 | 11.55 | 12.28 | 1.06 | 175.10 | 17.97 | 18.95 | 1.05 |
| B1.15C60V1.5S3 | 93.10 | 4.66 | 5.15 | 1.11 | 152.50 | 11.00 | 11.86 | 1.08 | 206.20 | 17.51 | 18.89 | 1.08 |
| B1.15C60V1.0S4 | 85.00 | 4.88 | 5.30 | 1.09 | 147.40 | 11.19 | 11.86 | 1.06 | 201.70 | 17.52 | 18.70 | 1.07 |
| B1.15C60V1.0S5 | 89.40 | 4.75 | 5.20 | 1.09 | 155.40 | 10.77 | 11.45 | 1.06 | 212.90 | 16.85 | 17.90 | 1.06 |
| B1.15C30V1.0S3 | 76.33 | 5.79 | 6.32 | 1.09 | 130.28 | 12.18 | 13.27 | 1.09 | 175.49 | 18.62 | 20.10 | 1.08 |
| Average value | | | | 1.12 | | | | 1.08 | | | | 1.08 |

Note: $\Delta_m$ is the deflection of a beam at the first cycle under the mth loading stage, and $\Delta_m'$ is the deflection of the beam after the third unloading–reloading cycle under the mth loading stage.

### 3.2.2. BFRP Reinforcement Ratio

Compared with other variables, the BFRP reinforcement ratio had the greatest influence on both the envelope and the load-displacement curves. The BFRP reinforcement ratio directly affects the failure modes of beams under bending. There was an obvious difference between the load-deflection curves of beams that failed by concrete crushing and those that failed by the rupture of BFRP bars. For the beams failed by BFRP bar ruptures, the crack width and height developed rapidly after cracking, which caused the slope of the load-deflection curves to decrease rapidly. More importantly, the peak load decreased with the increase in deflection. For beams B0.56C60V1.0S3 and B0.77C60V1.0S3, the peak load reached the maximum when the displacement was 36 mm, but the peak load decreased at the displacement of 42 mm, as shown in Figure 8a. Therefore, the beams failed by BFRP bars rupture exhibited poor ductility. The slope of the load-deflection curves of the beams with concrete crushing decreased gently after cracking. Moreover, the energy consumption of the beams that failed by concrete crushing was much higher than that of the beams that failed by BFRP bars rupture. The deflection of beams B0.77C60V1.0S3, B1.15C60V1.0S3, and B1.65C60V1.0S3 was 39.57%, 43.78%, and 62.95%, respectively, lower than that of beam B0.56C60V1.0S3 at the applied load of 110 kN.

### 3.2.3. Steel Fiber Volume Fraction and Shape

Remarkably, the envelope curve of the BFRP-RC beams with steel fibers was different from that of the BFRP-RC beam without steel fibers, as shown in Figure 8b,c. The slope of the first cycle envelope curve of the BFRP-RC beams with steel fibers decreased slowly after cracking, and the first cycle envelope curves were approximately trilinear. However, the first cycle envelope curve of the BFRP-RC beam without steel fibers is approximately bilinear. From Figure 8b, it can be seen that the envelope curve of beam B1.15C60 had the same features as the BFRP-RC beams with steel fibers before cracking, but the bridging effect of steel fibers after cracking limited the further development of crack width and height, which resulted in the stiffness of the beam decreased slowly and the slope of the first cycle envelope curve reduced slowly. The steel fiber volume fraction of beam B1.15C60V1.0S3 was 1.5% which was higher than that of beam B1.15C60, and the deflection of the former was reduced by 48.18% at 110 kN applied load compared with the latter. The increase in the number of fiber hook-ends was beneficial to improve the stiffness and thus reduce the deflection of the beam. As shown in Figure 8c, when the number of steel fiber hook-ends increased from 1 to 2 (i.e., from 3D to 5D), the deflection of the beam reduced by 11.56% at the applied load of 110 kN.

### 3.2.4. Concrete Strength

From Figure 8d, the envelope curves of beams with high strength concrete and low strength concrete exhibited the same features during both the elastic growth stage and the rising plastic stage. The load was shared by BFRP bars and concrete matrix before cracking, and the envelope curves demonstrated a linear increase manner until crack occurred. The stress and strain of the BFRP bars increase linearly. After cracks appeared, the first cycle envelope curve increased with the increase of deflection until failure, but the slope of the first cycle envelope curve decreased. This was because the width and height of the crack increased with the increase of deflection, resulting in the reduction of the stiffness of beams. High strength concrete had higher Young's modulus and tensile strength than low strength concrete, resulting in that the stiffness of beam B1.15C60V1.0S3 was larger than that of beam B1.15C30V1.0S3. Moreover, the area surrounded by the load-deflection curve of beam B1.15C60V1.0S3 was higher than that of beam B1.15C30V1.0S3, suggesting that increasing concrete strength is beneficial for increasing energy consumption, improving stiffness, and reducing deflection of the beam. Compared with beam B1.15C30V1.0S3 with low strength concrete grade 30, the energy dissipation of beam B1.15C60V1.0S3 with high strength concrete grade 60 increased by 2.67% before failure, but the deflection of the beam at 110 kN was reduced by 17.54%. Therefore, the deflection of FRP-RC beams can be effectively reduced by increasing concrete strength [21–23].

### 3.3. Residual Deflection

The residual deflection was defined as the irrecoverable deflection of a beam after the load was unloaded to 0 [36]. Figure 9 presents the load-residual deflection curves of beams. From Figure 9, it can be found that the residual deflection of all beams increased with the increase of the applied load and the number of unloading–reloading cycles under the same deflection. Moreover, the influence of the number of unloading–reloading cycles on the residual deflection became more significant under higher load. For beam B1.15C60V1.0S3, the residual deflection after the third unloading–reloading cycle was only 5.15% higher than that of the first loading when the deflection was 6 mm, while the residual deflection after the third unloading–reloading cycle was nearly 10% higher than that of the first loading when the deflection increased to 42 mm. The residual deflection of other beams demonstrated a similar trend. The reason was that the stiffness of a beam was larger at the initial stage of loading, the unloading–reloading cycle had a little cumulative effect on the internal damage of the beam, but the stiffness of the beam degraded rapidly at the later stage of loading, the damage accumulation of concrete and BFRP bars increased with the increase of a number of unloading–reloading cycles, resulting in larger residual deflection.

Figure 10 illustrates the load-residual deflection curves of the beams under the first unloading–reloading cycle. The influences of the four variables on the load-residual deflection curves of the beams are elaborated in Figure 10. The BFRP reinforcement ratio had the greatest influence on the load-residual deflection curves. The stress growth rate of beams with a high BFRP reinforcement ratio was lower than that of beams with a low BFRP reinforcement ratio after cracking. Therefore, beams with a low reinforcement ratio had a larger residual deflection. The residual deflection of B0.77C60V1.0S3, B1.15C60V1.0S3, and B1.65C60V1.0S3 under the 110 kN load was 40.31%, 62.61%, and 76.13%, respectively, lower than that of B0.56C60V1.0S3.

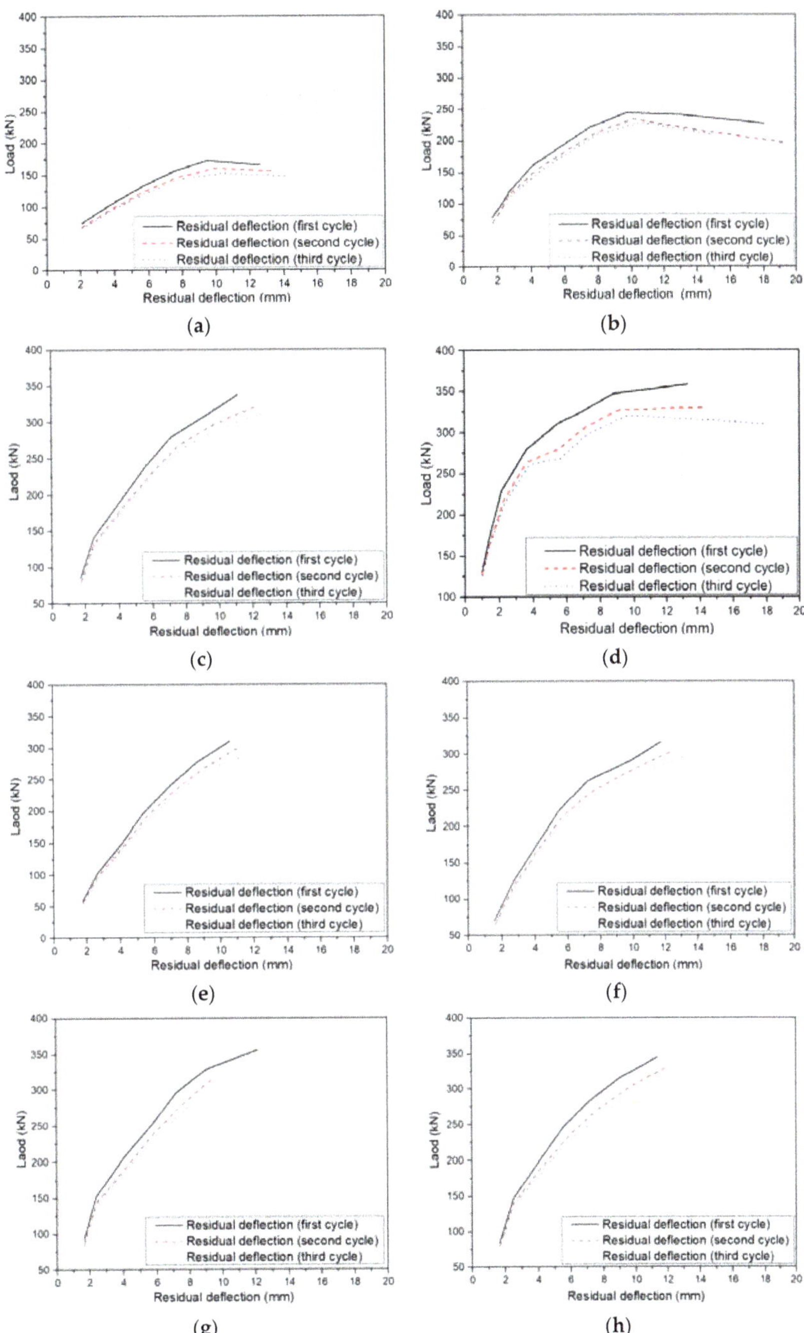

**Figure 9.** *Cont.*

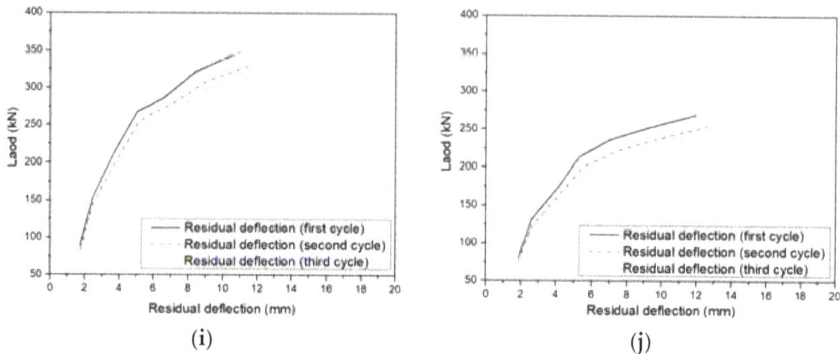

(i)          (j)

**Figure 9.** Load-residual deflection curves for beams: (**a**) B0.56C60V1.0S3; (**b**) B0.77C60V1.0S3; (**c**) B1.15C60V1.0S3; (**d**) B1.65C60V1.0S3; (**e**) B1.15C60; (**f**) B1.15C60V0.5S3; (**g**) B1.15C60V1.5S3; (**h**) B1.15C60V1.0S4; (**i**) B1.15C60V1.0S5; (**j**) B1.15C30V1.0S3.

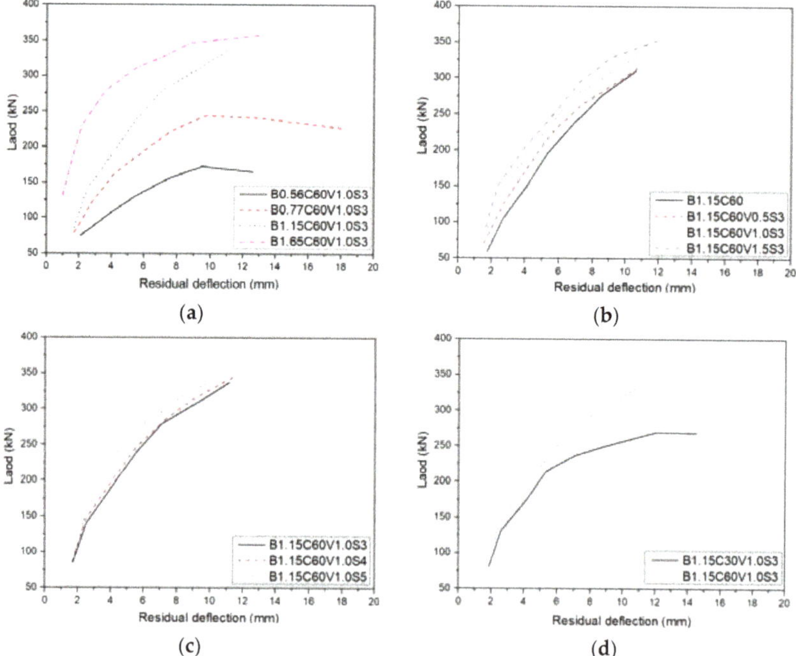

(a)          (b)

(c)          (d)

**Figure 10.** Load-residual deflection (first cycle) for beams with different: (**a**) BFRP reinforcement ratio; (**b**) steel fiber content; (**c**) steel fiber shape; (**d**) concrete strength.

Secondly, concrete strength and steel-fiber volume fraction also had significant effects on the load-residual deflection curves. From Figure 10a,b, the residual strength of the beam decreased with the increase of concrete strength and steel-fiber volume fraction. Compared with beam B1.15C30V1.0S3 with low-strength concrete grade 30, the residual deflection of beam B1.15C60V1.0S3 with high-strength concrete grade 60 at 110 kN reduced by increased by 5.56%. The increase of steel-fiber volume fraction increased the bridging action between concrete and steel fibers, which hindered the development of concrete cracking in the tensile zone, and then enhanced the stiffness of the beams, therefore reducing their residual deflection [36]. Compared with beam B1.15C60 without steel fibers, the residual deflection of beam B1.15C60V1.5S3 with a steel-fiber volume fraction of 1.5% at 110 kN applied load was reduced by 30.36%.

Finally, the steel fiber shape had the least influence on the load-residual deflection curves, as shown in Figure 10d. 3D, 4D, and 5D steel fibers all had the same fiber aspect ratio but different lengths, numbers of hook-ends, and tensile strength. The results indicated that all three types of steel fibers had a bond-slip failure, and no steel fibers were broken during testing. Longer fibers and more hook-ends are both beneficial for improving the bond between concrete and steel fibers, but the strength of steel fibers had little influence on the bond between concrete and fibers. Therefore, the residual deflection of the beams with 5D steel fibers was lower than that of beams with 3D steel fibers. However, the influence of steel fiber shape on the residual deflection was less obvious than the other three variables.

### 3.4. Stiffness Degradation

The deflection of the beams increased with the increase of the unloading–reloading cycles under the same applied load, which was called stiffness degradation. According to JGJ/T 101-2015 [37], the stiffness of a beam is expressed by secant stiffness $K_{ij}$, which can be calculated by the following equation.

$$K_{ij} = \frac{\left|+F_{ij}\right| + \left|-F'_{ij}\right|}{\left|+\Delta_{ij}\right| + \left|-\Delta'_{ij}\right|} \tag{4}$$

where $F_{ij}$ represents the peak load of the $j_{th}$ cycle under a displacement of $i_{th}$; $\Delta_{ij}$ represents the largest displacement of the $j_{th}$ cycle under a displacement of $i_{th}$; $F_{ij}'$ represents the minimum load of the $j_{th}$ cycle under a displacement of $i_{th}$; $\Delta_{ij}'$ represents the residual deflection of the $j_{th}$ cycle under a displacement of $i_{th}$; $j$ is the number of cycles under a displacement of $i_{th}$, where $j$ is less than or equal to 3 in this study.

As the loading mode was cyclic in this research, $F_{ij}'' = 0$, $\Delta_{ij}' = 0$. Therefore, the secant stiffness $K_{ij}$ can be simplified as the following equation:

$$K_{ij} = \frac{\left|+F_{ij}\right|}{\left|+\Delta_{ij}\right|} \tag{5}$$

Figure 11 depicts the stiffness–displacement curves of all beams. The stiffness of the beams decreased with the increase of displacement, and the stiffness degradation rate decreased with the increase of displacement. In particular, the stiffness degradation rate was the highest from the initiation of cracking to an actuator displacement of 6 mm. The stiffness of the beams remained unchanged before cracking. After cracking to an actuator displacement of 6 mm, the crack width and height increased rapidly, and the effective section of a beam decreased accordingly, leading to a higher rate of stiffness degradation. When the actuator displacement reached 6 mm, the crack height of a beam changed little, and the stiffness degradation was small. Noticeably, increasing the number of unloading–reloading cycles decreased the stiffness under the same deflection, but the stiffness degradation rate of beams decreased. The stiffness of the beams in the second cycle was 4.00% lower than in the first cycle under the same deflection, and their stiffness in the third cycle was 1.59% lower than in the second cycle. The main reason for this was that after the first cycle, new cracks appeared, and old cracks further developed, leading to rapid stiffness degradation. The peak load of the second cycle decreased under the same displacement, and no new cracks appeared, which had little effect on the stiffness of the beams.

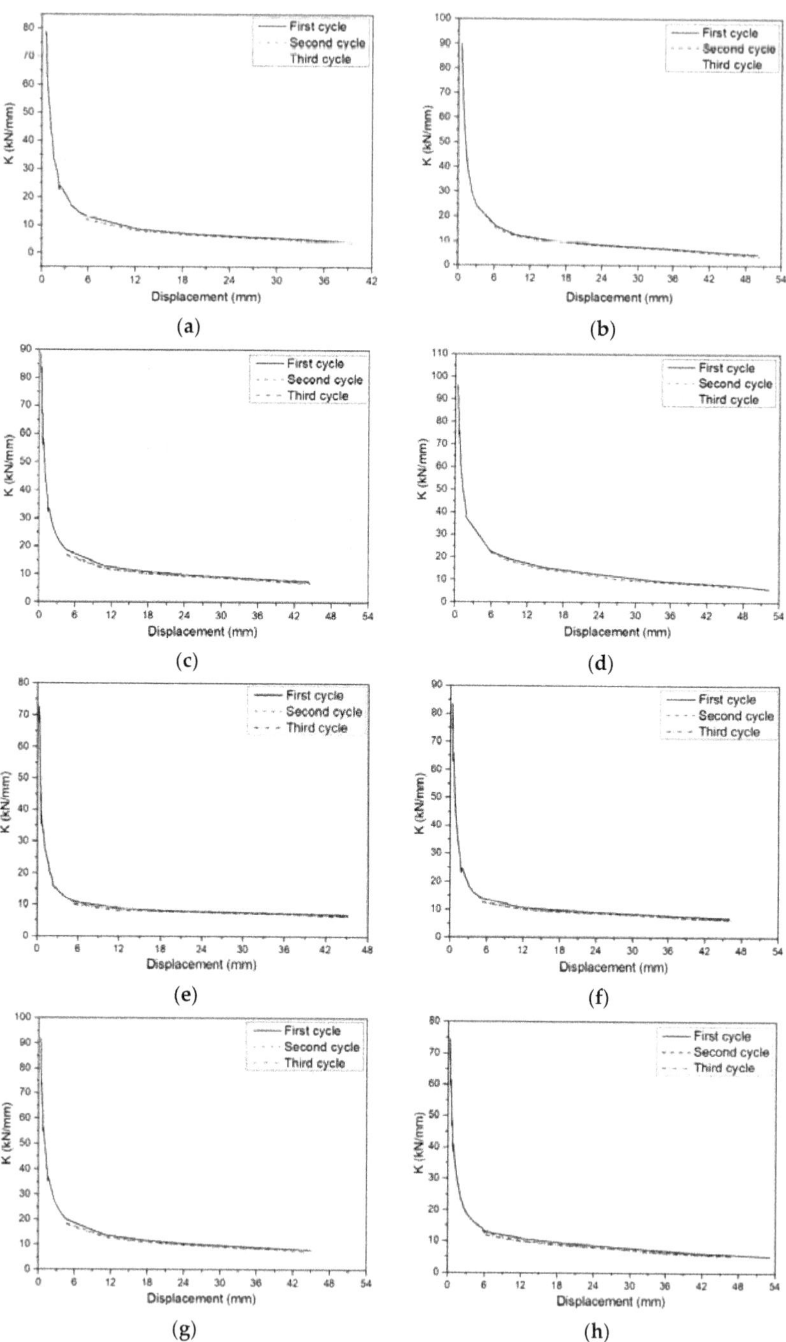

**Figure 11.** *Cont.*

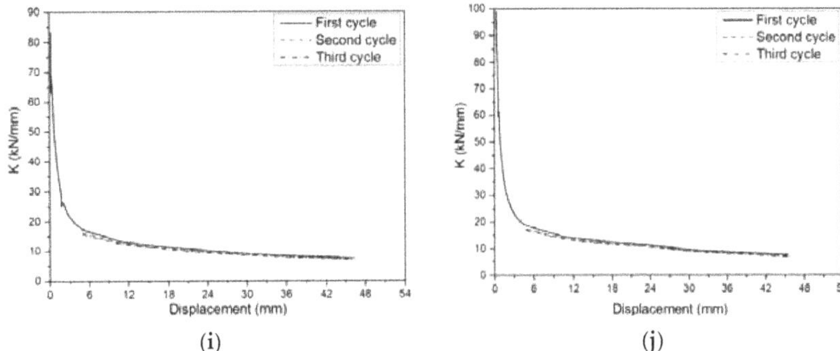

**Figure 11.** Stiffness–displacement curves: (**a**) B0.56C60V1.0S3; (**b**) B0.77C60V1.0S3; (**c**) B1.15C60V1.0S3; (**d**) B1.65C60V1.0S3; (**e**) B1.15C60; (**f**) B1.15C60V0.5S3; (**g**) B1.15C60V1.5S3; (**h**) B1.15C30V1.0S3; (**i**) B1.15C60V1.0S4; (**j**) B1.15C60V1.0S5.

Figure 12 illustrates stiffness–displacement curves of the beams in the first cycle under different variables. The stiffness had increased with the increase of the BFRP reinforcement ratio, but the stiffness degradation rate decreased. After cracking, the restraint force of the beams with a higher reinforcement ratio on crack width expansion was higher than that of the beams with a lower reinforcement ratio, so the stiffness degradation rate of beams with a higher reinforcement ratio was reduced. The increase of steel-fiber volume fraction and number of hook-ends helped to enhance the stiffness. From Figure 12b,c, it can be found that steel-fiber volume fraction and the number of hook-ends had a significant effect on the stiffness–displacement curve of the beams in the early stage of loading, but the effect became less significant in the later stage. Increasing volume fraction and number of hook-ends of steel fibers was beneficial for improving the tensile strength of concrete, and the random distribution of fibers helped to hinder the further development of cracks, thus reducing the deflection of the beams in the early stage of loading. The effects of fibers on deflection were reduced at the later stage of loading because most steel fibers in the tensile zone were pulled out at the ultimate failure. 5D steel fibers had higher tensile strength and more hook-ends than the 3D and 4D steel fibers. Therefore, the bond strength between concrete and fibers was higher than other steel fibers, which made the stiffness of beam B1.15C60V1.0S5 higher than that of beams B1.15C60V1.0S3 and B1.15C60V1.0S4. The effect of concrete strength on beam stiffness–displacement curves are depicted in Figure 12d. Increasing concrete strength can increase the stiffness of the beams, but it has little effect on the stiffness degradation rate.

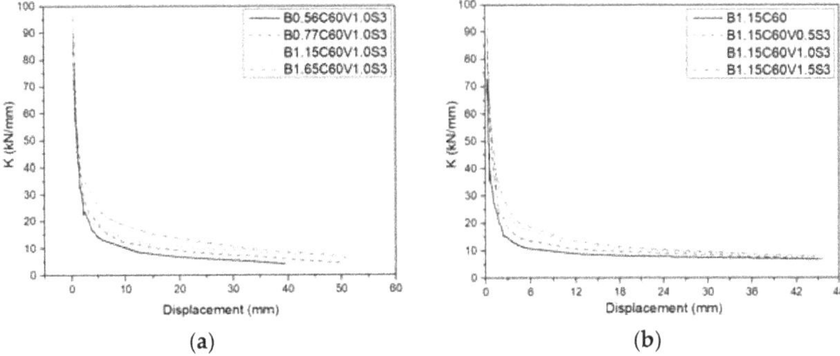

**Figure 12.** *Cont.*

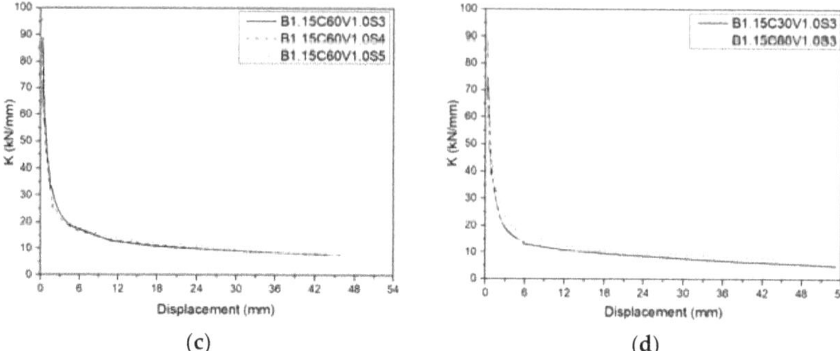

(c)                                       (d)

**Figure 12.** Stiffness–displacement curves of beams at the first loading–unloading cycle with respect to (**a**) reinforcement ratio; (**b**) steel-fiber volume fraction; (**c**) steel fiber shape; (**d**) concrete strength.

## 4. Experimental Results versus Model Prediction

FRP-RC beams usually possess larger deflection than RC beams due to the FRP bars having a lower Young's modulus. In this regard, the serviceability limit states usually control the structural design of FRP-RC beams. Controlling the deformation of FRP-RC beams under cyclic loading is particularly important for design. At present, most studies and design codes use the effective moment of inertia method to evaluate the deflection of FRP-RC beams under static loading, which is also used in this paper to predict and evaluate the deflections of the BFRP-RC beams with steel fibers under cyclic loading. Results from various analytical models/empirical equations were compared with experimental results from this research, through which the analytical model/empirical equations were evaluated for their appropriateness for calculating the deflection of the BFRP-RC beams with steel fibers under cyclic loading. Table 7 summarizes the comparisons between the experimental and theoretical results of the deflection of beams tested at a crack width of 0.5 mm.

### 4.1. Theoretical Calculation of Deflection of FRP-RC Beams

To simplify the analysis, the following assumptions were taken when evaluating the deflection of FRP-RC beams.

(1)    A beam section is homogeneous before concrete cracking, and the contribution of the BFRP bars to the total moment of inertia of a beam section is neglected. Therefore, the total moment of inertia ($I_g$) can be obtained by the following equation.

$$I_g = \frac{bh^3}{12} \tag{6}$$

(2)    After a crack is initiated in concrete, the contribution of the concrete in the tension zone is neglected. Therefore, the moment of inertia ($I_{cr}$) of the cracked beam section can be obtained by the following equation.

$$I_{cr} = \frac{b}{3}d^3k^3 + n_f A_f d^2 (1 - k)^2 \tag{7}$$

$$k = \sqrt{2\rho_f n_f + (\rho_f n_f)^2} - \rho_f n_f \tag{8}$$

$$n_f = \frac{E_f}{E_c} \tag{9}$$

where $d$ is the effective depth of the beam section, $k$ is the ratio of the depth of the neutral axis to the depth of reinforcement bars, $n_f$ is the ratio of Young's modulus of FRP bars to the modulus of elasticity of concrete, $E_c$ is the Young's modulus of concrete, $E_f$ is the Young's modulus of FRP bars, and $\rho_f$ is the FRP reinforcement ratio.

Currently, there are various calculation models for the effective moment of inertia ($I_e$) of an FRP-RC beam section. Bischoff [38,39] recommended that the effective moment of inertia ($I_{e,bischoff}$) of an FRP-RC beam section can be obtained by Equation (10).

$$I_{e,bischoff} = \frac{I_{cr}}{1 - \left(1 - \frac{I_{cr}}{I_g}\right)\left(\frac{M_{cr}}{M_a}\right)^2} \tag{10}$$

where $M_{cr}$ is the cracking moment, and $M_a$ is the applied moment.

According to Benmokrane et al. [40], the effective moment of inertia ($I_{e,benmokrane}$) of an FRP-RC beam section can be evaluated by Equation (11).

$$I_{e,benmokrane} = \left(\frac{M_{cr}}{M_a}\right)^3 \frac{I_g}{7} + 0.84\left[1 - \left(\frac{M_{cr}}{M_a}\right)^3\right]I_{cr} \leq I_g \tag{11}$$

Alsayed et al. [41] proposed that the effective moment of inertia ($I_{e,alsayed}$) of an FRP-RC beam section can be evaluated by Equation (12).

$$I_{e,alsayed} = \left(1.4 - \frac{2}{15}\left(\frac{M_a}{M_{cr}}\right)\right)I_{cr} \quad for \quad 1 < \frac{M_a}{M_{cr}} < 3$$
$$I_{e,alsayed} = I_{cr} \qquad\qquad for \qquad 3 < \frac{M_a}{M_{cr}} \tag{12}$$

Canadian ISIS [42] code recommends that the effective moment of inertia ($I_{ISIS}$) of an FRP-RC beam section can be evaluated by Equation (13).

$$I_e = \frac{I_g I_{cr}}{I_{cr} + \left[1 - 0.5\left(\frac{M_{cr}}{M_a}\right)^2\right](I_g - I_{cr})} \tag{13}$$

Combined with classical beam theory and the effective moment of inertia method, the mid-span deflection ($\Delta$) of an FRP-RC beam can be obtained by Equation (14).

$$\Delta = \frac{p l_a}{48 E_c I_e}(3l_o^2 - 4l_a^2) \tag{14}$$

where $p$ is the applied load, $l_a$ is the shear span, and $l_o$ is the clear span.

Using the various effective moment of inertia formulas in literature and design codes summarized above, combined with the deflection calculation method from the classical beam theory, the deflection of a BFRP-RC beam with steel fibers at 0.5 mm crack width can be obtained. However, none of the above analytical models for the effective moment of inertia considers the positive contribution of steel fibers to the moment of inertia of the section of a beam strengthened with BFRP bars. Indeed, for a BFRP-RC beam with steel fibers, the contribution of steel fibers in the concrete tensile zone cannot be neglected, because steel fibers reinforced concrete can bear large tensile stress after concrete cracking.

### 4.2. A New Model for the Deflection of the BFRP-RC Beams with Steel Fibers

As elaborated above, when calculating the deflections of the BFRP-RC beams with steel fibers, the effect of steel fibers on the deflection should be considered. However, the influences of steel fibers on the deflection of a BFRP-RC beam with steel fibers should be considered after concrete cracking [19]. The distance from the center of the mass of a fiber to the neutral axis of the beam cannot be calculated, resulting in the inability to obtain its area and moment of inertia. Rather, some scholars believe that the steel fibers in a beam section can be taken as a whole, which can obtain its area and moment of inertia [19]. The distribution and orientation of steel fibers dictate the concrete's performance, especially at the post-cracking stage. Generally, the distribution of steel fibers is described by the non-uniformity coefficient $\eta_v$, while the orientation is by the orientation coefficient $\eta_0$.

Zhu [19] recommends that the total area of steel fibers in an SFRC beam is obtained by the following equations:

$$A_{sf} = \eta_0 \eta_v bh\rho_{sf} = \eta bh\rho_{sf} \tag{15}$$

where $\eta$ is the steel fiber effective coefficient (in this research, $\eta$ was taken as 0.16).

Figure 13a depicts the gross section and transformed an uncracked section of a beam. Since the area moments of the compression and tension zones of the beam are equal, Equations (16) and (17) can be derived as follows:

$$\frac{bx_0^2}{2} + \frac{(n_{sf}-1)bx_0^2 A_{sf}}{2h} = \frac{b(h-x_0)^2}{2} + (n_{sf}-1)A_{sf}\frac{(h-x_0)^2}{2h} + (n_f-1)A_f(d-x_0) \tag{16}$$

$$x_0 = \frac{\frac{bh^2}{2} + (n_f-1)A_f d + \frac{(n_{sf}-1)A_{sf}h}{2h}}{bh + (n_f-1)A_f + (n_{sf}-1)A_{sf}} \tag{17}$$

$$n_{sf} = \frac{E_{sf}}{E_c} \tag{18}$$

The moment of inertia of the gross section ($I_g$) of a BFRP-RC beam with steel fibers is calculated by the following equation.

$$I_g = \frac{b}{3}\left[x_0^3 + (h-x_0)^3\right] + (n_f-1)A_f(d-x_0)^2 + \frac{(n_{sf}-1)A_{sf}}{3h}\left[x_0^3 + (h-x_0)^3\right] \tag{19}$$

Figure 13b describes the cracked and transformed cracked sections of a beam. Since the area moments of the compression and tension zones of the beam are equal, Equations (20) and (21) can be derived as follows:

$$\frac{bx_{cr}^2}{2} + \frac{(n_{sf}-1)bx_{cr}^2 A_{sf}}{2h} = n_{sf}A_{sf}(d-x_{cr}) + \frac{n_{sf}(h-x_{cr})^2 A_{sf}}{2h} \tag{20}$$

$$x_{cr} = \frac{-(n_{sf}A_{sf}+n_f A_f) + \sqrt{(n_{sf}A_{sf}+n_f A_f)^2 + 2(b-\frac{A_{sf}}{h})(\frac{n_{sf}}{2}hA_{sf}+n_f A_f d)}}{b - \frac{A_{sf}}{h}} \tag{21}$$

The moment of inertia of the cracked section ($I_{cr}$) of a BFRP-RC beam with steel fibers is calculated by the following equation.

$$I_{cr} = \frac{b}{3}x_{cr}^3 + n_f A_f(d-x_{cr})^2 + \frac{n_{sf}A_{sf}}{3h}(h-x_{cr})^3 \tag{22}$$

After the moment of inertia of the gross and cracked section of a BFRP-RC beam with steel fibers was obtained, the deflection of the beam can be calculated by introducing Equations (13) and (14).

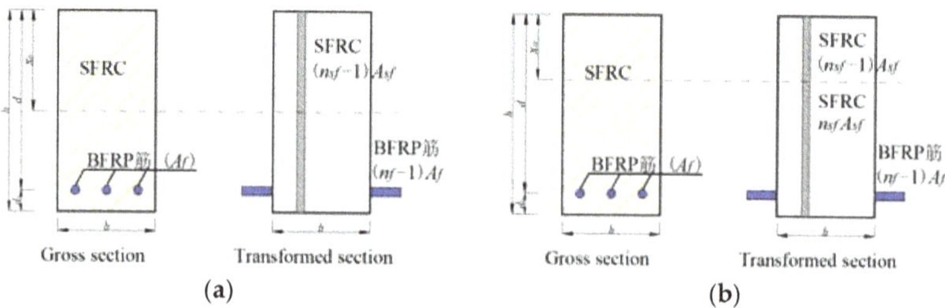


**Figure 13.** Sectional parameters of the gross and cracked sections: (**a**) gross section; (**b**) cracked section. "筋" = steel.

According to the loading regime adopted in this research, the deformation of the beam can be checked according to the static loading before the stroke of the actuator reached 6 mm, but the cyclic loading effect shall be considered after the actuator's stroke reached 6 mm. Table 9 lists the deflections of all beams investigated in this research when the stroke of the actuator reached 6 mm from experiment and calculation from various models in literature and design codes, as well as the new analytical model established in this study. From Table 9, the calculated deflections from the Bischoff, Benmokrane, Alsayed, and Canadian ISIS models are 13%~49% higher than the experimental value. While the average deflections calculated by the proposed analytical model in this paper are only 9% higher than the experimental ones, and the coefficient of variation is only 0.22, which is lower than the coefficient of variation of any other existing model investigated. Therefore, the analytical model proposed in this paper is more reliable and accurate for evaluating the deflection of the BFRP-RC beams with steel fibers.

**Table 9.** Deflections of the beams from the experiment and calculated from the proposed analytical momodelhen the stroke of the actuator reached 6 mm.

| Beams | $F_1$ (kN) | $\Delta_1$ (mm) | $\Delta_c$ (mm) | $\Delta_c/\Delta_1$ | $\Delta_{Benm.}$ (mm) | $\Delta_{Benm.}/\Delta_1$ | $\Delta_{Bisch.}$ (mm) | $\Delta_{Bisch.}/\Delta_1$ | $\Delta_{Alsa}$ (mm) | $\Delta_{Alsa}/\Delta_1$ | $\Delta_{ISIS}$ (mm) | $\Delta_{ISIS}/\Delta_1$ |
|---|---|---|---|---|---|---|---|---|---|---|---|---|
| B0.56C60V1.0S3 | 74.43 | 5.57 | 6.6 | 1.20 | 9.14 | 1.64 | 7.84 | 1.41 | 9.92 | 1.78 | 9.73 | 1.75 |
| B0.77C60V1.0S3 | 78.53 | 3.59 | 5.84 | 1.63 | 8.02 | 2.23 | 6.22 | 1.73 | 7.96 | 2.22 | 7.79 | 2.17 |
| B1.15C60V1.0S3 | 85.80 | 4.63 | 5.02 | 1.08 | 7.10 | 1.53 | 5.11 | 1.10 | 6.12 | 1.32 | 6.10 | 1.32 |
| B1.65C60V1.0S3 | 130.90 | 5.77 | 7.58 | 1.31 | 9.63 | 1.67 | 8.36 | 1.45 | 9.16 | 1.59 | 8.99 | 1.56 |
| B1.15C60 | 60.00 | 5.33 | 4.37 | 0.91 | 5.03 | 0.94 | 3.82 | 0.72 | 4.40 | 0.83 | 4.42 | 0.83 |
| B1.15C60V0.5S3 | 71.00 | 5.02 | 4.34 | 0.86 | 5.72 | 1.14 | 3.74 | 0.74 | 4.95 | 0.99 | 4.81 | 0.96 |
| B1.15C60V1.5S3 | 93.10 | 4.66 | 4.91 | 1.05 | 7.67 | 1.65 | 5.25 | 1.13 | 6.55 | 1.41 | 6.47 | 1.39 |
| B1.15C60V1.0S4 | 85.00 | 4.88 | 4.89 | 1.00 | 7.05 | 1.44 | 4.83 | 0.99 | 6.00 | 1.23 | 5.94 | 1.22 |
| B1.15C60V1.0S5 | 89.40 | 4.75 | 5.2 | 1.09 | 7.38 | 1.55 | 5.27 | 1.11 | 6.35 | 1.34 | 6.33 | 1.33 |
| B1.15C30V1.0S3 | 76.33 | 5.79 | 5.22 | 0.84 | 6.82 | 1.18 | 5.40 | 0.93 | 6.39 | 1.10 | 6.33 | 1.09 |
| Average value | | | | 1.09 | | 1.50 | | 1.13 | | 1.38 | | 1.36 |
| Coefficient of variation | | | | 0.22 | | 0.34 | | 0.30 | | 0.38 | | 0.37 |

Note: $F_1$ is the applied load on the beams when the stroke of the actuator reached 6 mm; $\Delta_1$ is the deflection of the beams when the stroke of the actuator reached 6 mm; $\Delta_c$ is the deflection of the beams calculated from the analytical model established in this paper; $\Delta_{Benm.}$ is the deflection of the beams calculated using Benmokrane's model; $\Delta_{Bisch.}$ is the deflection of the beams calculated using Bischoff's model; $\Delta_{Alsa}$ is the deflection of the beams calculated from Alsayed's model; $\Delta_{ISIS}$ is the deflection of the beams calculated using the Canadian ISIS model.

According to the loading regime adopted in this study, the cyclic loading effect shall be considered after the stroke of the actuator reached 6 mm. In sum, the deflection of a beam under a certain load can be calculated by using Equations (13), (14), (19) and (22). By introducing the calculated results into Equation (1), the deflection of the beam under cyclic loading can be calculated. Table 10 compares the calculated deflection and the actual deflection of the beams after three loading and unloading cycles. From Table 10, it can be seen that the ratio between the calculated value from the model to the counterpart from the experiment was 0.99, and the coefficient of variation was 0.16, suggesting that the analytical model proposed in this paper can accurately evaluate the deflection of the BFRP-RC beams with steel fibers under cyclic loading.

**Table 10.** Deflections of the beams were obtained from the experiment and calculated from the proposed analytical model after three loading and unloading cycles.

| Beams | $F_1$ (kN) | $\Delta_{1'}$ (mm) | $\Delta_{c1'}$ (mm) | $\Delta_{c1'}/\Delta_{1'}$ | $F_2$ (kN) | $\Delta_{2'}$ (mm) | $\Delta_{c2'}$ (mm) | $\Delta_{c2'}/\Delta_{2'}$ | $F_3$ (kN) | $\Delta_{3'}$ (mm) | $\Delta_{c3'}$ (mm) | $\Delta_{c3'}/\Delta_{3'}$ |
|---|---|---|---|---|---|---|---|---|---|---|---|---|
| B0.56C60V1.0S3 | 74.43 | 6.50 | 7.33 | 1.13 | 106.19 | 13.56 | 12.43 | 0.92 | 133.17 | 20.72 | 17.35 | 0.84 |
| B0.77C60V1.0S3 | 78.53 | 4.20 | 6.48 | 1.54 | 119.00 | 10.51 | 11.68 | 1.11 | 160.93 | 17.39 | 17.66 | 1.02 |
| B1.15C60V1.0S3 | 85.80 | 5.15 | 5.57 | 1.08 | 140.40 | 11.75 | 10.73 | 0.91 | 191.50 | 18.50 | 16.20 | 0.88 |
| B1.65C60V1.0S3 | 130.90 | 6.60 | 8.41 | 1.27 | 169.40 | 9.50 | 11.98 | 1.26 | 229.00 | 16.19 | 17.81 | 1.10 |
| B1.15C60 | 60.00 | 5.86 | 4.85 | 0.83 | 105.32 | 12.62 | 11.06 | 0.88 | 150.39 | 19.11 | 17.42 | 0.91 |
| B1.15C60V0.5S3 | 71.00 | 5.60 | 4.82 | 0.86 | 124.90 | 12.28 | 10.43 | 0.85 | 175.10 | 18.95 | 16.27 | 0.86 |
| B1.15C60V1.5S3 | 93.10 | 5.15 | 5.45 | 1.06 | 152.50 | 11.86 | 10.65 | 0.90 | 206.20 | 18.89 | 15.98 | 0.85 |
| B1.15C60V1.0S4 | 85.00 | 5.30 | 5.43 | 1.02 | 147.40 | 11.86 | 11.29 | 0.95 | 201.70 | 18.70 | 17.11 | 0.91 |
| B1.15C60V1.0S5 | 89.40 | 5.20 | 5.77 | 1.11 | 155.40 | 11.45 | 11.91 | 1.04 | 212.90 | 17.90 | 18.02 | 1.01 |
| B1.15C30V1.0S3 | 76.33 | 6.32 | 5.79 | 0.92 | 130.28 | 13.27 | 10.46 | 0.79 | 175.49 | 20.10 | 15.40 | 0.77 |
| Average value | | | | | | | 0.99 | | | | | |
| Coefficient of variation | | | | | | | 0.16 | | | | | |

Note: $\Delta_{cn'}$ is the deflection of the beams calculated from the analytical model established in this paper after three loading and unloading cycles.

## 5. Conclusions

The main purpose of this paper was to quantify the influences of short steel fibers on the flexural behaviors of the BFRP-RC beams. Ten beams, including nine BFRP-RC beams with steel fibers and one BFRP-RC beam without steel fibers, were tested via four-point bending under cyclic loading. To accurately calculate the deflection of the BFRP-RC beams under serviceability limit states, a modified analytical model for deflection of the BFRP-RC beams with steel fibers under cyclic loading was proposed and compared with the available deflection calculation models for FRP-RC beams without steel fibers. The following main conclusions can be drawn from the results of this research:

1.  The service load moment of the BFRP-RC beams with 1.5% by volume steel fibers was 103.3% higher than that of the beams without fibers, and the deflection and the residual deflection of the beams were reduced by 48.18% and 30.36% at the applied load of 100kN. Moreover, increasing the steel-fiber volume fraction can significantly enhance the stiffness of the BFRP-RC beams after cracking.
2.  Increasing the number of unloading–reloading cycles reduced the peak load and increased the residual deflection of the BFRP-RC beams under the same deflection. The deflection of the beams increased by 11% after the first stage of three loading and unloading cycles, while the deflection increased by only 8% after three unloading and reloading cycles in the second and third stages of loading.
3.  The BFRP reinforcement ratio had the greatest influence on the load-deflection curves, load-residual deflection curves, and stiffness–displacement curves of the BFRP-RC beams. Higher-strength concrete was beneficial in improving the stiffness of the beams and reducing their deflection. A higher BFRP reinforcement ratio was beneficial to improving the serviceability of the BFRP-RC beams, which is the controlling limit state for the structural design of the BFRP-RC beams with steel fibers.
4.  Combined with the influences of cyclic loading on the deflection, a new analytical method for evaluating the deflection of the BFRP-RC beams with steel fibers under cyclic loading was proposed in this research, which gives better results than any other available model in literature and design codes when compared with experimental results.

**Author Contributions:** Conceptualization, H.Z. and Z.L.; formal analysis, Z.L., S.C. and Q.C.; investigation, Q.C. and C.L.; data curation, Q.C.; writing—original draft preparation, Z.L.; writing—review and editing, Z.L. and X.Z. All authors have read and agreed to the published version of the manuscript.

**Funding:** This research was funded by the National Natural Science Foundation of China (No. 51578510), and the Science and Technology Research and Development Project of CSCEC (under grant No. CSCEC-2021-Z-24).

**Institutional Review Board Statement:** Not applicable.

**Informed Consent Statement:** Not applicable.

**Data Availability Statement:** The data presented in this study are available on request from the corresponding author.

**Conflicts of Interest:** The authors declare that they have no known competing financial interests or personal relationships that could have appeared to influence the work reported in this paper.

## References


1. Siddika, A.; Al Mamun, A.; Alyousef, R.; Amran, Y.M. Strengthening of reinforced concrete beams by using fiber-reinforced polymer composites: A review. *J. Build. Eng.* **2019**, *25*, 100798. [CrossRef]
2. Kadhim, M.M.; Jawdhari, A.R.; Altaee, M.J.; Adheem, A.H. Finite element modelling and parametric analysis of FRP strengthened RC beams under impact load. *J. Build. Eng.* **2020**, *32*, 101526. [CrossRef]
3. Sun, Z.; Fu, L.; Feng, D.-C.; Vatuloka, A.R.; Wei, Y.; Wu, G. Experimental study on the flexural behavior of concrete beams reinforced with bundled hybrid steel/FRP bars. *Eng. Struct.* **2019**, *197*, 109443. [CrossRef]
4. Mehany, S.; Mohamed, H.M.; Benmokrane, B. Contribution of lightweight self-consolidated concrete (LWSCC) to shear strength of beams reinforced with basalt FRP bars. *Eng. Struct.* **2021**, *231*, 111758. [CrossRef]
5. Tran, T.T.; Pham, T.M.; Hao, H. Effect of hybrid fibers on shear behaviour of geopolymer concrete beams reinforced by basalt fiber reinforced polymer (BFRP) bars without stirrups. *Compos. Struct.* **2020**, *243*, 112236. [CrossRef]
6. Barris, C.; Torres, L.; Vilanova, I.; Miàs, C.; Llorens, M. Experimental study on crack width and crack spacing for Glass-FRP reinforced concrete beams. *Eng. Struct.* **2017**, *131*, 231–242. [CrossRef]
7. Saleh, Z.; Goldston, M.; Remennikov, A.M.; Sheikh, M.N. Flexural design of GFRP bar reinforced concrete beams: An appraisal of code recommendations. *J. Build. Eng.* **2019**, *25*, 100794. [CrossRef]
8. Han, S.; Zhou, A.; Ou, J. Relationships between interfacial behavior and flexural performance of hybrid steel-FRP composite bars reinforced seawater sea-sand concrete beams. *Compos. Struct.* **2021**, *277*, 114672. [CrossRef]
9. Xiao, S.-H.; Lin, J.-X.; Li, L.-J.; Guo, Y.-C.; Zeng, J.-J.; Xie, Z.-H.; Wei, F.-F.; Li, M. Experimental study on flexural behavior of concrete beam reinforced with GFRP and steel-fiber composite bars. *J. Build. Eng.* **2021**, *2*, 103087. [CrossRef]
10. Khorasani, A.M.; Esfahani, M.R.; Sabzi, J. The effect of transverse and flexural reinforcement on deflection and cracking of GFRP bar reinforced concrete beams. *Compos. Part B Eng.* **2019**, *161*, 530–546. [CrossRef]
11. Dong, H.-L.; Zhou, W.; Wang, Z. Flexural performance of concrete beams reinforced with FRP bars grouted in corrugated sleeves. *Compos. Struct.* **2019**, *215*, 49–59. [CrossRef]
12. Zeng, W.; Ding, Y.; Zhang, Y.; Dehn, F. Effect of steel fiber on the crack permeability evolution and crack surface topography of concrete subjected to freeze-thaw damage. *Cem. Concr. Res.* **2020**, *138*, 106230. [CrossRef]
13. Jabbour, R.; Assaad, J.J.; Hamad, B. Cost-to-performance assessment of polyvinyl alcohol fibers in concrete structures. *Mech. Adv. Mater. Struct.* **2021**, 1–20. [CrossRef]
14. Chellapandian, M.; Mani, A.; Prakash, S.S. Effect of macro-synthetic structural fibers on the flexural behavior of concrete beams reinforced with different ratios of GFRP bars. *Compos. Struct.* **2020**, *254*, 112790. [CrossRef]
15. de Sá, F.R.; Silva, F.D.A.; Cardoso, D.C. Tensile and flexural performance of concrete members reinforced with polypropylene fibers and GFRP bars. *Compos. Struct.* **2020**, *253*, 112784. [CrossRef]
16. Liu, X.; Sun, Y.; Wu, T.; Liu, Y. Flexural cracks in steel fiber-reinforced lightweight aggregate concrete beams reinforced with FRP bars. *Compos. Struct.* **2020**, *253*, 112752. [CrossRef]
17. Abed, F.; AlHafiz, A.R. Effect of basalt fibers on the flexural behavior of concrete beams reinforced with BFRP bars. *Compos. Struct.* **2019**, *215*, 23–34. [CrossRef]
18. Ibrahim, S.S.; Kandasamy, S.; Pradeepkumar, S.; Bose, R.S.C. Effect of discrete steel fibres on strength and ductility of FRP laminated RC beams. *Ain Shams Eng. J.* **2020**, *12*, 1329–1337. [CrossRef]
19. Issa, M.S.; Metwally, I.M.; Elzeiny, S.M. Influence of fibers on flexural behavior and ductility of concrete beams reinforced with GFRP rebars. *Eng. Struct.* **2011**, *33*, 1754–1763. [CrossRef]
20. Hosseini, S.-A.; Nematzadeh, M.; Chastre, C. Prediction of shear behavior of steel fiber-reinforced rubberized concrete beams reinforced with glass fiber-reinforced polymer (GFRP) bars. *Compos. Struct.* **2021**, *256*, 113010. [CrossRef]
21. Dev, A.; Chellapandian, M.; Prakash, S.S.; Kawasaki, Y. Failure-mode analysis of macro-synthetic and hybrid fibre-reinforced concrete beams with GFRP bars using acoustic emission technique. *Constr. Build. Mater.* **2020**, *249*, 118737. [CrossRef]
22. Nie, X.; Fu, B.; Teng, J.; Bank, L.; Tian, Y. Shear Behavior of Reinforced Concrete Beams with GFRP Needles as Coarse Aggregate Partial Replacement: Full-Scale Experiments. In *Advances in Engineering Materials, Structures and Systems: Innovations, Mechanics and Applications*; CRC Press: Boca Raton, FL, USA, 2019; pp. 1548–1553. [CrossRef]

23. Junaid, M.T.; Elbana, A.; Altoubat, S.; Al-Sadoon, Z. Experimental study on the effect of matrix on the flexural behavior of beams reinforced with Glass Fiber Reinforced Polymer (GFRP) bars. *Compos. Struct.* **2019**, *222*, 110930. [CrossRef]
24. Zhu, H.; Cheng, S.; Gao, D.; Neaz, S.M.; Li, C. Flexural behavior of partially fiber-reinforced high-strength concrete beams reinforced with FRP bars. *Constr. Build. Mater.* **2018**, *161*, 587–597. [CrossRef]
25. Ge, W.; Song, W.; Ashour, A.; Lu, W.; Cao, D. Flexural performance of FRP/steel hybrid reinforced engineered cementitious composite beams. *J. Build. Eng.* **2020**, *31*, 101329. [CrossRef]
26. Al-Saawani, M.A.; El-Sayed, A.K.; Al-Negheimish, A.I. Effect of shear-span/depth ratio on debonding failures of FRP-strengthened RC beams. *J. Build. Eng.* **2020**, *32*, 101771. [CrossRef]
27. Zhu, H.; Li, Z.; Wen, C.; Cheng, S.; Wei, Y. Prediction model for the flexural strength of steel fiber reinforced concrete beams with fiber-reinforced polymer bars under repeated loading. *Compos. Struct.* **2020**, *250*, 112609. [CrossRef]
28. Li, Z.; Zhu, H.; Du, C.; Gao, D.; Yuan, J.; Wen, C. Experimental study on cracking behavior of steel fiber-reinforced concrete beams with BFRP bars under repeated loading. *Compos. Struct.* **2021**, *267*, 113878. [CrossRef]
29. Li, Z.; Zhu, H.; Zhen, X.; Wen, C.; Chen, G. Effects of steel fiber on the flexural behavior and ductility of concrete beams reinforced with BFRP rebars under repeated loading. *Compos. Struct.* **2021**, *270*, 114072. [CrossRef]
30. Cheng, L. Flexural fatigue analysis of a CFRP form reinforced concrete bridge deck. *Compos. Struct.* **2011**, *93*, 2895–2902. [CrossRef]
31. *GB/T 30022-2013*; The Method for Basic Mechanical Properties of Fiber Reinforced Polymer Bar. China National Standard: Beijing, China, 2013.
32. *JG/T 472-2015*; Steel Fiber Reinforced Concrete. China National Standard: Beijing, China, 2015.
33. *GB175-2007*; Common Portland Cement. China National Standard: Beijing, China, 2007.
34. *ACI 440. 1R-15*; Guide for the Design and Construction of Concrete Reinforced with FRP Bars. American Concrete Institute: Farmington Hills, MI, USA, 2015.
35. *CAN, CSA S806–12*; Canadian Standard, Design and Construction of Building Structures with Fiber-Reinforced Polymers. Canadian Standards Association: Toronto, ON, Canada, 2012.
36. Gao, D.; Gu, Z.; Wei, C.; Wu, C.; Pang, Y. Effects of fiber clustering on fatigue behavior ofs steel fiber reinforced concrete beams. *Constr. Build. Mater.* **2021**, *301*, 124070. [CrossRef]
37. *JGJ/T 101-2015*; Specification for Seismic Test of Building. China National Standard: Beijing, China, 2015.
38. Bischoff, P.H. Reevaluation of Deflection Prediction for Concrete Beams Reinforced with Steel and Fiber Reinforced Polymer Bars. *J. Struct. Eng.* **2005**, *131*, 752–767. [CrossRef]
39. Bischoff, P.H. Deflection Calculation of FRP Reinforced Concrete Beams Based on Modifications to the Existing Branson Equation. *J. Compos. Constr.* **2007**, *11*, 4–14. [CrossRef]
40. Benmokrane, B.; Chaallal, O.; Masmoudi, R. Flexural response of concrete beams reinforced with FRP reinforcing bars. *ACI Struct. J.* **1996**, *93*, 46–55.
41. Alsayed, S.; Al-Salloum, Y.; Almusallam, T. Performance of glass fiber reinforced plastic bars as a reinforcing material for concrete structures. *Compos. Part B Eng.* **2000**, *31*, 555–567. [CrossRef]
42. ISIS Canada Research Network. *Reinforcing Concrete Structures with Fibre Reinforced Polymers*; Design Manual No. 3; Canadian Standards Association: Toronto, ON, Canada, 2007.

*Article*

# Cationic Lignocellulose Nanofibers from Agricultural Waste as High-Performing Adsorbents for the Removal of Dissolved and Colloidal Substances

Liangyi Yao [1], Xiangyuan Zou [1], Shuqi Zhou [1], Hongxiang Zhu [1], Guoning Chen [2,3], Shuangfei Wang [1], Xiuyu Liu [2,3,4,*] and Yan Jiang [1,*]

[1]  School of Resources, Environment and Materials, College of Light Industry and Food Engineering, Guangxi University, Nanning 530004, China; 2015392082@st.gxu.edu.cn (L.Y.); zouxiangyuann@163.com (X.Z.); zsq990711@163.com (S.Z.); zhx@gxu.edu.cn (H.Z.); shuangfei_wang1964@163.com (S.W.)

[2]  School of Chemistry and Chemical Engineering, Guangxi University for Nationalities, Nanning 530006, China; chengn@bossco.cc

[3]  Guangxi Bossco Environmental Protection Technology Co., Ltd., Nanning 530007, China

[4]  Guangxi Key Laboratory of Chemistry and Engineering of Forest Products, Nanning 530006, China

*  Correspondence: xiuyu.liu@gxun.edu.cn (X.L.); jiangyan@gxu.edu.cn (Y.J.); Tel.: +86-0771-3267019 (Y.J.)

**Abstract:** The accumulation of dissolved and colloidal substances (DCS) in the increasingly closed paper circulating water system can seriously lower the productivity and safety of papermaking machines, and it has been a challenge to develop an adsorbent with low cost, high adsorption efficiency and large adsorption capacity for DCS removal. In this study, cationic lignocellulose nanofibers (CLCNF) were obtained by cationic modification of agricultural waste bagasse in deep eutectic solvents (DES) followed by mechanical defibrillation, and then CLCNF were employed as an adsorbent for DCS model contaminant polygalacturonic acid (PGA) removal. CLCNF was characterized by transmission electron microscopy, Fourier transform infrared, elemental analysis, X-ray diffraction, and thermogravimetric analysis. The analytical results confirmed the successful preparation of CLCNF with 4.6–7.9 nm diameters and 0.97–1.76 mmol/g quaternary ammonium groups. The effects of quaternary ammonium group contents, pH, contact time and initial concentration of PGA on the adsorption were investigated in a batch adsorption study. According to the results, the cationic modification significantly enhanced the adsorption of PGA by CLCNF and the adsorption performance increased with the increase of the quaternary ammonium group contents. The adsorption of PGA on CLCNF followed the pseudo-second-order and the fitted Langmuir isotherm model. The adsorption showed fast initial kinetics and the experimental maximum adsorption capacity was 1054 mg/g, which is much higher than PGA adsorbents previously reported in the literature. Therefore, CLCNF with high cationic group content developed in this paper is a promising adsorbent for DCS removal.

**Keywords:** lignocellulose nanofibers; adsorption; deep eutectic solvents; cationization; dissolved and colloidal substances removal

---

**Citation:** Yao, L.; Zou, X.; Zhou, S.; Zhu, H.; Chen, G.; Wang, S.; Liu, X.; Jiang, Y. Cationic Lignocellulose Nanofibers from Agricultural Waste as High-Performing Adsorbents for the Removal of Dissolved and Colloidal Substances. *Polymers* **2022**, *14*, 910. https://doi.org/10.3390/polym14050910

Academic Editors: Wei Wu, Hao-Yang Mi, Chongxing Huang, Hui Zhao and Tao Liu

Received: 4 February 2022
Accepted: 22 February 2022
Published: 24 February 2022

**Publisher's Note:** MDPI stays neutral with regard to jurisdictional claims in published maps and institutional affiliations.

## 1. Introduction

The paper industry needs to reduce the consumption of freshwater resources due to the requirements of environmental protection. It is of great significance to increase the utilization of recycled white water and purposely convert it into a totally effluent-free papermaking process [1]. This will lead to an accumulation of pollutants in the system as the white water reuse rate increases. The accumulated pollutants in the water recycling system are called dissolved and colloidal substances (DCS) [2]. The composition of DCS, which mainly comes from pulp, filler, recycled water and the chemicals added during the papermaking process, is very complex [3]. DCS are also known as "anionic waste" because they are generally negatively charged substances in water [4]. Polygalacturonic acid (PGA) is one of the main sources of anionic waste in white water [5,6]. PGA was often selected as a model contaminant of DCS due to its relatively high content in DCS [7]. Excessive buildup of the DCS in process water stream may decrease the functioning of the paper machine and increase corrosion, foaming, pitch, precipitation, scaling, consumption of chemicals and the poor physical properties of the paper produced [2,8]. Therefore, it is necessary to remove these harmful substances or reduce their negative impact in order to obtain a completely closed loop water system.

The traditional method of controlling and reducing DCS in paper mills is to use cationic polyelectrolytes for neutralization, but this leads to a high consumption of hazardous chemicals [9]. As technology continues to improve, some new environmentally friendly methods have been invented. For example, membrane filtration treatment [10], dissolved air flotation [11], membrane reactor [12] and biological enzymes [13] have been proposed during recent years, but adaptability and costs have limited their application. Apart from these methods, adsorption methods gained increasing attention due to the advantages of simple and safe operations, low cost, no secondary pollution and the ability to treat water with high concentrations of waste. In current times, some adsorbents have been designed to reduce DCS content in white water through electrostatic interactions [1,8,14]. However, the current technology is still unsatisfactory, so it is necessary to find new adsorbents with larger capacity, higher adsorption efficiency and lower cost.

The vision has shifted to bio-based adsorbents, such as cellulosic nanomaterials, a promising class of adsorbents in the field of environmental remediation. Cellulosic nanomaterials are generally obtained from pretreated cellulose followed by nanofibrillation. So far, cellulosic nanomaterials have been used to adsorb and remove many types of water contaminants, such as natural organic matter [15], dye [16], heavy metals [17,18], fluoride [19], pharmaceutical agents [20] and viruses [21]. Cellulosic nanomaterials have proven to be an ideal sorbent for water contaminants. In contrast to other materials, they have the characteristics of high specific surface area, versatile surface chemistry, environmental inertness and renewability [22,23]. Conventional adsorbents usually have limited low adsorption efficiency and adsorption capacity due to the limited surface area or active sites for adsorption [24]. When the size of adsorbents is reduced to nanoscale, high specific surface area [25] and short intraparticle diffusion distance are expected to improve the situation. At the same time, its strong potential for surface chemical modification [26] means that a large number of active sites can be added. Cellulosic nanomaterials are renewable, widely sourced and environmentally inert biomaterials, and therefore pose little threat to the environment. However, cellulosic nanomaterials have not yet been investigated as a DCS adsorbent.

In general, cellulosic nanomaterials can reduce cost and improve performance by adjusting raw materials and pretreatment methods. In most cases, cellulosic nanomaterials used for water purification are obtained from purified cellulose sources, i.e., cellulose fibers where noncellulosic components (mainly lignin) have been removed [17]. This usually requires a complex and hazardous bleaching process. Therefore, a more beneficial way to produce water purification nanomaterials is directly from lignocellulosic raw materials without or with mild chemical treatment while achieving full lignocellulose utilization. Moreover, there is a preference for agricultural by-products rather than wood as lignocellulosic raw materials due to the lack of forest resources. Over 32 billion kilograms of high volume, low value and underutilized lignocellulosic biomaterial are produced from agricultural by-products annually, creating significant disposal problems [27]. In terms of pretreatment methods, deep eutectic solvents (DES) have been heralded as the most promising environmentally benign solvents to replace volatile organic solvents due to their almost null toxicity and total biodegradability [28]. DES are a fluid obtained by simply mixing two or three cheap and safe components with lower melting point than any of the original components [29]. Recently, deep eutectic solvents (DES) have been used as pretreatment mediums for production of functionalized lignocellulosic nanofibers [30,31].

The current study aims to use a green and simple strategy to prepare functionalized nanofiber adsorbents with high adsorption efficiency, large adsorption capacity and low cost from bagasse for the efficient removal of DCS model contaminant PGA. Concretely, cationic lignocellulose nanofibers (CLCNF) with different contents of quaternary ammonium groups have been prepared via cationic modification of bagasse in a DES followed by mechanical disintegration. DES is composed of aqueous tetraalkylammonium hydroxide and 1,3-dimethylurea, and Glycidyltrimethylammonium chloride was chosen as the cationization agent. The structure of CLCNF has been characterized using transmission electron microscopy, Fourier transform infrared, elemental analysis, X-ray diffraction, and thermogravimetric analysis. On the other hand, the effects of quaternary ammonium group contents, pH, contact time and initial concentration of PGA on the adsorption were investigated in a batch adsorption study. Moreover, the kinetics of adsorption and adsorption isotherms were performed to analyze the adsorption mechanism and predict adsorption capacity.

## 2. Materials and Methods

### 2.1. Materials

Between sixty to eighty mesh powder of sugarcane bagasse (cellulose: 43 wt%; hemicellulose: 30 wt%; lignin: 24 wt%) was collected after grinding and sieving. It was washed with water and ethanol alternately, then dried at 60 °C for 24 h before being used.

Tetraethylammonium hydroxide solution (TEAOH, 35 wt% in water) was obtained from TCI (Shanghai, China). 1,3-dimethylurea (1,3-DMU) and the cationization agent glycidyltrimethylammonium chloride (GTAC) were purchased from the Aladdin Industrial Corporation (Shanghai, China). The cationization agent was formulated into an 80 wt% solution for later use. Polygalacturonic acid (PGA) with a molecular weight between 25,000–50,000, sodium tetraborate and 3-Phenylphenol were purchased from Shanghai Macklin Biochemical Co., Ltd. (Shanghai, China). The stock solution of PGA (1 g/L) was made by dissolving PGA in pH 11 water adjusted with NaOH and then adjusted to pH 7 with HCl. 0.15% solution of 3-Phenylphenol in 0.5% NaOH and sodium tetraborate 0.0125 M in concentrated sulphuric acid were used as color rendering agents in adsorption experiments. To adjust the pH, 0.1 M HCl or NaOH was used. Deionized water was used throughout the experiments.

### 2.2. Preparation of CLCNF

#### 2.2.1. Cationization of Bagasse

The simplified reaction scheme of the cationization of bagasse is illustrated in Scheme 1. Specifically, 1,3-DMU and TEAOH were mixed in a 2:1 (1,3-DMU: TEAOH) molar ratio at room temperature (24 °C) to obtain 90 g of transparent aqueous deep eutectic solvent (DES). Then, 10 g of sugarcane bagasse was added followed by the addition of GTAC solution (80 wt% in water). The dry weight of GTAC was 10 g, 20 g and 30 g, respectively, and the control group did not use the cationization agent. The reaction was mixed by mechanical stirring at room temperature for 8 h. At the end of the reaction, excessive water was added to terminate this reaction. The mixture was centrifuged for 20 min using a rotation speed of 4000 rpm. The supernatant was discarded and the sample was rediluted with water and then centrifuged again, and so on until the pH of the supernatant was neutral, indicating that the agents involved in the reaction were almost removed. These samples were collected and stored at 4 °C.

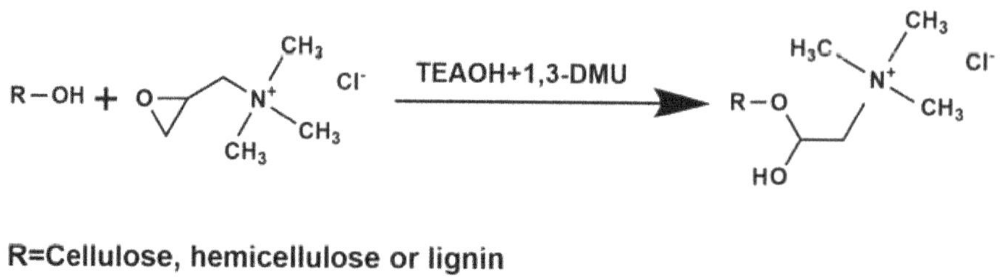

**R=Cellulose, hemicellulose or lignin**

**Scheme 1.** The cationization reaction between hydroxyl groups of lignocellulose and GTAC (DES as a reaction medium).

### 2.2.2. Disintegration of Cationic Bagasse into Nanofibers

Mechanical defibrillationdecomposition of cationic bagasse was performed with a high-pressure homogenizer (AH-PILOT 2018, ATS Nano Technology Co., Ltd., Suzhou, China). The cationic bagasse was prepared as a 1 wt% suspension and subsequently decomposed in the high-pressure homogenizer at a pressure of 1000 bar for 30 min. The cationic bagasse directly entered the chamber of the high-pressure homogenizer and was easily decomposed to obtain cationic lignocellulose nanofibers (CLCNF). The nanofiber samples obtained with GTAC dosages of 10 g, 20 g and 30 g were named CLCNF-1, CLCNF-2 and CLCNF-3, respectively. The control group required pre-mechanical pulverization by shearing to enter the chamber of the high-pressure homogenizer for disintegration, otherwise it would block the narrow chamber. The nanofiber sample of the control group was named LCNF.

### 2.3. Characterizations

### 2.3.1. Transmission Electron Microscopy (TEM)

The analysis of the size and the morphology of CLCNF and LCNF were performed by transmission electron microscopy (JEM-1400plus, JEOL, Tokyo, Japan). The samples were highly diluted to around 0.005 wt% by deionized water and then added dropwise to carbon coated copper grids. Finally, phosphotungstic acid solution (2 wt%) was applied as a negative stain before imaging. The average width of the nanofibers was measured using ImageJ software. The width of each nanofiber sample with standard errors was calculated based on over 100 individual nanofibers.

### 2.3.2. Light Transmittance

Light transmittance of the nanofiber suspensions was measured using a UV-Vis spectrophotometer (Specord 50 Plus, Analytik Jena AG, Jena, Germany). Well-dispersed CLCNF and LCNF suspensions (0.1 wt%) were placed in a cuvette with an optical path of 10 mm and the percent transmittance in the 400–800 nm wavelength range was recorded.

### 2.3.3. Fourier Transform Infrared (FTIR)

The chemical structure changes of pristine bagasse, CLCNF and LCNF were characterized using a Fourier transform infrared spectrometer (Tensor II, Bruker, Ettlingen, Germany). FTIR spectra was collected in the wavenumber range from 400 to 4000 $cm^{-1}$ with a resolution of 2 $cm^{-1}$.

### 2.3.4. Elemental Analysis

The nitrogen content of CLCNF and LCNF was analyzed using an elemental analyser (Elemantar Vario EL cube, Elementar, Hanau, Germany). All samples were freeze-dried prior to analysis. Each of the introduced quaternary ammonium group contains one nitrogen, so the quaternary ammonium group of the nanofiber samples are directly related to the nitrogen content.

### 2.3.5. Surface Charge Density

The surface charge densities of CLCNF and LCNF were determined using the polyelectrolyte titration method through a particle charge detector (PCD-05, BTG Mütek, Heidenheim an der Brenz, Germany). 10 mL of nanofiber suspension (0.01 wt% in water) was titrated with 0.001 N anion standard solution (Polyanetholesulfonic acid sodium) and 0.001 N cation standard solution (Polydadmac), and the surface charge density was calculated based on the consumption of standard solution.

### 2.3.6. Zeta-Potential

The zeta-potential of CLCNF and LCNF was determined using a Zetasizer NanoZS instrument (Zetasizer Nano ZS90, Malvern Instruments Limited, Worcestershire, UK). CLCNF and LCNF were prepared at the same consistency of 0.1 wt% and adjusted to different pH for zeta-potential measurements. First, the zeta-potential of all samples was measured at pH 7. Then, the zeta-potential of CLCNF-3 and LCNF in the pH range of 3–11 (i.e., pH = 3, 5, 7, 9 and 11) was reported.

### 2.3.7. Nanofibers Yield

The nanofiber yield of CLCNF and LCNF was determined and calculated using centrifugation. Briefly, the nanofiber suspension with a concentration of 0.2 wt% was centrifuged at a rate of 4000 rpm for 20 min to separate nanofibers (in supernatant) and non-nanofiber parts (in sediment). After carefully discarding the supernatant, the sediment was dried at 104 °C. A nanofiber yield of CLCNF and LCNF was calculated according to the following formula:

$$Nanofibers\ yield = \frac{W_L - W_s}{W_L} \times 100\% \tag{1}$$

where $W_L$ is the weight of dried lignocellulose in the suspension before centrifugation and $W_S$ is the weight of dried sediment after centrifugation.

### 2.3.8. X-ray Diffraction (XRD)

The crystal structures of pristine bagasse, CLCNF and LCNF were investigated using an X-ray diffractometer (D8 Discover, Bruker AXS GmbH, Karlsruhe, Germany). Cu Kα radiation was generated at 40 kV and 30 mA. The scanning range is 5–50°, and the scanning speed is 5°/min. The crystallinity index (*CrI*) was estimated according to the patterns using the following equation [32]:

$$CrI = \frac{I_{200} - I_{am}}{I_{200}} \times 100\% \tag{2}$$

where $I_{200}$ was the maximum intensity of the peak at 2θ between 22° and 23°, and $I_{am}$ was the minimum intensity of the amorphous cellulose at 2θ between 18° and 19°.

### 2.3.9. Thermogravimetric Analysis (TGA)

Thermogravimetric analysis of pristine bagasse, CLCNF and LCNF was performed using a TGA-DSC/DTA analyzer (STA 449 F5, NETZSCH-Gerätebau GmbH, Selb, Germany) under a nitrogen atmosphere at a constant rate of 30 mL/min. Approximately 5 mg of dry sample was placed in an aluminum oxide pan and thermally degraded by heating from 30 °C to 850 °C at a rate of 10 °C/min.

### 2.3.10. Viscosity

The viscosity of CLCNF-3 suspension (0.5 wt%) was measured using the rotational rheometer (Haake MARS 4, Thermo Fisher Scientific, Waltham, MA, USA). The measurements were conducted at 25 °C and at a shear rate of 0.01–1000 s$^{-1}$.

### 2.4. Adsorption Studies

The adsorption experiments were performed in batch mode, and the performance of the nanofiber adsorbents were researched in terms of the different quaternary ammonium groups' content of nanofibers (LCNF, CLCNF-1, CLCNF-2, and CLCNF-3), pH (3–11), initial concentration of PGA (400–800 mg/L), and adsorption time (0–6 h).

In batch adsorption experiments, 5 g of nanofiber suspensions at 0.5 wt%was mixed with PGA stock solution (1 g/L) and the total volume was adjusted as 50 mL with deionized water. Then the suspensions were mixed at 300 rpm by a magnetic stirrer. In adsorption experiments of various variables, the adsorption time was 6 h excluding the time series, and the initial concentration of PGA was 400 mg/L excluding the initial concentration series, and the pH of the solution was adjusted to 7 excluding the pH series. Next, the supernatant obtained after centrifuging the mixture was filtered using a 0.22 μm membrane syringe filter. The change in the concentration of PGA in the solution before and after adsorption was analyzed by a UV-Vis spectrophotometer (Specord 50 Plus, Analytik Jena AG, Jena, Germany). Briefly, in a similar way to a previously published method [33], 1 mL of PGA-containing sample and 5 mL of the sulphuric/tetraborate solution were mixed in a water-ice bath. The solution was reacted in a boiling water bath for 8 min, immediately cooled in a water-ice bath, and 0.1 mL of 3-Phenylphenol solution was added and turned to pink after mixing. The absorbance of the solution was measured in 524 nm by the UV-Vis spectrophotometer, and the PGA concentration was determined by comparison to a stable calibration curve ($R^2 = 0.9993$).

## 3. Results and Discussion

In this study, sugarcane bagasse was directly cationized using a deep eutectic solvent (DES) as a reaction medium. Afterwards, three cationic lignocellulose nanofiber (CLCNF) samples with different positive charge contents and molar quantities of quaternary ammonium groups were prepared via mechanical disintegration, respectively denoted CLCNF-N where N increases with the surface charge content (i.e., CLCNF-1, CLCNF-2 and CLCNF-3). LCNF, as the reference sample, was also prepared by mechanical disintegration with DES-based pretreatment but without cationic modification. In this reaction, Glycidyltrimethylammonium chloride (GTAC) was chosen as a cationization agent, as it grafted quaternary ammonium groups to bagasse fibers and promoted mechanical decomposition. DES was composed of aqueous tetraalkylammonium hydroxide and 1,3-dimethylurea (1,3-DMU), providing an alkaline condition to allow cationic modification. This DES has been proven to be a green solvent because of the low toxicity and biodegradability of its components [34]. With the applied cationization method (Scheme 1), the hydroxyl groups of cellulose, lignin and hemicellulose are deprotonated under the alkaline conditions provided by DES, and the active hydroxyl groups react with GTAC to generate quaternary ammonium groups. DES acted as a swelling agent, which is confirmed in the optical microscope images of bagasse and precursors of CLCNF and LCNF (no mechanical disintegration) in Figure S1. Robust lignocellulosic structures swelled and dissociated after DES-based treatment, and this dissociation phenomenon was significantly enhanced after GTAC usage increased.

### 3.1. Adsorbent Characterization

TEM images confirmed the nanoscaled structure of all the prepared nanofibril samples. Specifically, LCNF has an average width of 25.3 ± 6.7 nm (Figure 1A). Compared with LCNF, the CLCNF samples have well-individualized structures with a homogeneous size distribution (Figure 1B–D). This was more directly reflected in the width of CLCNF samples. As shown in Figure 1E, the average width for CLCNF-1, CLCNF-2 and CLCNF-3 was 7.9 ± 1.7, 5.5 ± 1.0 and 4.6 ± 0.8 nm, respectively. These results suggest that the cationization of bagasse could facilitate the mechanical disintegration process, thus resulting in nanofibers with small and homogeneous widths. Such a positive effect for mechanical disintegration is especially remarkable when upgrading the amount of GTAC in DES. Additionally, the precursor of LCNF needs to be mechanically pretreated for avoiding the blockage of the chamber for the high-pressure homogenizer. However, this is not the case with the cationized bagasse fibers, which can readily pass through the chamber of the high-pressure homogenizer without the occurrence of clogging. It is presumed that the charged groups on the cationized bagasse fibers can create electrostatic repulsion between fibirls and thus enhance the penetration of water into the fibers to create osmotic pressure. Undoubtedly, such effects would promote the mechanical defibrillation process of fibers [26,35].

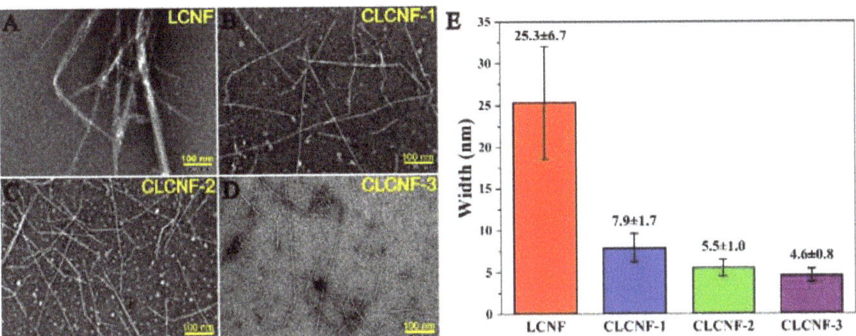


**Figure 1.** TEM images of LCNF (**A**), CLCNF-1 (**B**), CLCNF-2 (**C**), CLCNF-3 (**D**) and their corresponding widths (**E**). Scale bar is 100 nm.

The suspension transmittance is one of the main methods to indirectly evaluate the degree of fibrillation [36]. The light transmittance measurement results of the nanofiber suspensions are shown in Figure 2A. It was shown that the percentage transmittance in the UV–visible range of LCNF suspension was the lowest because of its huge fiber size. In contrast, the percentage transmittance of the CLCNF suspensions was significantly higher than the LCNF suspension and the CLCNF-3 suspension exhibited the highest transmittance. This is because the light transmittance of the suspension is related to the light-scattering phenomenon, and the light scattering is proportional to the cross section area of the particles [37]. A higher transmittance means a smaller fiber size and a higher degree of fibrillation. This result corroborates what was observed in the TEM images and it is also consistent with the visual observations of the nanofiber suspensions in Figure 2B.

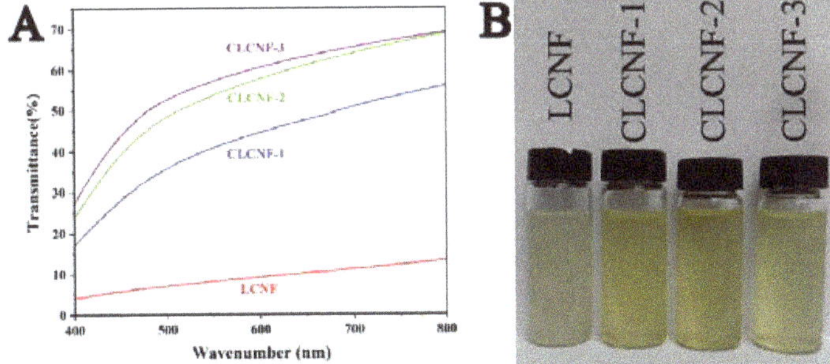


**Figure 2.** Light transmittance (**A**) and photograph (**B**) of LCNF and various CLCNF suspensions having a concentration of 0.1 wt%.

In order to confirm the successful cationic modification of bagasse, a Fourier transform infrared (FTIR) experiment was carried out. Figure 3 illustrates the FTIR spectra of bagasse, LCNF and various CLCNF samples. It was shown that the new peak at 1484 cm$^{-1}$ corresponding to the C-N stretching vibration appeared in all CLCNF samples but not in the other samples. And the epoxy ether vibrations that belong to GTAC at 1261 cm$^{-1}$ were not present in all CLCNF samples, indicating the epoxide ring opening reaction of bagasse with GTAC [38]. This confirms the successful introduction of quaternary ammonium groups on bagasse. Another significant chemical structure change is the disappearance of the characteristic peak at 1733 cm$^{-1}$ attributed to the C=O bond after DES treatment. This is because highly alkaline conditions of DES caused the breakage of naturally occurring ester bonds in bagasse, specifically the acetyl group of hemicellulose and the ester linkage of the carboxylic groups in the ferulic and p-coumaric acids of lignin/hemicellulose [32,39]. In addition, the broadening of the hydroxyl peak around 3300 cm$^{-1}$ indicates that chemical modification altered the hydrogen bonding pattern of bagasse and made the bagasse more susceptible to moisture uptake from air. This will be further confirmed in the results of the thermogravimetric analysis later.

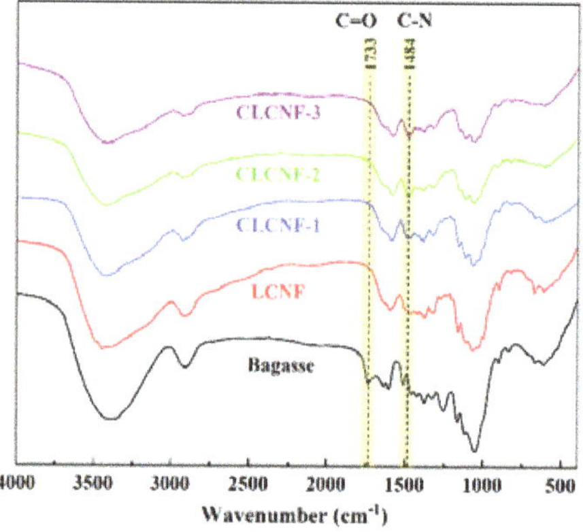


**Figure 3.** FTIR spectra of bagasse, LCNF and various CLCNF.

The specific number of quaternary ammonium groups grafted to bagasse was measured by elemental analysis (based on the weight percentages of N) and was also confirmed by polyelectrolyte titration. The results of elemental analysis and polyelectrolyte titration are listed in Table 1. The result shows that the quaternary ammonium group contents and surface charge densities were 0.97–1.76 mmol/g and 0.87–1.85 meq/g, respectively. The contents of the quaternary ammonium group in the same sample varied slightly due to the analyzing methods [40]. These results show that the increase of GTAC content in the reaction mixture improved the reactivity. The charged group content is a key performance indicator for nanofiber adsorbents in water purification applications, and it may directly affect the purification effect of contaminants [41]. CLCNF-3 was selected as the nanofiber with the highest cationic group content in this study. This was a fairly high value when compared with previous reports. In the literature, cationized CNF with an ammonium content of 0.134 mmol/g were obtained using 3-chloro-2-hydroxypropyl trimethylammonium chloride as the cationization agent [42]. Cationic CNF with a quaternary ammonium group content of 1.2 mmol/g were obtained from cellulose pulp etherified with a quaternary ammonium salt in water [15]. Cationic wood nanofiber with a cationic group content around 1.5 mmol/g were obtained using four different aqueous solvents containing TEAOH with different carbamides and GTAC as the cationization agent [32]. Surface quaternized cellulose nanofibrils with a trimethylammonium chloride content of 2.31 mmol/g were obtained from wood pulp [41]. In addition, LCNF were detected to contain trace amounts of N elements, which may be due to incomplete washing of TEAOH during centrifugal washing. Zeta-potential and surface charge densities both reflected the charge on the nanofiber surface. The zeta-potential and surface charge density were detected as negative for LCNF, but positive for CLCNF. A negative zeta-potential corresponds to a negatively charged surface, and vice versa. The results again confirmed that the presence of GTAC successfully introduced positively charged quaternary ammonium groups to the fibers.

**Table 1.** Nitrogen content, cationic group content, surface charge density, zeta-potential and nanofibers yield of LCNF and various CLCNF.

| Sample | N (%) | Cationic Group Content (mmol/g) | Surface Charge Density (meq/g) | Zeta-Potential (mV) | Nanofibers Yield (%) |
|---|---|---|---|---|---|
| LCNF | 0.08 | 0 | $-0.11 \pm 0.02$ | $-13.6 \pm 0.3$ | 8.49 |
| CLCNF-1 | 1.36 | 0.97 | $0.87 \pm 0.02$ | $26.7 \pm 0.9$ | 76.49 |
| CLCNF-2 | 1.49 | 1.06 | $0.97 \pm 0.01$ | $35.7 \pm 0.9$ | 88.59 |
| CLCNF-3 | 2.46 | 1.76 | $1.85 \pm 0.06$ | $44.7 \pm 0.8$ | 92.50 |

In this study, nanofibers were separated and quantified by means of centrifugation. After centrifuging the nanofiber suspensions, the larger-sized fibers will sink to the bottom and the smaller-sized fibers will remain suspended in the upper layer. The nanofiber yield refers to the percentage of small size fibers with diameters in nanoscale in the total fiber weight [43]. It showed that the nanofiber yield of LCNF was extremely low at only 8.49%. The increase in the amount of GTAC caused the value of the nanofiber yield to increase linearly to at least 76.49%, with a maximum value of 92.50%. This again confirms the results observed in the TEM images.

Zeta-potential is an important and useful parameter to describe the electric potential in the solid/liquid interfacial layer of a material in aqueous solution [44]. The zeta-potential as a function of pH for LCNF and CLCNF-3 water suspensions at consistencies of 0.1 wt% was shown in Figure 4. In this way, more information on the surface charge states of representative samples LCNF and CLCNF-3 was obtained. In the studied pH range (3–11), the zeta-potential of CLCNF-3 was always positive, while LCNF was always negative. The absolute value of the zeta-potential of LCNF became larger as the pH in water increased, with values ranging from $-6.49$ to $-19.20$ mV. This was attributed to the negatively charged surface groups (i.e., hydroxyl groups) on LCNF and the interactions of protons and sodium ions with them at different pH. The zeta-potential of CLCNF-3 fluctuates in the range of 21.17 to 44.73 mV. These values were always positively attributed to the presence of quaternary ammonium functional groups on the surface of the nanofibers. The zeta-potential of CLCNF-3 was maximum at neutral pH, and a pH that is too high or too low will decrease the zeta-potential. This is similar to the phenomenon observed for another cationized cellulose nanofiber [42]. The state of the nanofiber surface charge will most likely affect its adsorption performance for anionic contaminants in water [41,45].

CLCNF-3 is predicted to be the most outstanding adsorbent due to it containing the highest surface charge density of all samples. Therefore, the rheological properties of CLCNF-3 were investigated. The rheological properties were obtained from the viscosity measurements of CLCNF-3 suspension and are shown in Figure 5. The results show that the viscosity of the CLCNF-3 suspension decreases with increasing shear rate, suggesting the typical shear thinning behavior of the nanofiber suspension [40]. This indicates that the CLCNF-3 suspension is a pseudoplastic fluid and its diffusion in water was unimpeded. This would facilitate the capture of contaminants in the wastewater by CLCNF-3. The literature reports that some samples containing the naturally hydrophobic non-derivated lignin were difficult to disperse in water [46,47]. This would have raised concerns about the diffusivity of CLCNF-3 in water affecting water-based applications but did not occur.

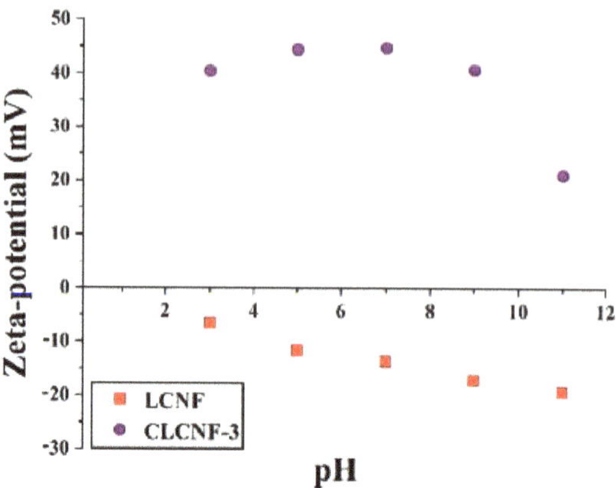

**Figure 4.** Zeta-potential of LCNF and CLCNF-3 water suspensions at consistencies of 0.1 wt% at different pH values.

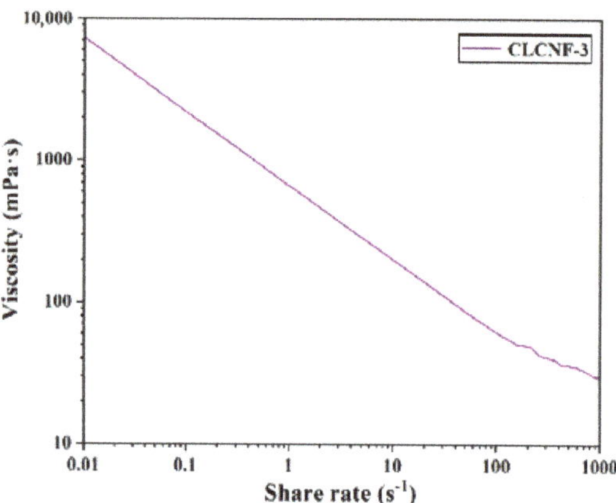

**Figure 5.** Viscosities of CLCNF-3 water suspensions at consistencies of 0.5 wt%.

The crystal structure and the crystallinity index (*CrI*) of bagasse, LCNF and various CLCNF was investigated by X-ray diffraction (Figure 6). All samples exhibit a sharp high peak around $2\theta = 22.2°$ in XRD graphs (Figure 6A), which corresponds the (0 0 2) lattice plane of cellulose I. The results indicate that all samples contain a large amount of native cellulose I, and the cellulose I crystal structure was basically retained after modification [48]. The *CrI* of various samples are shown in Figure 6B. In general, mechanical disintegration led to the destruction of the crystalline region of cellulose and the depolymerization of cellulose, resulting in a decrease in crystallinity [38]. Chemical treatment can dissolve the amorphous fraction of cellulose [49], hemicellulose [38] and lignin [50], which in turn may lead to an increase in crystallinity. The *CrI* values of bagasse and LCNF were 50.6% and 48.8%, respectively. It may be due to the combined effect of mechanical defibrillation and chemical treatment that the *CrI* of LCNF is slightly lower than that of bagasse. On the other hand, the *CrI* values of all CLCNF samples were significantly lower compared to that of bagasse, and CLCNF-3 showed the lowest *CrI* value at 26.5%. This is mainly due to the fact that the charged quaternary ammonium groups on CLCNF made a strong electrostatic repulsion between fibers making the mechanical disintegration effect amplified. Secondly, the quaternary ammonium groups on CLCNF were detected as an amorphous region that negatively affects *CrI* [40].

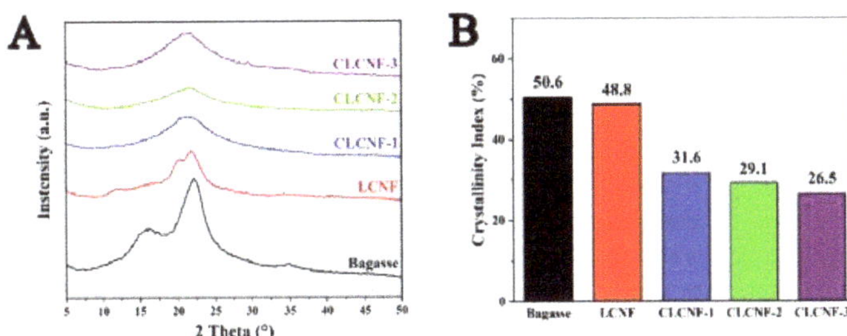

**Figure 6.** X-ray diffractogram (**A**) and calculated crystallinity indices (**B**) of bagasse, LCNF and various CLCNF.

The TGA and DTG curves are presented in Figure 7, and they were used to assess the effect of DES treatment and cationic modification on the thermal degradation behavior. Bagasse exhibited a typical thermal degradation profile of lignocellulose. Its DTG curve shows two main peaks, a shoulder peak at 200~310 °C attributed to hemicellulose decomposition, and the most prominent peak at 310~395 °C, which is attributable to cellulose decomposition [51]. The TGA curves of the LCNF and bagasse almost overlap, indicating that they have similar thermal degradation behaviors. A different thermal degradation behavior was observed for CLCNF compared to bagasse. CLCNF lost more mass in the low temperature region from 30 to 160 °C. Bagasse lost about 8.8% of mass in water form and CLCNF lost 10.6% or more. This was attributed to the grafting of CLCNF with extremely hydrophilic quaternary ammonium groups. At the same time, the onset decomposition temperature ($T_{onset}$) and the maximum decomposition temperature ($T_{max}$) of CLCNF were significantly lower than that of bagasse. The $T_{onset}$ of bagasse was 313.8 °C, and the $T_{onset}$ of CLCNF was reduced by 6.3 to 24.5 °C in comparison. The $T_{max}$ of bagasse was 356.1 °C, and the $T_{max}$ of CLCNF was reduced by 20.6 to 37.8 °C in comparison. This is attributed to the fact that the quaternary ammonium groups on CLCNF contain a volatile component ($NCH_2(CH_3)_2$), making it unstable and thus susceptible to degradation [32]. Secondly, the decrease in *CrI* also negatively affected the degradation temperature [52]. Clearly, DES treatment has little effect on the pyrolysis behavior of bagasse, but cationic modification greatly reduced its thermal stability.

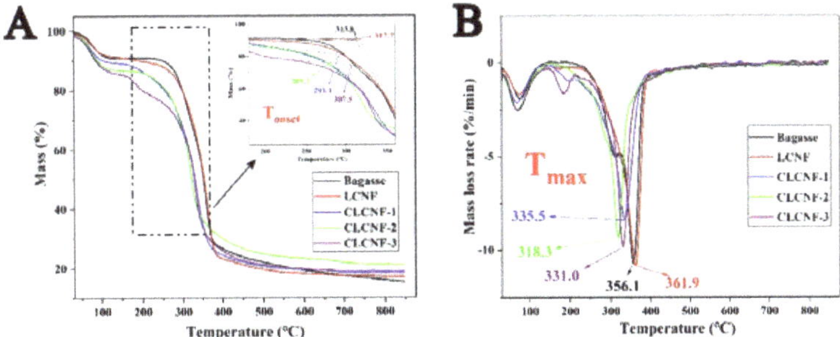

**Figure 7.** TGA (**A**) and DTG (**B**) curves of bagasse, LCNF and various CLCNF.

### 3.2. Adsorption Studies

#### 3.2.1. Different Quaternary Ammonium Groups Content of Nanofibers Effect on PGA Adsorption

The relationship between the quaternary ammonium groups content and the ability of lignocellulose nanofiber adsorbents to adsorb PGA is presented in Figure 8. The non-modified LCNF reference displayed a very low adsorption removal rate of about 7.25%. Since PGA is a negatively charged contaminant in water, and LCNF with no quaternary

By fitting pseudo-first-order (PFO) and pseudo-second-order (PSO) kinetic models to the experimental data (Figure 10B,C), this allows further description and understanding of the adsorption process. The equations of these two models are as follows [42]:

PFO:

$$log(Q_e - Q_t) = logQ_e - \frac{K_1}{2.303}t \tag{3}$$

PSO:

$$\frac{t}{Q_t} = \frac{1}{K_2 Q_e^2} + \frac{t}{Q_e} \tag{4}$$

where $Q_t$ (mg/g) and $Q_e$ (mg/g) represent the amounts adsorbed at time $t$ (min) and at equilibrium, respectively, and $K_1$ (min$^{-1}$) and $K_2$ [mg/(g·min)] represents the rate coefficients for the PFO and PSO kinetic models, respectively.

The kinetic parameters calculated from the fitted equations in these figures along with the regression coefficients are listed in Table 2. The $R^2$ value of the PSO rate equation (0.9999) is higher than that of the PFO rate equation (0.9727), indicating that the process of PGA adsorption by CLCNF-3 is more consistent with PSO kinetics. The excellent applicability of the PSO model to the adsorption kinetics of PGA onto CLCNF-3 implies that the rate-controlling step of adsorption is the chemisorption between adsorbent and adsorbate.

**Table 2.** Parameters of pseudo-first-order, pseudo-second-order and intraparticle diffusion for PGA adsorption.

| Pseudo-First-Order | | Pseudo-Second-Order | | Intraparticle Diffusion | |
|---|---|---|---|---|---|
| Parameter | Value | Parameter | Value | Parameter | Value |
| $K_1$ (min$^{-1}$) | 0.01566 | $K_2$ [mg/(g·min)] | 0.00121 | $K_i$ [mg/(g·min$^{0.5}$)] | 16.4826 |
| $Q_e$ (mg/g) | 775.36 | $Q_e$ (mg/g) | 769.23 | $C$ | 595.49 |
| $R^2$ | 0.9727 | $R^2$ | 0.9999 | $R^2$ | 0.2091 |

Since neither PFO nor PSO kinetic describe the process of diffusion, the Weber-Morris intraparticle diffusion model was also used to explain the adsorption process. The Weber-Morris intraparticle diffusion equation is as follows [45]:

$$Q_t = K_i t^{0.5} + C \tag{5}$$

where $K_i$ [mg/(g·min$^{0.5}$)] is the rate constant of intraparticle diffusion; $C$ is the constant related to the thickness of the boundary layer, which is in direct ratio to the effect of the boundary layer.

The fitted curve of the Weber-Morris intraparticle diffusion model is shown in Figure 10D, and the parameters are listed in Table 2. According to this model, the plots of $Q_t$ versus $t^{0.5}$ must pass through the origin and yield a straight line. However, in this study, the plots of $Q_t$ versus $t^{0.5}$ were not linear over the whole time range, and the fit was very low. This result shows that the Weber Morris intraparticle diffusion model was not suitable for predicting the adsorption kinetics of PGA onto CLCNF-3 over the whole range. This suggests that intraparticle diffusion is not the rate-limiting step in the adsorption process [53].

### 3.2.4. Adsorption Isotherm

As the concentration of PGA in wastewater is often fluctuating, the adsorption capacity of the adsorbent at different concentrations was investigated. Figure 11 illustrates the adsorption of PGA as a function of initial concentration. Figure 11A shows the reduction in the percentage adsorption removals of PGA onto CLCNF-3 with increasing initial concentrations of PGA. This reduction can be attributed to the finite number of active sites available on CLCNF-3 that were occupied by adsorbed PGA, subsequently leading to a decrease in PGA adsorption. On the other hand, it was found that the maximum adsorption capacity of CLCNF-3 increased with increasing initial concentration until the initial concentration of PGA was increased to 550 mg/L. This is because a higher concentration provides a driving force to overcome all the resistances of PGA between the aqueous and solid phases, thus increasing adsorption; moreover, as the initial concentration increased, so did the number of collisions between PGA and CLCNF-3, thus improving the adsorption process.

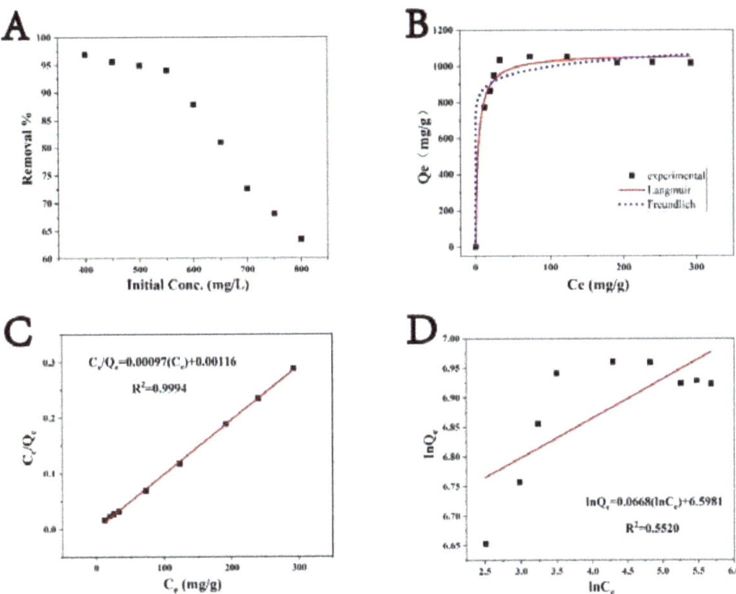

**Figure 11.** (**A**) Effects of initial concentration on the adsorption of PGA onto CLCNF-3 (pH = 7; adsorbent dose, 0.5 g/L; stirring time, 6 h; concentration range, 400–800 mg/L); (**B–D**) Fits to (**B**) nonlinear adsorption isotherms, (**C**) the Langmuir isotherm, and (**D**) the Freundlich isotherm.

The experimental data were fitted to the Langmuir model [54], which applies to monolayer adsorption, and the Freundlich model [55], which applies to multilayer adsorption, in order to identify the mechanisms of contaminant removal (Figure 11B–D). The respective equations for these models are:

Langmuir model:

$$\frac{C_e}{Q_e} = \frac{C_e}{Q_m} + \frac{1}{K_L Q_m} \tag{6}$$

Freundlich model:

$$lnQ_e = lnk_F + \frac{lnC_e}{n} \tag{7}$$

where $C_e$ and $Q_e$ are the PGA solution concentration (mg/L) and adsorption capacity (mg/g) at equilibrium, respectively, and $Q_m$ is the theoretical saturation capacity (mg/g). $K_L$ and $K_F$ are the Langmuir adsorption constant (L/g) and the Freundlich adsorption constant (L/mg), respectively, and $n$ is the heterogeneity factor for the adsorption.

The fitting parameters are presented in Table 3. The $R^2$ value for the Langmuir model is 0.9994 compared to only 0.5520 for the Freundlich model, demonstrating that the adsorption process is more consistent with monolayer adsorption without lateral interactions between adsorbed molecules. Moreover, the maximum experimental adsorption value (i.e., $Q_e$ = 1054 mg/g) is very close to the calculated value of $Q_m$, the maximum theoretical adsorption value of PGA uptake for CLCNF-3.

**Table 3.** Parameters of the Langmuir and Freundlich Adsorption Isotherms for PGA adsorption.

| Langmuir Isotherm | | Freundlich Isotherm | |
|---|---|---|---|
| Constant | Value | Constant | Value |
| $K_L$ (L/mg) | 0.84 | $K_F$ (mg/L) | 733.69 |
| $Q_m$ (mg/g) | 1026.17 | n | 14.9768 |
| $R^2$ | 0.9994 | $R^2$ | 0.5520 |

The value for $K_L$ can be calculated from the Langmuir equation which can then be used to determine the separation factor $R_L$ using:

$$R_L = \frac{1}{1 + K_L C_0} \tag{8}$$

The relationship between the initial concentration $C_0$ and the separation factor $R_L$ is shown in Figure S2. $R_L$ can be used to determine the favorability and feasibility of an adsorption process. An $R_L$ value between 0 and 1 is favorable for adsorption and values greater than 1 are adverse for adsorption [56]. The $R_L$ values for CLCNF-3 are all between 0 and 1 indicating that the adsorption of PGA by CLCNF-3 is a favorable process. It should be noted that as the initial concentration increased from 400 to 800 mg/L, the $R_L$ value gradually decreased from 0.0030 to 0.0015, indicating that higher concentrations promote PGA adsorption by CLCNF-3.

## 4. Conclusions

In conclusion, cationic lignocellulose nanofibers (CLCNF) with nanoscale diameters and quaternary ammonium group contents in the range of 0.97–1.76 mmol/g were successfully prepared by cationic modification of bagasse in deep eutectic solvents (DES) followed by mechanical defibrillation. Such lignocellulosic nanomaterials showed excellent adsorption performance in removing dissolved and colloidal substances (DCS) model contaminant polygalacturonic acid (PGA). The cationic modification in DES changed the surface charge of bagasse from negative to positive and obtained a large number of active adsorption sites, so CLCNF can easily adsorb the negatively charged PGA in water. Electrostatic interactions were considered as the main mechanism for capturing PGA by CLCNF in water. The kinetic process of PGA adsorption by CLCNF can be predicted by a pseudo-second-order model and the Langmuir model fitted the data well, indicating monolayer adsorption. The CLCNF adsorbent in current research has the advantages of low cost, high efficiency and large adsorption capacity, which can meet the new demand of DCS adsorbents. It is expected to be used for the adsorption of other anionic contaminants in water.

**Supplementary Materials:** The following supporting information can be downloaded at: https://www.mdpi.com/article/10.3390/polym14050910/s1, Figure S1: Optical microscope images of bagasse (A), the precursors of LCNF, CLCNF-1, CLCNF-2, and CLCNF-3 (B–E); Figure S2: Separation factor of PGA adsorption by CLCNF-3.

**Author Contributions:** Methodology, L.Y.; formal analysis, L.Y., X.L. and Y.J.; investigation, X.Z., L.Y., Y.J., H.Z. and G.C.; data creation, X.Z. and S.Z.; writing original draft preparation, L.Y.; writing review and editing, X.L. and Y.J.; supervision, X.L.; project administration, S.W. All authors have read and agreed to the published version of the manuscript.

**Funding:** The research was supported by the China Postdoctoral Science Foundation (2020M683209), the Natural Science Foundation of Guangxi (2020GXNSFBA159009, 2021GXNSFBA075043, 2018GXNS-FAA138126), the Opening Project of Guangxi Key Laboratory of Clean Pulp & Papermaking and Pollution Control (2019ZR04, 2019KF28), the Scientific Research Foundation of Guangxi University for Nationalities (2019KJQD10), Guangxi Ba-Gui Scholars Program, the Guangxi Major Projects of Science and Technology (GXMPSTAB21196064) and the Specific Research Project of Guangxi for Research Bases and Talents (AD18126002).

**Institutional Review Board Statement:** Not applicable.

**Informed Consent Statement:** Not applicable.

**Data Availability Statement:** Not applicable.

**Conflicts of Interest:** The authors declare that they have no conflict of interest.


## References

1. Guyard, A.; Daneault, C.; Chabot, B. Use of modified silica nanoparticle for fixation and elimination of colloidal and dissolved substances from white water. *Nord. Pulp Pap. Res. J.* **2006**, *21*, 620–628. [CrossRef]
2. Liu, K.; Li, X.-F.; Li, X.-M.; He, B.-H.; Zhao, G.-L. Lowering the cationic demand caused by PGA in papermaking by solute adsorption and immobilized pectinase on chitosan beads. *Carbohydr. Polym.* **2010**, *82*, 648–652. [CrossRef]
3. Petzold, G.; Petzold-Welcke, K.; Qi, H.; Stengel, K.; Schwarz, S.; Heinze, T. The removal of stickies with modified starch and chitosan—Highly cationic and hydrophobic types compared with unmodified ones. *Carbohydr. Polym.* **2012**, *90*, 1712–1718. [CrossRef] [PubMed]
4. Xiao, H.; He, B.; Qian, L.; Li, J. Cationic polystyrene spheres for removal of anionic contaminants in white water of papermaking. *J. Appl. Polym. Sci.* **2015**, *132*. [CrossRef]
5. Boegh, K.H.; Garver, T.M.; Henry, D.; Yuan, H.; Hill, G.S. New methods for on-line analysis of dissolved substances in white-water—It's important to maintain an equilibrium. *Pulp Pap.-Can.* **2001**, *102*, 40–45.
6. He, Z.B.; Ni, Y.H.; Zhang, E. Further understanding on the cationic demand of dissolved substances during peroxide bleaching of a spruce TMP. *J. Wood Chem. Technol.* **2004**, *24*, 153–168. [CrossRef]
7. Xiao, H.; He, B.; Li, J. Adsorption of polygalacturonic acid on crosslinked polystyrene spheres with cationic polyelectrolyte. *RSC Adv.* **2016**, *6*, 11522–11527. [CrossRef]
8. Xiao, H.; He, B.; Li, J. Immobilization of cationic polyelectrolyte on polystyrene spheres and adsorption for model whitewater contaminants. *J. Appl. Polym. Sci.* **2015**, *132*. [CrossRef]
9. Ho, Y.S.; Mckay, G. The kinetics of sorption of divalent metal ions onto sphagnum moss peat. *Water Res.* **2000**, *34*, 735–742. [CrossRef]
10. Nuortila-Jokinen, J.; Mänttäri, M.; Huuhilo, T.; Kallioinen, M.; Nyström, M. Water circuit closure with membrane technology in the pulp and paper industry. *Water Sci. Technol.* **2004**, *50*, 217–227. [CrossRef]
11. Miranda, R.; Negro, C.; Blanco, A. Internal Treatment of Process Waters in Paper Production by Dissolved Air Flotation with Newly Developed Chemicals. 1. Laboratory Tests. *Ind. Eng. Chem. Res.* **2009**, *48*, 2199–2205. [CrossRef]
12. Lin, H.; Liao, B.-Q.; Chen, J.; Gao, W.; Wang, L.; Wang, F.; Lu, X. New insights into membrane fouling in a submerged anaerobic membrane bioreactor based on characterization of cake sludge and bulk sludge. *Bioresour. Technol.* **2011**, *102*, 2373–2379. [CrossRef]
13. Liu, K.; Zhao, G.; He, B.; Chen, L.; Huang, L. Immobilization of pectinase and lipase on macroporous resin coated with chitosan for treatment of whitewater from papermaking. *Bioresour. Technol.* **2012**, *123*, 616–619. [CrossRef]
14. Vallerand, R.; Daneault, C.; Chabot, B. Adsorption of Model Whitewater Contaminants on Ion-Exchange Resins. *J. Pulp Pap. Sci.* **2010**, *36*, 22–27.
15. Sehaqui, H.; Perez de Larraya, U.; Tingaut, P.; Zimmermann, T. Humic acid adsorption onto cationic cellulose nanofibers for bioinspired removal of copper(ii) and a positively charged dye. *Soft Matter* **2015**, *11*, 5294–5300. [CrossRef]
16. Dassanayake, R.S.; Acharya, S.; Abidi, N. Recent Advances in Biopolymer-Based Dye Removal Technologies. *Molecules* **2021**, *26*, 4697. [CrossRef]
17. Sirviö, J.A.; Visanko, M. Lignin-rich sulfated wood nanofibers as high-performing adsorbents for the removal of lead and copper from water. *J. Hazard. Mater.* **2020**, *383*, 121174. [CrossRef]
18. Lehtonen, J.; Hassinen, J.; Kumar, A.A.; Johansson, L.-S.; Mäenpää, R.; Pahimanolis, N.; Pradeep, T.; Ikkala, O.; Rojas, O.J. Phosphorylated cellulose nanofibers exhibit exceptional capacity for uranium capture. *Cellulose* **2020**, *27*, 10719–10732. [CrossRef]
19. Sehaqui, H.; Mautner, A.; Perez de Larraya, U.; Pfenninger, N.; Tingaut, P.; Zimmermann, T. Cationic cellulose nanofibers from waste pulp residues and their nitrate, fluoride, sulphate and phosphate adsorption properties. *Carbohydr. Polym.* **2016**, *135*, 334–340. [CrossRef]
20. Selkälä, T.; Suopajärvi, T.; Sirviö, J.A.; Luukkonen, T.; Lorite, G.S.; Kalliola, S.; Sillanpää, M.; Liimatainen, H. Rapid uptake of pharmaceutical salbutamol from aqueous solutions with anionic cellulose nanofibrils: The importance of pH and colloidal stability in the interaction with ionizable pollutants. *Chem. Eng. J.* **2018**, *350*, 378–385. [CrossRef]
21. Sato, A.; Wang, R.; Ma, H.; Hsiao, B.S.; Chu, B. Novel nanofibrous scaffolds for water filtration with bacteria and virus removal capability. *J. Electron Microsc.* **2011**, *60*, 201–209. [CrossRef]
22. Abouzeid, R.E.; Khiari, R.; El-Wakil, N.; Dufresne, A. Current State and New Trends in the Use of Cellulose Nanomaterials for Wastewater Treatment. *Biomacromolecules* **2019**, *20*, 573–597. [CrossRef] [PubMed]
23. Voisin, H.; Bergström, L.; Liu, P.; Mathew, A.P. Nanocellulose-Based Materials for Water Purification. *Nanomaterials* **2017**, *7*, 57. [CrossRef] [PubMed]

24. Anjum, M.; Miandad, R.; Waqas, M.; Gehany, F.; Barakat, M.A. Remediation of wastewater using various nano-materials. *Arab. J. Chem.* **2019**, *12*, 4897–4919. [CrossRef]

25. Sehaqui, H.; Zhou, Q.; Ikkala, O.; Berglund, L.A. Strong and tough cellulose nanopaper with high specific surface area and porosity. *Biomacromolecules* **2011**, *12*, 3638–3644. [CrossRef] [PubMed]

26. Rol, F.; Belgacem, M.N.; Gandini, A.; Bras, J. Recent advances in surface-modified cellulose nanofibrils. *Prog. Polym. Sci.* **2019**, *88*, 241–264. [CrossRef]

27. Reddy, D.H.K.; Seshaiah, K.; Reddy, A.V.R.; Lee, S.M. Optimization of Cd(II), Cu(II) and Ni(II) biosorption by chemically modified Moringa oleifera leaves powder. *Carbohydr. Polym.* **2012**, *88*, 1077–1086. [CrossRef]

28. Yang, Z.; Asoh, T.-A.; Uyama, H. Cationic functionalization of cellulose monoliths using a urea-choline based deep eutectic solvent and their applications. *Polym. Degrad. Stab.* **2019**, *160*, 126–135. [CrossRef]

29. Zhang, Q.; De Oliveira Vigier, K.; Royer, S.; Jérôme, F. Deep eutectic solvents: Syntheses, properties and applications. *Chem. Soc. Rev.* **2012**, *41*, 7108–7146. [CrossRef]

30. Sirviö, J.A.; Isokoski, E.; Kantola, A.M.; Komulainen, S.; Ämmälä, A. Mechanochemical and thermal succinylation of softwood sawdust in presence of deep eutectic solvent to produce lignin-containing wood nanofibers. *Cellulose* **2021**, *28*, 6881–6898. [CrossRef]

31. Yu, W.; Wang, C.; Yi, Y.; Wang, H.; Yang, Y.; Zeng, L.; Tan, Z. Direct pretreatment of raw ramie fibers using an acidic deep eutectic solvent to produce cellulose nanofibrils in high purity. *Cellulose* **2021**, *28*, 175–188. [CrossRef]

32. Segal, L.; Creely, J.J.; Martin, A.E.; Conrad, C.M. An Empirical Method for Estimating the Degree of Crystallinity of Native Cellulose Using the X-ray Diffractometer. *Text. Res. J.* **1959**, *29*, 786–794. [CrossRef]

33. Ibarz, A.; Pagán, A.; Tribaldo, F.; Pagán, J. Improvement in the measurement of spectrophotometric data in the m-hydroxydiphenyl pectin determination methods. *Food Control* **2006**, *17*, 890–893. [CrossRef]

34. Sirviö, J.A.; Ismail, M.Y.; Zhang, K.; Tejesvi, M.V.; Ämmälä, A. Transparent lignin-containing wood nanofiber films with UV-blocking, oxygen barrier, and anti-microbial properties. *J. Mater. Chem. A* **2020**, *8*, 7935–7946. [CrossRef]

35. Sirviö, J.A.; Ukkola, J.; Liimatainen, H. Direct sulfation of cellulose fibers using a reactive deep eutectic solvent to produce highly charged cellulose nanofibers. *Cellulose* **2019**, *26*, 2303–2316. [CrossRef]

36. Uetani, K.; Kasuya, K.; Koga, H.; Nogi, M. Direct determination of the degree of fibrillation of wood pulps by distribution analysis of pixel-resolved optical retardation. *Carbohydr. Polym.* **2021**, *254*, 117460. [CrossRef]

37. Saito, T.; Nishiyama, Y.; Putaux, J.-L.; Vignon, M.; Isogai, A. Homogeneous Suspensions of Individualized Microfibrils from TEMPO-Catalyzed Oxidation of Native Cellulose. *Biomacromolecules* **2006**, *7*, 1687–1691. [CrossRef]

38. Wang, Y.; Xie, W. Synthesis of cationic starch with a high degree of substitution in an ionic liquid. *Carbohydr. Polym.* **2010**, *80*, 1172–1177. [CrossRef]

39. Hafrén, J.; Oosterveld-Hut, H.M.J. Fluorescence lifetime imaging microscopy study of wood fibers. *J. Wood Sci.* **2009**, *55*, 236–239. [CrossRef]

40. Li, P.; Sirviö, J.A.; Hong, S.; Ämmälä, A.; Liimatainen, H. Preparation of flame-retardant lignin-containing wood nanofibers using a high-consistency mechano-chemical pretreatment. *Chem. Eng. J.* **2019**, *375*, 122050. [CrossRef]

41. Pei, A.; Butchosa, N.; Berglund, L.A.; Zhou, Q. Surface quaternized cellulose nanofibrils with high water absorbency and adsorption capacity for anionic dyes. *Soft Matter* **2013**, *9*, 2047–2055. [CrossRef]

42. Muqeet, M.; Malik, H.; Mahar, R.B.; Ahmed, F.; Khatri, Z.; Carlson, K. Cationization of Cellulose Nanofibers for the Removal of Sulfate Ions from Aqueous Solutions. *Ind. Eng. Chem. Res.* **2017**, *56*, 14078–14088. [CrossRef]

43. Jiang, Y.; Liu, X.; Yang, Q.; Song, X.; Qin, C.; Wang, S.; Li, K. Effects of residual lignin on mechanical defibrillation process of cellulosic fiber for producing lignocellulose nanofibrils. *Cellulose* **2018**, *25*, 6479–6494. [CrossRef]

44. Cho, D.; Lee, S.; Frey, M.W. Characterizing zeta potential of functional nanofibers in a microfluidic device. *J. Colloid Interface Sci.* **2012**, *372*, 252–260. [CrossRef]

45. Zhao, J.; Wang, L.; Xiao, J.; Tao, M.; Zhang, W. Removal of anionic azo dyes from aqueous solutions by quaternary ammonium salt-functionalized fibers with adjustable surface microenvironments. *React. Funct. Polym.* **2020**, *154*, 104684. [CrossRef]

46. Gordobil, O.; Herrera, R.; Llano-Ponte, R.; Labidi, J. Esterified organosolv lignin as hydrophobic agent for use on wood products. *Prog. Org. Coat.* **2017**, *103*, 143–151. [CrossRef]

47. Visanko, M.; Sirviö, J.A.; Piltonen, P.; Sliz, R.; Liimatainen, H.; Illikainen, M. Mechanical fabrication of high-strength and redispersible wood nanofibers from unbleached groundwood pulp. *Cellulose* **2017**, *24*, 4173–4187. [CrossRef]

48. Qing, Y.; Sabo, R.; Zhu, J.Y.; Agarwal, U.; Cai, Z.; Wu, Y. A comparative study of cellulose nanofibrils disintegrated via multiple processing approaches. *Carbohydr. Polym.* **2013**, *97*, 226–234. [CrossRef]

49. Sirviö, J.A.; Anttila, A.-K.; Pirttilä, A.M.; Liimatainen, H.; Kilpeläinen, I.; Niinimäki, J.; Hormi, O. Cationic wood cellulose films with high strength and bacterial anti-adhesive properties. *Cellulose* **2014**, *21*, 3573–3583. [CrossRef]

50. Francisco, M.; van den Bruinhorst, A.; Kroon, M.C. New natural and renewable low transition temperature mixtures (LTTMs): Screening as solvents for lignocellulosic biomass processing. *Green Chem.* **2012**, *14*, 2153–2157. [CrossRef]

51. Wan, C.; Lu, Y.; Jiao, Y.; Jin, C.; Sun, Q.; Li, J. Ultralight and hydrophobic nanofibrillated cellulose aerogels from coconut shell with ultrastrong adsorption properties. *J. Appl. Polym. Sci.* **2015**, *132*. [CrossRef]

52. Hideno, A. Comparison of the Thermal Degradation Properties of Crystalline and Amorphous Cellulose, as well as Treated Lignocellulosic Biomass. *BioResources* **2016**, *11*, 6309–6319. [CrossRef]

53. Duman, O.; Tunç, S.; Gürkan Polat, T. Adsorptive removal of triarylmethane dye (Basic Red 9) from aqueous solution by sepiolite as effective and low-cost adsorbent. *Microporous Mesoporous Mater.* **2015**, *210*, 176–184. [CrossRef]
54. Jin, L.; Li, W.; Xu, Q.; Sun, Q. Amino-functionalized nanocrystalline cellulose as an adsorbent for anionic dyes. *Cellulose* **2015**, *22*, 2443–2456. [CrossRef]
55. Lu, M.; Lü, X.; Xu, X.; Guan, X. Thermodynamics and kinetics of bacterial cellulose adsorbing persistent pollutant from aqueous solutions. *Chem. Res. Chin. Univ.* **2015**, *31*, 298–302. [CrossRef]
56. Zhao, R.; Li, X.; Sun, B.; Li, Y.; Li, Y.; Yang, R.; Wang, C. Branched polyethylenimine grafted electrospun polyacrylonitrile fiber membrane: A novel and effective adsorbent for Cr(vi) remediation in wastewater. *J. Mater. Chem. A* **2017**, *5*, 1133–1144. [CrossRef]

*Article*

# Preparation of Rosin-Based Composite Membranes and Study of Their Dencichine Adsorption Properties

**Long Li** [1,2,3,4], **Xiuyu Liu** [1,2,3,4], **Lanfu Li** [1], **Sentao Wei** [1] **and Qin Huang** [1,2,3,4,*]

[1]  School of Chemistry and Chemical Engineering, Guangxi Minzu University, Nanning 530006, China; lilong19980227@163.com (L.L.); xiuyu.liu@gxun.edu.cn (X.L.); a17876072393@163.com (L.L.); a1184306866@163.com (S.W.)

[2]  Key Laboratory of Chemistry and Engineering of Forest Products, State Ethnic Affairs Commission, Nanning 530006, China

[3]  Guangxi Key Laboratory of Chemistry and Engineering of Forest Products, Nanning 530006, China

[4]  Guangxi Collaborative Innovation Center for Chemistry and Engineering of Forest Products, Guangxi Minzu University, Nanning 530006, China

*  Correspondence: huangqin@gxun.edu.cn

**Abstract:** In this work, rosin-based composite membranes (RCMs) were developed as selective sorbents for the preparation of dencichine for the first time. The rosin-based polymer microspheres (RPMs) were synthesized using 4-ethylpyridine as a functional monomer and ethylene glycol maleic rosinate acrylate as a crosslinking. RCMs were prepared by spinning the RPMs onto the membranes by electrostatic spinning technology. The optimization of various parameters that affect RCMs was carried out, such as the ratio concentration and voltage intensity of electrospinning membrane. The RCMs were characterized by SEM, TGA and FT-IR. The performances of RCMs were assessed, which included adsorption isotherms, selective recognition and adsorption kinetics. The adsorption of dencichine on RCMs followed pseudo-second-order and adapted Langmuir–Freundlich isotherm model. As for the RCMs, the fast adsorption stage appeared within the first 45 min, and the experimental maximum adsorption capacity was 1.056 mg/g, which is much higher than the previous dencichine adsorbents reported in the literature. The initial decomposition temperature of RCMs is 297 °C, the tensile strength is 2.15 MPa and the elongation at break is 215.1%. The RCMs have good thermal stability and mechanical properties. These results indicated that RCMs are a tremendously promising adsorbent for enriching and purifying dencichine from the notoginseng extracts.

**Keywords:** rosin-based composite membranes; dencichine; electrostatic spinning technology; notoginseng extracts

**Citation:** Li, L.; Liu, X.; Li, L.; Wei, S.; Huang, Q. Preparation of Rosin-Based Composite Membranes and Study of Their Dencichine Adsorption Properties. *Polymers* **2022**, *14*, 2161. https://doi.org/10.3390/polym14112161

Academic Editors: Wei Wu, Hao-Yang Mi, Chongxing Huang, Hui Zhao and Tao Liu

Received: 20 April 2022
Accepted: 11 May 2022
Published: 26 May 2022

**Publisher's Note:** MDPI stays neutral with regard to jurisdictional claims in published maps and institutional affiliations.

## 1. Introduction

Pharmacologically active natural products have gained unprecedented popularity in recent decades [1]. They have made great contributions historically to drug development, and many of them have had profound effects on our lives. Dencichine (b-N-oxalyl-L-a, b-diaminopropionic acid, b-ODAP), isolated from the roots of panax notoginseng, has a high medicinal value [2]. Dencichine has been reported to show beneficial effects against numerous diseases, such as dispersing stasis and hemostasis, improving platelet number, relieving swelling and pain, kidney diseases, neuroprotection, lowering blood glucose [2–5] and rheumatic diseases [6]. Natural ingredients in plants are characterized by low concentration [7], the existence of multi-component mixtures, structural diversity and similarity, which pose great challenges to the separation and purification of natural ingredients [8]. Therefore, it is essential to explore efficient extraction and purification processes for dencichine. According to several literature reviews, various methods have been developed to carry ou the separation, determination and enrichment of dencichine in panax notoginseng, such as colorimetry [9], high-performance liquid chromatography (HPLC) [10], gas chromatography–mass spectrometry (GC/MS) [11], liquid chromatographic–tandem mass spectrometric (LC/MS) [12], ultra-high-performance liquid chromatography–tandem mass spectrometry (UPLC-MS/MS) [13], water-methanol method [14] and ultrasonic/microwave-assisted method [15]. The instrumental method employed a specific type of specialization and sophistication of the instruments. The separation method is not highly efficient and profitless to mass production. At the same time, the separated products need to be purified to remove harmful solvents such as acetonitrile. Water-methanol extraction is a traditional and easy-to-operate extraction approach, although inefficient. The ultrasonic/microwave-assisted method in the extraction process has a low separation effect on the structural analogs of dencichine. Molecularly imprinted technology is an easy and efficient method for the preparation of tailor-made polymeric materials with molecular recognition abilities [16]. However, the imprinting of dencichine is problematic and easy to cause template molecule waste. An efficient, low-cost material for dencichine extraction and purification is currently lacking.

In recent years, adsorption-based methods have been widely used in the separation and enrichment of bioactive compounds from many natural products. For example, RPM separation of total alkaloids from coptidis [17] and chitosan membrane purification of artemisinin [18]. The adsorption effect depends on the type and quality of the adsorption material. At present, many kinds of adsorption materials have been developed and deployed for the purification and separation of natural products. The commonly used adsorbents are activated carbon, iron oxide, silica gel, starch, adsorption resin, clay minerals and composite membranes. They use the porous matrices [19–21] as an adsorption platform and have good application prospects in the mass transfer and adsorption process. Among them, the composite membranes [22–25] have excellent thermal stability, high adsorption capacities and stable three-dimensional structure. In addition, they are used repeatedly, so they are widely used for isolating and the enrichment of active ingredients from natural products, such as half terpene [26], flavonoids [27], alkaloids [17] and other active compounds. Poly-ethersulfone (PES) [28], poly-sulfone (PSF) [29], poly-vinylidene fluoride (PVDF) [30,31], poly-vinylpyrrolidone (PVP) [32], poly-acrylonitrile (PAN) [33], poly-vinyl alcohol (PVA) [34] and natural product cellulose [35–38] are widely used as the polymeric membranes. With people paying more attention to the environment, porous chitosan, cellulose, rosin and other natural polymers appear in people's field of vision, and the use of natural adsorption materials is becoming more and more prevalent. The rosin-based crosslinking agent has excellent rigidity and contains three double bonds, non-toxic, non-carcinogenic, high abundance, low cost, environmental protection and other advantages. The polymer materials prepared from rosin have the advantages of degradation, high mechanical strength and excellent luster. Rosin-based polymers have been generally implemented not only in traditional fields such as coatings [39] and adhesives [40], but also in emerging fields such as energy [41], environment [42], drug delivery [43] and drug analysis. There are several methods for the preparation of composite membranes, including the casting flow method [44], freeze-drying approach [45] and electrostatic spinning et al. The composite membranes prepared by electrospinning have unique properties such as high specific surface area and uniform nanofiber structure. Electrospinning is currently applied in various applications including electrochemistry, natural product extraction, medicine, the environment and batteries.

In this work, RCMs were developed as excellent adsorbents for the preparation of dencichine for the first time. The preparation of RPMs was carried out using ethylene glycol maleic rosinate acrylate as a crosslinking by precipitation polymerization, and the RCMs were further prepared by electrospinning. Furthermore, we also study the feasibility of RCMs as effective sorptive materials for the dissociation and enrichment of dencichine. The RCMs were characterized by SEM, TGA and FT-IR. Then, the RCMs were evaluated for their sorbent performance of dencichine from notoginseng extracts. The adsorption kinetics of dencichine on RCMs was studied, and the adsorption mechanism was analyzed in detail. Studies have shown that the membranes have excellent application prospects for the extraction of dencichine.

## 2. Materials and Methods

### 2.1. Materials

Dencichine was provided from Chengdu Plant standard pure Biotechnology Co., Ltd. (Chengdu, China). 4-ethylpyridine and 2,2′-azobisisobutyronitrile (AIBN) were provided from Shanghai Aladdin Biochemical Technology Co., Ltd. (Shanghai, China). N, N′-methylenebis (acrylamide) and acetic acid were provided from Shanghai Maclin Biochemical Technology Co., Ltd. (Shanghai, China). Methyl alcohol was obtained from Chengdu Cologne Chemicals Co., Ltd. (Chengdu, China). Glycine-DL-leucine (GL) and glycyl-L-phenylalanine (GP) were provided from Aladdin Chemistry Co. Ltd. (Shanghai, China). HPLC-grade acetonitrile and methanol were provided from Agilent Technologies (Shanghai, China) Co., Ltd. Ethylene glycol maleic rosinate acrylate (EGMRA) was purchased by Wuzhou Sun Shine Forestry & Chemicals Co., Ltd. (Wuzhou, China) [17].

### 2.2. Preparation of RPMs

The functional monomers 4-VP (2.4 mmol) and the EGMEA (0.48 mmol) and MBA (9.6 mmol) were dissolved in 100 mL of methanol in a 250 mL flask. Subsequently, 0.1260 g AIBN was added to the mixed solutions and the organic phase was formed by ultrasound. Then, the mixture was heated followed by mechanical agitation (50 rpm), and heat-polymerized at 70 °C for 11 h. The unreacted monomers were removed by methanol extraction for 48 h. The RPMs were dried under vacuo at 50 °C for 12 h.

### 2.3. Preparation of RCMs

The RCMs were prepared via the electrospinning method. First, the mixed solution with contents of 10 wt% (PAN) and 1 wt% (RPMs) was prepared by adding PAN (2.0000 g) and RPMs (0.2000 g) into DMSO (20 mL) under continuous magnetic stirring at 85 °C for 2.5 h. The mixed solution was extracted 10 mL by using a 10 mL disposable syringe and then the composite membranes were prepared by electrostatic spinning. Electrospinning was carried out at a temperature of 25 ± 5 °C, humidity of 30 ± 5%, the fixed voltage of 20 KV, feeding rate of 0.1 mL/h and a distance of 16 cm. Electrospinning was stopped after 10 h. Finally, RCMs were obtained after methanol extraction for 24 h and vacuum dried at 60 °C for 12 h. The scheme for preparing the RCMs was represented in Figure 1.

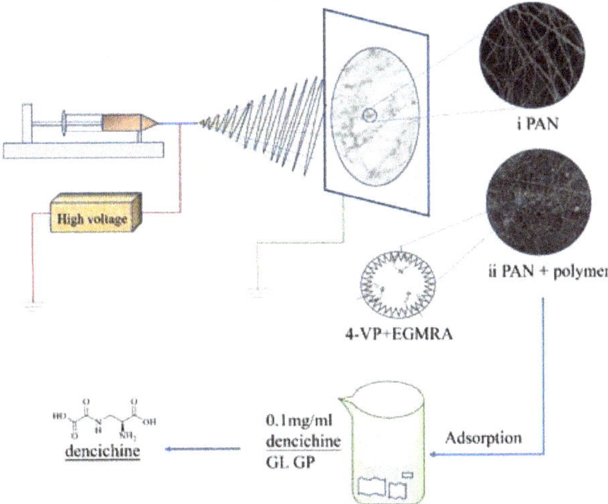


**Figure 1.** The scheme for preparing the RCMs.

### 2.4. Characterization of RCMs and RPMs

The RCMs and RPMs were investigated by FT-IR (MagnA-IR550, Thermo Fisher Scientific, Waltham, MA, USA) in the range 4000–400 cm$^{-1}$. The analysis of the size and morphology of the RCMs and RPMs were performed using field-emission scanning electron microscopy (SEM, Supra 55 Sapphire, Carl Zeiss, Jena, Germany)). Thermogravimetric analysis of the RCMs and RPMs performed using TGA-DSC/DTA analyzer (STA 449 F5, NETZSCH-Gerätebau GmbH, Selb, Germany). The RCMs and RPMs were weighed by analytical balance with an accuracy of 0.1 mg (Practum124-1cn, Sartorius AG, Göttingen, Germany). The zeta potentials of the RCMs and RPMs were measured using a Laser Nanoparticle Size and Zeta Potential Analyzer (Zetasizer Nano, Malvern, UK). The RCM and RPM pore and specific surface area were measured at 77 k using surface area and pore size analyzer (ASAP2020, Micromeritics, Norcross, GA, USA). The mechanical properties of RCM were tested by an electronic universal testing machine (JDL-10000N, Yangzhou Tianfa test Machinery Co., Ltd., Yangzhou, China).

#### 2.4.1. Scanning Electron Microscopy (SEM)

The surface morphology of the RCMs and RPMs was determined by SEM analysis (SEM, Supra 55 Sapphire, Carl Zeiss Germany, Oberkochen, Germany). The samples were evenly coated on the conductive adhesive of the sample sheet and then sprayed with gold for 0.5 h. The surface morphology of the RCMs and RPMs after the samples were sprayed with gold was observed by SEM under low vacuum conditions.

#### 2.4.2. Thermogravimetric Analysis (TGA)

The thermal stability of the RCMs and RPMs was determined by thermogravimetric analysis (TGA) (STA 449 F5, NETZSCH-Gerätebau GmbH, Selb, Germany). The sample was heated from 30 °C to 800 °C for thermal degradation under nitrogen protection at a rate of 10 °C/min.

#### 2.4.3. Dynamic Mechanical Analyzer

The samples were cut to 1 cm in width and 4 cm in length, and the stress–strain curve of the RCMs was measured at a lifting rate of 1 mm/min by the electronic universal testing machine (JDL-10000N, Yangzhou Tianfa test Machinery Co., Ltd., Yangzhou, China). The tensile strength of the RCMs was calculated based on the following Equation (1) [46]:

$$\sigma_b = \frac{P}{A_0} = \frac{P}{bd} \tag{1}$$

where $\sigma_b$ represents the tensile strength, P represents the maximum tensile load, A represents the cross sectional area of the sample, b represents the width and d represents thickness.

The elongation at the break of RCMs was calculated based on the following Equation (2) [46]:

$$\delta = \frac{\Delta L_b}{L_0} \tag{2}$$

where $\delta$ represents the elongation at the break, $\Delta L_b$ represents the increase in length at the breaking point and $L_0$ represents the original length.

### 2.5. HPLC Analysis

All analysis were performed on an Agilent Series 1260 (Agilent Technologies, La Jolla, CA, USA) system, equipped with an autosampler, a quaternary pump, a diode-array detector and a column compartment, controlled by Agilent1260 LC software. Separation was achieved on a ZORBAX SB-C18 analytical column (4.6 × 250 mm, 5 µm, USA) [47]. The mobile phase was 0.05% $H_3PO_4$ aqueous solution and acetonitrile, and the ratio was 95:5. The detection wavelength was 213 nm, the mobile phase flow rate was 1 mL/min and the injected sample volume was 10 µL. The temperature remained at 25 °C.

*2.6. Static Adsorption*

2.6.1. Standard Curve of Dencichine

The content of examined dencichine was determined by HPLC. Dencichine solution with concentration of 0, 0.06, 0.08, 0.10, 0.12, 0.14, 0.16 mg/mL was prepared. The contents of dencichine at different concentrations were analyzed by HPLC (Agilent Technologies, La Jolla, CA, USA) at 213 nm. The standard curve is represented by the following fitting Equation (3).

$$A = 4071.577C - 0.1569 \tag{3}$$

where A represents the absorption peak area of dencichine solution at 213 nm, C represents the concentration of the different dencichine standards (mg/mL) and the correlation coefficient ($R^2$) for this equation was 0.9995.

2.6.2. Static Adsorption of Dencichine on RCMs and RPMs

The RPMs were accurately weighed ($0.02 \pm 0.0002$ g) and placed into round 50 mL conical flasks. After adding 20 mL of dencichine solution, it was adsorbed for 5 h in an 80 rpm air constant temperature oscillator at 25 °C. The content of dencichine after adsorption was determined by HPLC. The adsorption amount of the RPMs on dencichine was calculated based on the following Equation (4) [17].

$$Q_e = \frac{(C_0 - C_e) \times V}{W} \tag{4}$$

Here, $Q_e$ represents the amount adsorbed (mg/g), $C_e$ represents the equilibrium solution concentration (mg/mL), $C_0$ represents the initial concentration (mg/mL), V represents adsorbed solution volume (mL) and W represents the mass of the RPMs (g).

Similarly, the RCMs were accurately weighed ($0.15 \pm 0.0001$ g) and placed into round 50 mL conical flasks. After adding 20 mL of dencichine solution, it was adsorbed for 5 h in an 80 rpm air constant temperature oscillator at 25 °C. The content of dencichine after adsorption was determined by HPLC. The dencichine adsorption amount of the RCMs was calculated based on Equation (4).

2.6.3. Adsorption of Dencichine on Different Mass RCMs and RPMs

The RPMs (10, 15, 20, 25, 30, 35, 40 mg) were weighed with an analytical balance and placed into round 50 mL conical flasks. After adding 20 mL of dencichine solution, it was adsorbed for 5 h in an 80 rpm air constant temperature oscillator at 25 °C. The RCMs (75, 100, 125, 150, 175 mg) were weighed with an analytical balance. After adding 20 mL of dencichine solution, it was adsorbed for 5 h in an 80 rpm air constant temperature oscillator at 25 °C. The content of dencichine after adsorption was determined by HPLC. The dencichine adsorption amount of RCMs and RPMs was calculated based on Equation (4).

2.6.4. Adsorption Kinetics of Dencichine on RCMs and RPMs

The adsorption process of dencichine was studied with the optimal experimental mass. The RCMs and RPMs were used to adsorb dencichine (0.1 mg/mL) in an 80 rpm air constant temperature oscillator at 25 °C. Samples of 0.5 mL were absorbed with 1 mL syringe at 1, 2, 3, 4, 5, 6, 7, 8, 9, 11, 13, 15, 20, 25, 30, 45, 60, 75, 90, 105, 120, 135, 150, 180, 240, 300 min, respectively, and the content of dencichine was determined by HPLC. Subsequently, the adsorption kinetic curve according to the relationship between time and adsorption amount was plotted.

2.6.5. Adsorption Isotherm and Thermodynamics of Dencichine on the RCMs and RPMs

To evaluate the adsorption isotherm of dencichine on RCMs and RPMs, the adsorption was carried out by the addition of 150 mg RCMs or 20mg RPMs into 20 mL of dencichine solution with different concentrations (0.08, 0.10, 0.12, 0.14, 0.16, 0.18 mg/mL). The adsorption equilibrium time was determined by adsorption kinetics experiments. Following the experiment, the adsorption content of RCMs for dencichine in each sample was measured.

To evaluate the thermodynamics of dencichine on the RCMs and RPMs, a series of adsorption tests were carried out, including different temperatures 15, 25, 35, 45 and 55 °C.

### 2.6.6. Adsorption of Dencichine on the RCMs and RPMs at Different PH

To evaluate the influence of PH on the adsorption performance of dencichine on RCMs and RPMs, a series of adsorption tests were carried out at different pH values (1, 3, 5, 6, 7, 8, 9, 11). The adsorption tests were carried out at 25 °C, 80 rpm with 0.1 mg/mL of dencichine.

### 2.7. Selective Adsorption on the RCMs and RPMs

To evaluate the selective adsorption on the RCMs and RPMs, the adsorption tests were carried out at different solutions (0.10 mg/mL dencichine, 0.10 mg/mL glycine-DL-leucine (GL), 0.10 mg/mL glycyl-L-phenylalanine (GP)). An aqueous GL solution with concentration of 0, 0.06, 0.08, 0.10, 0.12, 0.14, 0.16 mg/mL was prepared. The concentrations of different solutions were measured at 200 nm (GL) using HPLC (Agilent Technologies, La Jolla, CA, USA). The standard curve was represented by the following Equation (5).

$$A_2 = 4251.53C_2 - 0.4393 \tag{5}$$

Here, $A_2$ represents the absorption peak area of GL solution at 200 nm, $C_2$ represents the concentration of different GL standards (mg/mL) and the correlation coefficient (R2) for this equation was 0.9999.

An aqueous GP solution with a concentration of of 0, 0.06, 0.08, 0.10, 0.12, 0.14, 0.16 mg/mL was prepared. The concentrations of different solutions were measured at 210 nm (GP) using HPLC (Agilent Technologies, La Jolla, CA, USA). The standard curve is represented by the following Equation (6).

$$A_3 = 7873.23C_3 - 4.6747 \tag{6}$$

Here, $A_3$ represents the absorption peak area of GP solution at 210 nm, $C_3$ represents the concentration of different GP standards (mg/mL) and the correlation coefficient ($R^2$) for this equation was 0.9999.

## 3. Results and Discussion

### 3.1. Characterization of the RCMs and RPMs

SEM was used to study the surface morphology of RCMs and RPMs, and the results are represented in Figure 2. As can be seen in Figure 2a,b, the RPMs are found to be spherical or nearly spherical objects, and the surfaces of the microspheres had a porous, large surface area. These pores are formed as a result of the diffusion of methanol from the particle to the surface during the polymerization of the polymer. This interconnected porous network provides accessibility and active sites for dencichine adsorption, thus facilitating adsorption. In Figure 2b, it can be seen that the RCMs and the RPMs form a three-dimensional network structure, and the polymers are encased in the membranes. As can be seen from Figure 2c,d, the diameter of the composite membrane fiber prepared by electrostatic spinning is at the nanometer level, with a range of 322.9 ± 73.49 nm. The RCMs present random distributions and are very uniform with dense structure and interconnected large pores. Therefore, electrospinning can provide a rigid frame for the RPMs, which is conducive to the further recycling of the RPMs.

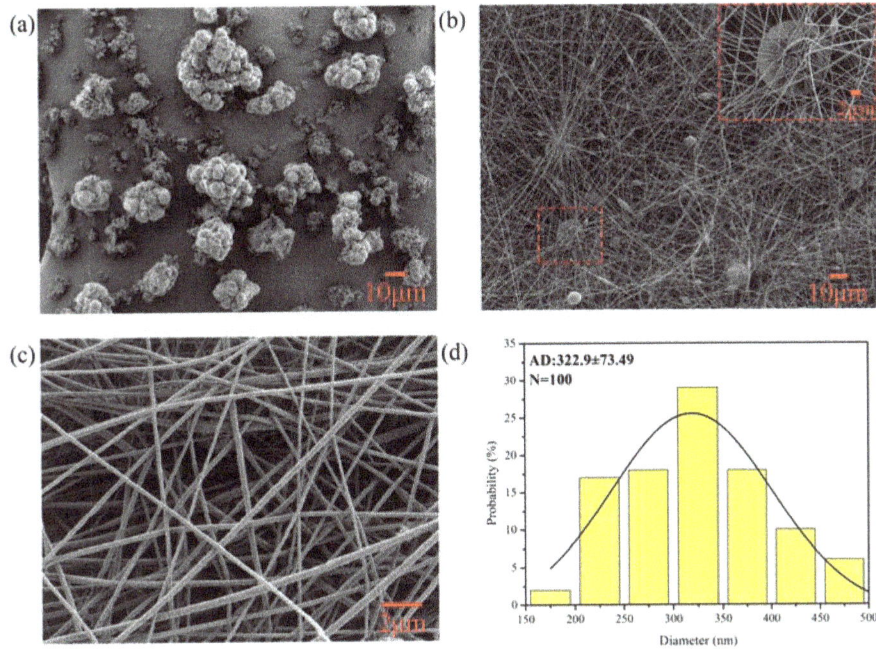

**Figure 2.** (**a**) SEM images of RPMs; (**b**,**c**) SEM images of RCMs; (**d**) Average particle size and particle size distribution images of RCMs.

The structure of the RCMs and RPMs was investigated by FT-IR spectroscopy, and the results are represented in Figure 3. In Figure 3a, the RPMs display a stretching vibration peak of -NH occurs at 3308 cm$^{-1}$, a stretching vibration peak of -C=O occurs at 1653 cm$^{-1}$, a -C=C stretching vibration absorption peak is observed at 1521 cm$^{-1}$ and a -COO- stretching vibration absorption peak is observed at 2359 cm$^{-1}$, indicating the successful preparation of RPMs. In the RCM curve, the stretching vibration absorption peak of -C≡N is observed at 2243 cm$^{-1}$, indicating the presence of PAN. Compared with RPMs, the presence of PAN leads to a blue shift in -C=O and a red shift in -C=C. In addition, the FT-IR spectra of the RCMs and RPMs samples are similar, indicating the successful preparation of RCMs.

N$_2$ adsorption–desorption experiments were carried out to study the pore volumes [48], pore size distributions, average pore diameters and specific surface areas of the RPMs, and the results are shown in Figure 3b. The pore volume and surface area of the RPMs are $2.756 \times 10^{-3}$ cm$^3$/g and $6.1776 \pm 0.1204$ m$^2$/g, respectively. The RPMs have a high specific surface area, which was favorable for the adsorption and extraction of analytes.

TGA curves were presented in Figure 3c, and they were used to describe the thermal stability of the RCMs and RPMs. The RPMs start to decompose at 360 °C, the temperature of the fastest decomposition rate occurs at 381 °C and the maximum decomposition temperature (Tmax) is 462 °C. The decomposition process is divided into two stages, including dehydration in the low temperature zone (100–360 °C) and decomposition in the high temperature zone (381–462 °C). The initial decomposition temperature for the RCMs is 297 °C, the temperature of the fastest decomposition rate occurred at 306 °C, and the Tmax is 333 °C. Decomposition occurs in two stages, including dehydration (100–297 °C) and decomposition (297–333 °C). The second stage of decomposition is due to the breakdown of the PAN in the RCMs. Compared with RCMs, the RPMs have different thermal degradation behavior. The RPMs lost more mass in the high-temperature region from 360 to 500 °C. The TGA results demonstrate that the RCMs and RPMs have excellent thermal stability.

The mechanical properties stability of the membrane long-term stability is one of the significant parameters. Figure 3e shows the stress–strain curves of the as-prepared membranes measured by a dynamic mechanical analyzer. In Figure 3e, it can be seen that the RCMs exhibited excellent mechanical stability with the tensile strength of 2.15 MPa, along with the elongation at the break of 215.1%. The stress–strain curve results demonstrate that the RCMs had excellent mechanical properties.

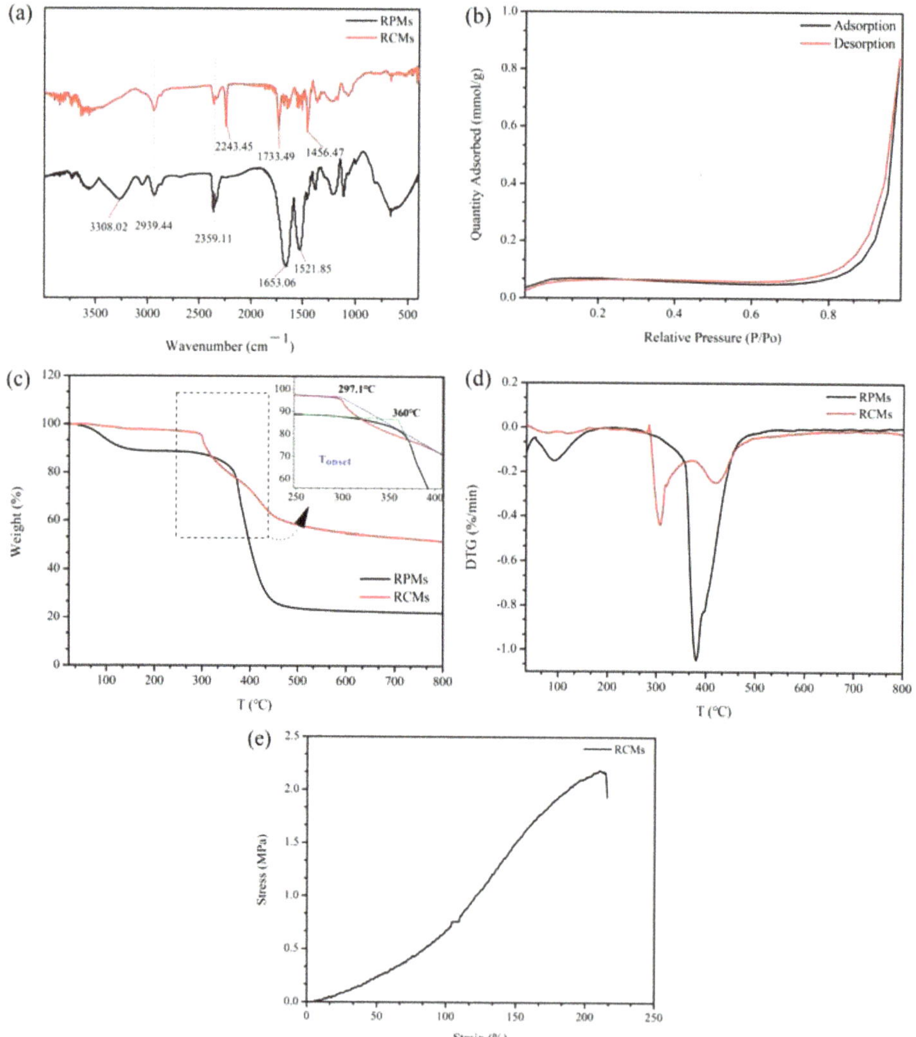



**Figure 3.** (**a**) FI-IR images of RCMs and RPMs; (**b**) $N_2$ adsorption–desorption of RPMs; (**c**) TGA of RCMs and RPMs; (**d**) DTG of RCMs and RPMs (**e**) Stress–strain curves of RCMs.

### 3.2. Optimization Preparation Conditions of the RCMs

The results of the investigation into the preparation conditions of the RCMs are represented in Figure 4. As can be seen in Figure 4a, the adsorption amount of dencichine increases as the RPM concentration increases. When the RPM concentration exceeds 1 wt%, the adsorption amount decreases with the increase in the RPM concentration. The sediment volume is also increased, for a certain PAN concentration, when the polymer concentration increases. Thus, the optimal concentration of the RPMs is 1.0 wt%. In Figure 4b, it can be seen that the adsorption amount of dencichine increases with the increase in the PAN concentration. When the concentration exceeds 10 wt%, the adsorption amount decreases with the increase in the PAN concentration. This may be because the spinning solution with low PAN content has low adhesion, poor spinning effect and poor adsorption amount of the dencichine. PAN content continues to increase, the spinning effect is excellent, and the preparation of the RCMs has also increased the adsorption amount. However, as the content of PAN continues to increase, the spinning solution surface tension increases, the droplet formation of jet flow in the electric field tension is difficult, even blocking the syringe needle which affects the spinning, and the preparation of the morphology of the RCMs will become worse, reducing the adsorption amount. In Figure 4c, it can be seen that the adsorption amount of dencichine increased as the voltage increased. When the concentration exceeds 20 KV, the adsorption amount decreases with the increased voltage. When the voltage concentration is increased, membranes thickness is also increased. With the further increase in the electric field, the drop gradually stays in the electric field for a shorter time and the radius of the RCM center circle decreases. This may be due to the RPMs being wrapped in membranes, thus reduce reducing the adsorption amount.

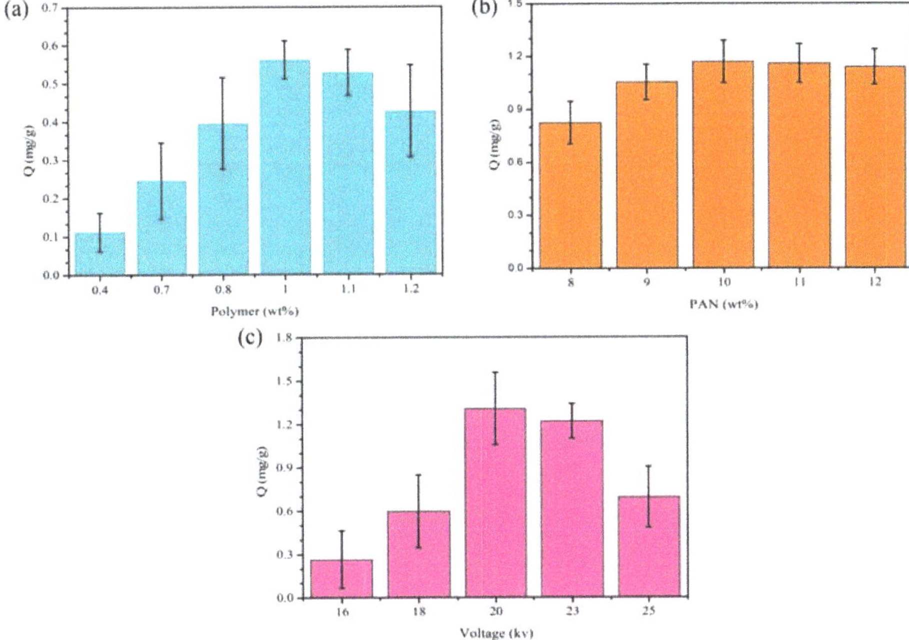

**Figure 4.** Optimization preparation conditions of the RCMs: (**a**) Polymer content; (**b**) PAN content; (**c**) Voltage adsorption amount.

### 3.3. Adsorption of Dencichine on Different Mass RCMs and RPMs

The adsorption of dencichine (0.1 mg/mL) on different mass RCMs and RPMs is represented in Figure 5. As can be seen in Figure 5a, the adsorption amount of dencichine is increased by increasing the mass of the RPMs. When the mass exceeds 20 mg, the adsorption amount decreases with the increase in the RPM mass. In Figure 5b, it can be seen that the adsorption amount of dencichine increases with the increase in the RCM mass. When the mass exceeds 150 mg, the adsorption amount decreases with the increase in the RCM mass. Under the same concentration conditions, more adsorption sites are provided to dencichine at a small increase, which raises the effective contact area and the amount of dencichine adsorption. With the increase in mass, the RCMs and RPMs had inadequate adsorption of dencichine, and the adsorption amount decreased. In other words, the adsorption of the sorption system stays correlated with the availability of adsorption sites on the surface of the adsorbent and the concentration of the dencichine solution.

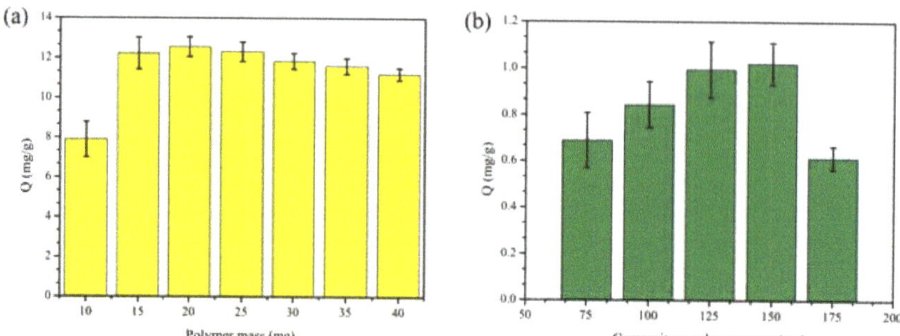

**Figure 5.** Adsorption of dencichine on different mass RCMs and RPMs; (**a**) RPM adsorption amount; (**b**) RCM adsorption amount.

### 3.4. Adsorption Kinetics of Dencichine on the RCMs and RPMs

The adsorption kinetics curves of dencichine on the RCMs and RPMs at initial concentrations (0.1 mg/mL) are represented in Figure 6a. The adsorption of dencichine on the RCMs and RPMs showed excellent characteristics of the adsorption kinetics, the adsorption capacity increased with the increase in the adsorption time, and the adsorption rate decreased gradually with increasing adsorption time. As for the RPMs, the fast adsorption stage appeared within the first 15 min, while the slow adsorption stage appeared at 15 to 60 min and the adsorption equilibrium appeared after 150 min. As for the RCMs, the fast adsorption stage appeared within the first 45 min, while the slow adsorption stage appeared at 45 to 90 min and the adsorption equilibrium appeared after 240 min. Compared with RPMs, the RCMs have different adsorption kinetics behavior. This is because dencichine molecules are adsorbed on the 4-VP surface of the RPMs and RCMs during the initial stages. Then, over time, it becomes increasingly difficult for the dencichine to enter the RCMs.

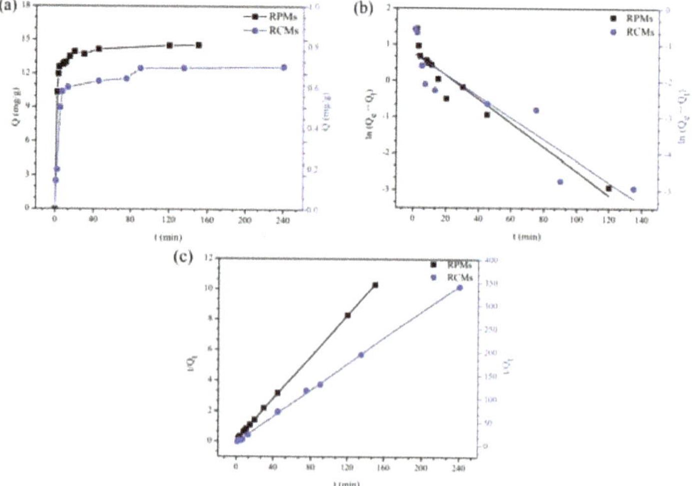

**Figure 6.** Adsorption kinetics of dencichine on the RCMs and RPMs. (**a**) Adsorption kinetics curves; (**b**) Pseudo-first-order adsorption kinetics model; (**c**) Pseudo-second-order adsorption kinetic.

To determine the mass transfer mechanisms and rate controlling, adsorption kinetics of dencichine onto the RCMs and RPMs are evaluated using fitting pseudo-second-order (PSO) and pseudo-first-order (PFO) models [49–53].

PFO adsorption kinetics models:

$$\ln(Q_e - Q_t) = \ln Q_e - \frac{K_1}{2.303}t \tag{7}$$

PSO adsorption kinetics models:

$$\frac{t}{Q_t} = \frac{1}{K_2 Q_e^2} + \frac{1}{Q_e}t \tag{8}$$

where $Q_e$ represents the adsorbed amount at equilibrium (mg/g), $Q_t$ represents the amounts adsorbed at time t (mg/g), $K_1$ represents the PFO adsorption kinetics models rate constant $(min^{-1})$ and $K_2$ represents the PSO adsorption kinetics models rate constant (g/(mg min)).

The corresponding kinetic parameters calculated by Origin are shown in Table 1. The PFO kinetic model is based on the assumption that adsorption controls diffusion and the PSO kinetic model assumes that the adsorption rate is controlled by the chemisorption process.

**Table 1.** Kinetic data of PFO kinetic model and PSO kinetic model.

| Samples | PFO Kinetic | | PSO Kinetic | |
|---|---|---|---|---|
| | $K_1 \, (min^{-1})$ | $R^2$ | $K_2 \, (g \, mg^{-1} \, min^{-1})$ | $R^2$ |
| RCMs | 0.0678 | 0.8235 | 0.3662 | 0.9992 |
| RPMs | 0.0758 | 0.9082 | 0.0603 | 0.9999 |

Figure 6b shows the relationship between ln $(Q_e - Q_t)$ and time (t), and Figure 6c shows the relationship between $t/Q_t$ and time t. It can be seen from the kinetics parameters of both adsorbents presented that the coefficient of determination of PFO kinetics $R^2$ (the RPMs) and $R^2$ (the RCMs) are 0.9082 and 0.8235, respectively, and that of PSO kinetics $R^2$ (the RPMs) and $R^2$ (the RCMs) are 0.9999 and 0.9992. The results indicate that the PSO kinetics model fits well with experimental data, and the $R^2$ values of the PSO kinetics model are higher than that of the PFO kinetics model. This phenomenon also indicates that chemisorption is a dominant role in the adsorption process.

### 3.5. Adsorption Isotherm and Thermodynamics of Dencichine on the RCMs and RPMs

The adsorption isotherms of dencichine on the RCMs and RPMs at (298 K) with dencichine concentrations of 80, 100, 120, 140, 160 and 180 µg/mL are shown in Figure 7a. As can be seen in Figure 7, the adsorption process of dencichine on RCMs and RPMs was obviously affected by the initial concentration. The dencichine adsorption amount for RCMs and RPMs increased with increasing dencichine concentration.

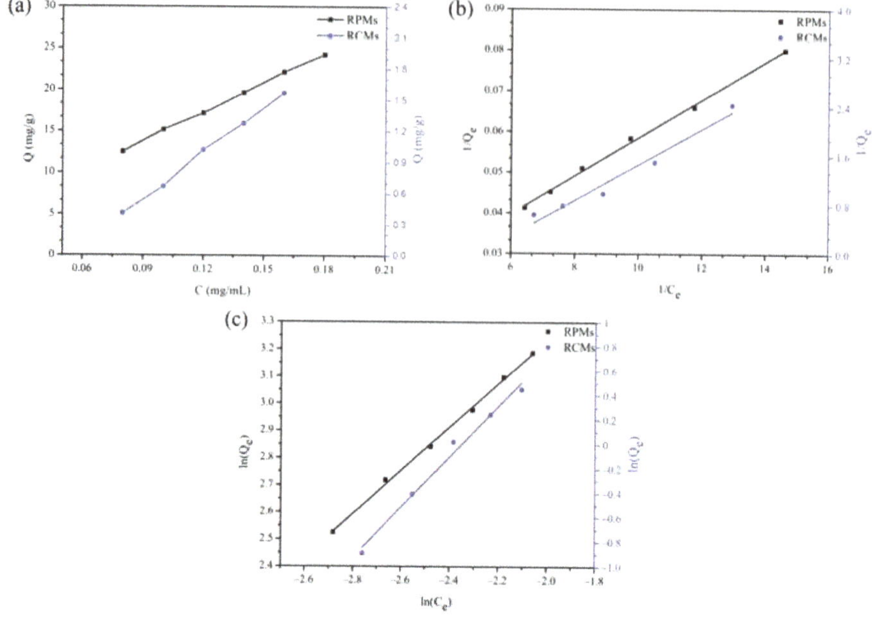

**Figure 7.** Adsorption thermodynamics of dencichine on RCMs and RPMs. (**a**) Adsorption isotherms; (**b**) Langmuir isotherm model; (**c**) Freundlich isotherm model.

In order to analyze the adsorption mechanism, fitting Langmuir Freundlich isotherm models to the experimental data (Figure 7b,c) is helpful and allows further understanding of the adsorption mechanism. The equations of these two models are as follows [54–56].
Langmuir isotherm equation:

$$\frac{1}{Q_e} = \frac{1}{Q_m} + \frac{1}{K_3 Q_m} \times \frac{1}{C_e} \tag{9}$$

Freundlich isotherm equation:

$$\ln Q_e = \ln k_4 + \frac{1}{n}\ln C_e \tag{10}$$

where $C_e$ represents the concentration of dencichine at equilibrium (mg/mL), $Q_e$ represents the dencichine adsorption amount for the RCMs and RPMs at equilibrium (mg/g), $1/n$ is the dimensionless Freundlich constant, $Q_m$ represents the saturation adsorption capacities of monolayer coverage (mg/g), $K_3$ represents the Langmuir constant (mL/mg) and $K_4$ represent the Freundlich constant (mg/mL).

The Freundlich and Langmuir isotherms for the adsorption of dencichine on the RCMs and RPMs are represented in Figure 7b,c, and the fitting data is shown in Table 2. From the adsorption isotherms data, it is observed that the correlation coefficient ($R^2$) for the adsorption of dencichine on the RPMs adsorption has a higher value for the Freundlich equation (0.9976) than the Langmuir (0.9968), indicating that the Freundlich model is more suitable for the adsorption process of dencichine on RPMs. Overall, $0 < 1/n < 1$ indicates that the adsorption process easily occurs and has excellent adsorption capacity, $1/n$ (0.7988) in the Freundlich equation can be seen as a reflection of the easy adsorption behavior. The Langmuir model (0.9570) fits the adsorption data less than the Freundlich model (0.9842) for the adsorption of dencichine on the RCMs. In total, $1/n$ (2.062) in the Freundlich equation can be seen as a reflection of the adsorption behavior. These results indicated that the adsorption of dencichine on the RCMs was multilayer adsorption. According to the prediction of the Langmuir isothermal model, the maximum adsorption capacity of RPMs for dencichine at 25 °C is 85.11 mg/g. These results further proved the application prospect of RCMs in the separation of dencichine.

**Table 2.** Parameters of Langmuir adsorption model and Freundlich adsorption model.

| Samples | Langmuir Isotherm | | | Freundlich Isotherm | | |
|---|---|---|---|---|---|---|
| | $K_3$ (mL·mg$^{-1}$) | $R^2$ | $Q_m$ (mg·g$^{-1}$) | $K_4$ (mL·mg$^{-1}$) | $R^2$ | 1/n |
| RCMs | −4.990 | 0.9570 | 0.6864 | 84.19 | 0.9842 | 2.062 |
| RPMs | 2.511 | 0.9968 | 85.11 | 106.8 | 0.9976 | 0.7988 |

To understand the effect of temperature on the adsorption amount of dencichine on the RCMs and RPMs, the results are discussed for the different temperatures, and the results are represented in Figure 8. As shown in this figure, the adsorption process of dencichine on the RCMs and RPMs is affected by temperature. At low temperatures, the dencichine adsorption capacities for the RCMs and RPMs increased with increasing temperature. The contact probability of the RCMs of dencichine increases with the increase in temperature. The RCMs and RPMs have temperature sensitivity because of the hydrogen bonding interaction with the dencichine molecules, and the hydrogen bond is destroyed gradually with the increase in temperature. With the increase in temperature, the adsorption capacity of the RCMs decreased, indicating that high temperature is not conducive to the progress of the adsorption process. This phenomenon proves that chemisorption is dominant in the adsorption process.

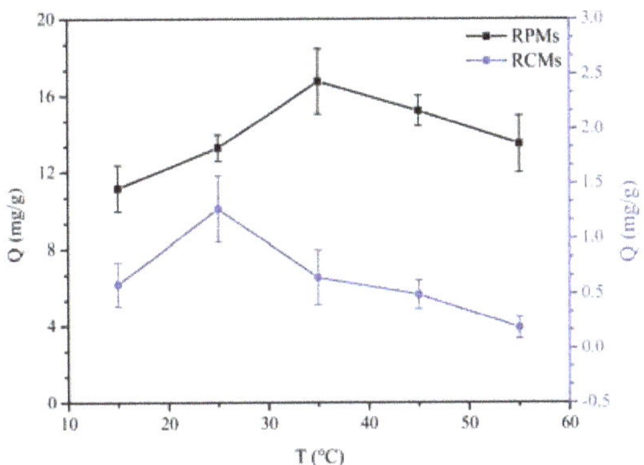


**Figure 8.** Adsorption thermodynamics of dencichine on RCMs and RPMs.

### 3.6. Adsorption of Dencichine on the RCMs and RPMs at Different PH Levels

To understand the effect of PH on the adsorption amount of dencichine on the RCMs and RPMs, the results were discussed for the different PH levels, and the results are shown in Figure 9. The adsorption of dencichine on the RCMs and RPMs at the different PH levels of 1, 3, 5, 6, 7, 8, 9 and 11 are shown in Figure 9a.

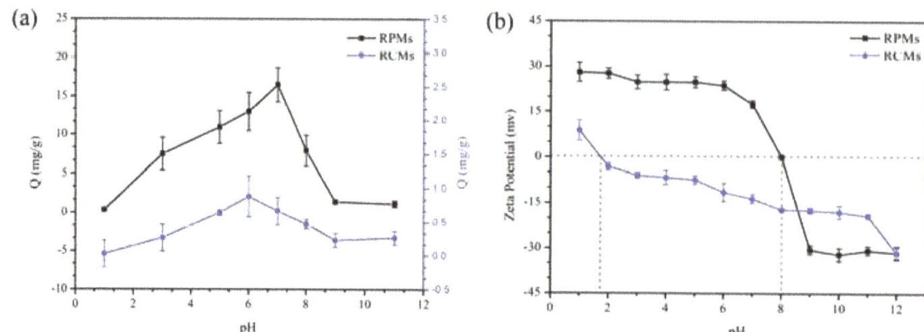

**Figure 9.** (**a**) Adsorption of dencichine on RCMs and RPMs in different pH conditions; (**b**) The zeta potential of RCM and RPM adsorption amounts.

As shown in this Figure 9a, the maximum dencichine adsorption amount of the RPMs is reached when the pH is 6, and the RCMs are reached when the pH is 7. The results are due to the fact that dencichine is acidic and 4-ethylpyridine is basic, and proves that a weakly acidic or neutral condition could favor the adsorption process. The zeta potentials of the RCMs and RPMs at various pH levels were measured, and the results are presented in Figure 9b. The points of zero charges are at the pH values: 1.7 for RCMs and 8.0 for RPMs, the zeta potential of the RCMs and RPMs decreased as PH increased, and RCMs have more surface negative charge than RPMs. The results indicate that the amino group in PAN reduces the surface charge. Therefore, the adsorption of dencichine by RCMs may be affected by electrostatic interactions. These results further prove that RCMs have exceptional promise for the separation and enrichment of dencichine.

### 3.7. Selective Adsorption on RCMs and RPMs

The selectivity study of RCMs and RPMs was evaluated by using dencichine, two analogues including GL and GP, compared with the adsorption of dencichine in a single solution and the mixed solution (Table 3). Table 3 illustrated the data obtained from the selectivity experiment for both RCMs and RPMs, concerning the adsorption quantity.

**Table 3.** Preferential adsorption of dencichine on RCMs and RPMs.

| Material | The Single Adsorption (mg/g) | | | The Compound Adsorption (mg/g) | | |
|---|---|---|---|---|---|---|
| | Dencichine | GL | GP | Dencichine | GL | GP |
| RPMs | 15.57 | 1.148 | 0.5734 | 13.79 | 1.261 | 0.3456 |
| PCMs | 1.056 | 0.0260 | 0.1841 | 0.8625 | 0.2930 | 0.4494 |

In the single solution of three substances, the adsorption amount for dencichine on the RPMs and RCMs is 15.57 mg/g and 1.056 mg/g, respectively. The adsorption amount is better than those for the two analogues. Similarly, in the mixed solution of three substances, the adsorption amount for dencichine on the RPMs and RCMs was 13.79 mg/g and 0.8625 mg/g, respectively, which is significantly higher than those for the two analogues. This phenomenon illustrates a high discrimination property of RCMs and RPMs between dencichine and analogues. These results further proved the application prospect of RCMs in the separation of dencichine.

## 4. Conclusions

In the present work, RPMs were synthesized using 4-VP as a functional monomer and EGMRA as a crosslinker and RCMs were further prepared by electrospinning. Their physicochemical properties and chemical structures were analyzed and characterized. It was observed that RCMs have a negative charge on the surface and have an excellent

*Polymers* **2022**, *14*, 2161

adsorption effect on denichine in an aqueous solution and the RCMs showed excellent mechanical and thermodynamic properties. The support of the adsorption isotherm model for Langmuir-Freundlich indicated that the adsorption of denichine and its analogues by RCMs was multilayers adsorption, and the kinetic adsorption suggested that chemisorption was the main adsorption mechanism. In the meantime, the RCMs showed excellent water solubility and were highly discriminating against denichine and its analogues, and were conducive to the extraction and purification of a water-soluble bioactive component from natural products. This synthetic method is green, efficient, eco-friendly and cost effective. Therefore, our study provides a novel, efficient and green polymer for the concentration and the purification of denichine and will open up a new avenue for the purification of denichine.

**Author Contributions:** Methodology, L.L. (Long Li); formal analysis, L.L. (Long Li), X.L. and Q.H.; investigation, L.L. (Long Li), L.L. (Lanfu Li), Q.H. and S.W.; data creation, L.L. (Lanfu Li) and X.L.; writing—original draft preparation, L.L. (Long Li); writing—review and editing, L.L. (Long Li), X.L. and Q.H.; supervision, Q.H.; project administration, Q.H. and X.L. All authors have read and agreed to the published version of the manuscript.

**Funding:** This research received no external funding.

**Informed Consent Statement:** Informed consent was obtained from all subjects involved in the study.

**Acknowledgments:** The research was supported by the National Natural Science Foundation of China (31860192), the Natural Science Foundation of Guangxi (2020GXNSFBA159009), the China Postdoctoral Science Foundation (2020M683209), the Opening Project of Guangxi Key Laboratory of Clean Pulp & Papermaking and Pollution Control (2019KF28) and the Scientific Research Foundation of Guangxi Minzu University (2019KJQD10).

**Conflicts of Interest:** The authors declare no conflict of interest.

## References

1. Li, J.; Chase, H.A. Development of adsorptive (non-ionic) macroporous resins and their uses in the purification of pharmacologically-active natural products from plant sources. *Nat. Prod. Rep.* **2010**, *27*, 1493–1510. [CrossRef] [PubMed]
2. Huang, C.; Cheng, L.; Feng, X.; Li, X.; Wang, L. Denichine ameliorates renal injury by improving oxidative stress, apoptosis and fibrosis in diabetic rats. *Life Sci.* **2020**, *258*, 118146. [CrossRef] [PubMed]
3. Xie, H.; Ji, W.; Liu, D.; Liu, W.; Wang, D.; Lv, R.; Wang, X. Surface molecularly imprinted polymers with dummy templates for the separation of denichine from *Panax notoginseng*. *RSC Adv.* **2015**, *5*, 48885–48892. [CrossRef]
4. Ng, T.B. Pharmacological activity of sanchi ginseng (*Panax notoginseng*). *J. Pharm. Pharmacol.* **2006**, *58*, 1007–1019. [CrossRef]
5. Jie, L.; Pengcheng, Q.; Qiaoyan, H. Denichine ameliorates kidney injury in induced type II diabetic nephropathy via the TGF-beta/Smad signalling pathway. *Eur. J. Clin. Pharmacol.* **2017**, *812*, 196–205. [CrossRef]
6. Cang, D.; Zou, G.; Yang, C.; Shen, X.; Li, F.; Wu, Y.; Ji, B. Denichine prevents ovariectomy-induced bone loss and inhibits osteoclastogenesis by inhibiting RANKL-associated NF-kappaB and MAPK signaling pathways. *J. Pharmacol. Sci.* **2021**, *146*, 206–215. [CrossRef]
7. Milke, L.; Kallscheuer, N.; Kappelmann, J.; Marienhagen, J. Tailoring Corynebacterium glutamicum towards increased malonyl-CoA availability for efficient synthesis of the plant pentaketide noreugenin. *Microb. Cell Factories* **2019**, *18*, 71. [CrossRef]
8. Liu, X.; Wang, Y.; Chen, Y.; Xu, S.; Gong, Q.; Zhao, C.; Cao, J.; Sun, C. Characterization of a Flavonoid 3′/5′/7-O-Methyltransferase from Citrus reticulata and Evaluation of the In Vitro Cytotoxicity of Its Methylated Products. *Molecules* **2020**, *25*, 858. [CrossRef]
9. Liu, B.; Su, K.; Wang, J.; Wang, J.; Xin, Z.; Li, F.; Fu, Y. Corynoline Exhibits Anti-inflammatory Effects in Lipopolysaccharide (LPS)-Stimulated Human Umbilical Vein Endothelial Cells through Activating Nrf2. *Inflammation* **2018**, *41*, 1640–1647. [CrossRef]
10. Zhu, J.; Zhou, X.; Zheng, H.; Li, Z. Enantioselective determination of denichine in rabbit plasma by high-performance liquid chromatography–electrospray mass spectrometry. *J. Chromatogr. B-Anal. Technol. Biomed. Life Sci.* **2006**, *840*, 124–131. [CrossRef]
11. Xie, G.X.; Qiu, Y.P.; Qiu, M.F.; Gao, X.F.; Liu, Y.M.; Jia, W. Analysis of denichine in *Panax notoginseng* by gas chromatography-mass spectrometry with ethyl chloroformate derivatization. *J. Pharm. Biomed. Anal.* **2007**, *43*, 920–925. [CrossRef] [PubMed]
12. Zhang, Y.; Chen, X.; Li, X.; Zhong, D. Development of a liquid chromatographic–tandem mass spectrometric method with precolumn derivatization for the determination of denichine in rat plasma. *Anal. Chim. Acta* **2006**, *566*, 200–206. [CrossRef]
13. Dong, J.; Yin, Z.; Su, L.; Yu, M.; Wang, M.; Li, L.; Mao, C.; Lu, T. Comparative pharmacokinetic analysis of raw and steamed *Panax notoginseng* roots in rats by UPLC-MS/MS for simultaneously quantifying seven saponins. *Pharm. Biol.* **2021**, *59*, 651–659. [CrossRef] [PubMed]
14. Ma, L.J.; Liu, F.; Zhong, Z.F.; Wan, J.B. Comparative study on chemical components and anti-inflammatory effects of *Panax notoginseng* flower extracted by water and methanol. *J. Sep. Sci.* **2017**, *40*, 4730–4739. [CrossRef] [PubMed]

15. Shen, S.; Zhou, C.; Zeng, Y.; Zhang, H.; Hossen, A.; Dai, J.; Li, S.; Qin, W.; Liu, Y. Structures, physicochemical and bioactive properties of polysaccharides extracted from *Panax notoginseng* using ultrasonic/microwave-assisted extraction. *LWT—Food Sci. Technol.* **2022**, *154*, 112446. [CrossRef]

16. Qin, S.; Jin, F.; Gao, L.; Su, L.; Li, Y.; Han, S.; Wang, P. Determination of sulfamerazine in aquatic products by molecularly imprinted capillary electrochromatography. *R. Soc. Open Sci.* **2019**, *6*, 190119. [CrossRef]

17. Li, P.; Qin, L.; Wang, T.; Dai, L.; Li, H.; Jiang, J.; Zhou, J.; Li, H.; Cheng, X.; Lei, F. Preparation and adsorption characteristics of rosin-based polymer microspheres for berberine hydrochloride and separation of total alkaloids from coptidis rhizoma. *Chem. Eng. J.* **2019**, *392*, 123707. [CrossRef]

18. Cao, Y.; Zhang, Y.; Zhang, Y.; Wang, L.; Lv, L.; Ma, X.; Zeng, S.; Wang, H. Biodegradable functional chitosan membrane for enhancement of artemisinin purification. *Carbohydr. Polym.* **2020**, *246*, 116590. [CrossRef]

19. Losada-Perez, P.; Polat, O.; Parikh, A.N.; Seker, E.; Renner, F.U. Engineering the interface between lipid membranes and nanoporous gold: A study by quartz crystal microbalance with dissipation monitoring. *Biointerphases* **2018**, *13*, 011002. [CrossRef]

20. McClements, J.; Seumo Tchekwagep, P.M.; Vilela Strapazon, A.L. Immobilization of Molecularly Imprinted Polymer Nanoparticles onto Surfaces Using Different Strategies: Evaluating the Influence of the Functionalized Interface on the Performance of a Thermal Assay for the Detection of the Cardiac Biomarker Troponin I. *ACS Appl. Mater. Interfaces* **2021**, *13*, 27868–27879. [CrossRef]

21. Wackers, G.; Vandenryt, T.; Cornelis, P. Array formatting of the heat-transfer method (HTM) for the detection of small organic molecules by molecularly imprinted polymers. *Sensors* **2014**, *14*, 11016–11030. [CrossRef]

22. Mousa, H.M.; Alenezi, J.F.; Mohamed, I.M.A.; Yasin, A.S.; Hashem, A.-F.M.; Abdal-hay, A. Synthesis of $TiO_2$@ZnO heterojunction for dye photodegradation and wastewater treatment. *J. Alloys Compd.* **2021**, *886*, 161169. [CrossRef]

23. Mousa, H.M.; Fahmy, H.S.; Abouzeid, R.; Abdel-Jaber, G.T.; Ali, W.Y. Polyvinylidene fluoride-cellulose nanocrystals hybrid nanofiber membrane for energy harvesting and oil-water separation applications. *Mater. Lett.* **2022**, *306*, 130965. [CrossRef]

24. Mousa, H.M.; Alfadhel, H.; Ateia, M.; Abdel-Jaber, G.T. Polysulfone-iron acetate/polyamide nanocomposite membrane for oil-water separation. *Environ. Nanotechnol. Monit. Manag.* **2020**, *14*, 100314. [CrossRef]

25. Mousa, H.M.; Alfadhel, H.; Nasr, E.A. Engineering and Characterization of Antibacterial Coaxial Nanofiber Membranes for Oil/Water Separation. *Polymers* **2020**, *12*, 2597. [CrossRef]

26. Fan, J.-P.; Cheng, Y.-T.; Zhang, X.-H.; Xiao, Z.-P.; Liao, D.-D.; Chen, H.-P.; Huang, K.; Peng, H.-L. Preparation of a novel mixed non-covalent and semi-covalent molecularly imprinted membrane with hierarchical pores for separation of genistein in Radix Puerariae Lobatae. *React. Funct. Polym.* **2019**, *146*, 104439. [CrossRef]

27. Zhang, H.; Li, Y.; Zheng, D.; Cao, S.; Chen, L.; Huang, L.; Xiao, H. Bio-inspired construction of cellulose-based molecular imprinting membrane with selective recognition surface for paclitaxel separation. *Appl. Surf. Sci.* **2018**, *466*, 244–253. [CrossRef]

28. Ahmad, A.L.; Abdulkarim, A.A.; Ooi, B.S.; Ismail, S. Recent development in additives modifications of polyethersulfone membrane for flux enhancement. *Chem. Eng. J.* **2013**, *223*, 246–267. [CrossRef]

29. Saljoughi, E.; Mousavi, S. Preparation and characterization of novel polysulfone nanofiltration membranes for removal of cadmium from contaminated water. *Sep. Purif. Technol.* **2012**, *90*, 22–30. [CrossRef]

30. Yuliwati, E.; Ismail, A. Effect of additives concentration on the surface properties and performance of PVDF ultrafiltration membranes for refinery produced wastewater treatment. *Desalination* **2011**, *273*, 226–234. [CrossRef]

31. Liu, F.; Hashim, N.A.; Liu, Y.; Moghareh Abed, M.R.; Li, K. Progress in the production and modification of PVDF membranes. *J. Membr. Sci.* **2011**, *375*, 1–27. [CrossRef]

32. Jiang, J.; Shen, Y.; Yu, D.; Yang, T.; Wu, M.; Yang, L.; Petru, M. Porous Film Coating Enabled by Polyvinyl Pyrrolidone (PVP) for Enhanced Air Permeability of Fabrics: The Effect of PVP Molecule Weight and Dosage. *Polymers* **2020**, *12*, 2961. [CrossRef]

33. Li, H.; Mu, P.; Li, J.; Wang, Q. Inverse desert beetle-like ZIF-8/PAN composite nanofibrous membrane for highly efficient separation of oil-in-water emulsions. *J. Mater. Chem. A* **2021**, *9*, 4167–4175. [CrossRef]

34. Abral, H.; Atmajaya, A.; Mahardika, M.; Hafizulhaq, F.; Handayani, D.; Sapuan, S.M.; Ilyas, R.A. Effect of ultrasonication duration of polyvinyl alcohol (PVA) gel on characterizations of PVA film. *J. Mater. Res. Technol.* **2020**, *9*, 2477–2486. [CrossRef]

35. Jiang, Y.; Liu, X.; Yang, Q.; Song, X.; Qin, C.; Wang, S.; Li, K. Effects of residual lignin on composition, structure and properties of mechanically defibrillated cellulose fibrils and films. *Cellulose* **2019**, *26*, 1577–1593. [CrossRef]

36. Liu, X.; Jiang, Y.; Qin, C.; Yang, S.; Song, X.; Wang, S.; Li, K. Enzyme-assisted mechanical grinding for cellulose nanofibers from bagasse: Energy consumption and nanofiber characteristics. *Cellulose* **2018**, *25*, 7065–7078. [CrossRef]

37. Liu, X.; Jiang, Y.; Wang, L.; Song, X.; Qin, C.; Wang, S. Tuning of size and properties of cellulose nanofibers isolated from sugarcane bagasse by endoglucanase-assisted mechanical grinding. *Ind. Crops Prod.* **2020**, *146*, 112201. [CrossRef]

38. Song, X.; Yang, S.; Liu, X.; Wu, M.; Li, Y.; Wang, S. Transparent and Water-Resistant Composites Prepared from Acrylic Resins ABPE-10 and Acetylated Nanofibrillated Cellulose as Flexible Organic Light-Emitting Device Substrate. *Nanomaterials* **2018**, *8*, 648. [CrossRef]

39. Si, H.; Liu, H.; Shang, S.; Song, J.; Liao, S.; Wang, D.; Song, Z. Preparation and properties of maleopimaric acid-based polyester polyol dispersion for two-component waterborne polyurethane coating. *Prog. Org. Coat.* **2016**, *90*, 309–316. [CrossRef]

40. Liu, B.; Nie, J.; He, Y. From rosin to high adhesive polyurethane acrylate: Synthesis and properties. *Int. J. Adhes. Adhes.* **2016**, *66*, 99–103. [CrossRef]

41. Choi, S.J.; Yim, T.; Cho, W.; Mun, J.; Jo, Y.N.; Kim, K.J.; Jeong, G.; Kim, T.-H.; Kim, Y.-J. Rosin-Embedded Poly(acrylic acid) Binder for Silicon/Graphite Negative Electrode. *ACS Sustain. Chem. Eng.* **2016**, *4*, 6362–6370. [CrossRef]

42. Liu, S.; Ding, Y.; Li, P.; Diao, K.; Tan, X.; Lei, F.; Zhan, Y.; Li, Q.; Huang, B.; Huang, Z. Adsorption of the anionic dye Congo red from aqueous solution onto natural zeolites modified with N,N-dimethyl dehydroabietylamine oxide. *Chem. Eng. J.* **2014**, *248*, 135–144. [CrossRef]
43. Yadav, B.K.; Gidwani, B.; Vyas, A. Rosin: Recent advances and potential applications in novel drug delivery system. *J. Bioact. Compat. Polym.* **2015**, *31*, 111–126. [CrossRef]
44. Chen, P.; Wang, X.; Kong, J.; Hu, X. A Facile Route to Fabricate CS/GO Composite Film for the Application of Therapeutic Contact Lenses. *Adv. Mater. Sci. Eng.* **2020**, *2020*, 8476025. [CrossRef]
45. Wei, S.; Yu, Q.; Fan, Z.; Liu, S.; Chi, Z.; Chen, X.; Zhang, Y.; Xu, J. Fabricating high thermal conductivity rGO/polyimide nanocomposite films via a freeze-drying approach. *RSC Adv.* **2018**, *8*, 22169–22176. [CrossRef] [PubMed]
46. Yang, F.; Ma, J.; Zhu, Q.; Qin, J. Fluorescent and mechanical properties of UiO-66/PA composite membrane. *Colloids Surf. A Physicochem. Eng. Asp.* **2021**, *627*, 127083. [CrossRef]
47. Li, X.; Li, H.; Ma, W.; Guo, Z.; Li, X.; Li, X.; Zhang, Q. Determination of patulin in apple juice by single-drop liquid-liquid-liquid microextraction coupled with liquid chromatography-mass spectrometry. *Food Chem.* **2018**, *257*, 1–6. [CrossRef]
48. Zhang, Z.; Scherer, G.W. Evaluation of drying methods by nitrogen adsorption. *Cem. Concr. Res.* **2019**, *120*, 13–26. [CrossRef]
49. Arami, M.; Limaee, N.; Mahmoodi, N. Evaluation of the adsorption kinetics and equilibrium for the potential removal of acid dyes using a biosorbent. *Chem. Eng. J.* **2008**, *139*, 2–10. [CrossRef]
50. Chang, Q.; Lin, W.; Ying, W.C. Impacts of amount of impregnated iron in granular activated carbon on arsenate adsorption capacities and kinetics. *Water Environ. Res.* **2012**, *84*, 514–520. [CrossRef]
51. Guo, L.; Li, G.; Liu, J.; Meng, Y.; Xing, G. Nonlinear Analysis of the Kinetics and Equilibrium for Adsorptive Removal of Cd(II) by Starch Phosphate. *J. Dispers. Sci. Technol.* **2012**, *33*, 403–409. [CrossRef]
52. Din, M.I.; Mirza, M.L.; Ata, S.; Athar, M.; Mohsin, I.U. Thermodynamics of Biosorption for Removal of Co(II) Ions by an Efficient and Ecofriendly Biosorbent (Saccharum bengalense): Kinetics and Isotherm Modeling. *J. Chem.* **2012**, *2013*, 528542.
53. Abukhadra, M.R.; El-Meligy, M.A.; El-Sherbeeny, A.M. Evaluation and characterization of Egyptian ferruginous kaolinite as adsorbent and heterogeneous catalyst for effective removal of safranin-O cationic dye from water. *Arab. J. Geosci.* **2020**, *13*, 169. [CrossRef]
54. Azizi, S.; Shahri, M.M.; Mohamad, R. Green Synthesis of Zinc Oxide Nanoparticles for Enhanced Adsorption of Lead Ions from Aqueous Solutions: Equilibrium, Kinetic and Thermodynamic Studies. *Molecules* **2017**, *22*, 831. [CrossRef]
55. Moon, H.G.; Jung, Y.; Shin, B.; Song, Y.G.; Kim, J.H.; Lee, T.; Lee, S.; Jun, S.C.; Kaner, R.B.; Kang, C.; et al. On-Chip Chemiresistive Sensor Array for On-Road NO x Monitoring with Quantification. *Adv. Sci.* **2020**, *7*, 2002014. [CrossRef]
56. HunBok Jung, M.; YanZheng, A.C. A field, laboratory and modeling study of reactive transport of groundwater arsenic in a coastal aquifer. *Environ. Sci. Technol.* **2009**, *43*, 5333–5338. [CrossRef]

MDPI

St. Alban-Anlage 66

4052 Basel

Switzerland

Tel. +41 61 683 77 34

Fax +41 61 302 89 18

www.mdpi.com

*Polymers* Editorial Office

E-mail: polymers@mdpi.com

www.mdpi.com/journal/polymers